Marius Durea and Radu Strugariu
An Introduction to Nonlinear Optimization Theory

Marius Durea and Radu Strugariu

An Introduction to Nonlinear Optimization Theory

Managing Editor: Aleksandra Nowacka-Leverton

Associate Editor: Vicentiu Radulescu

Language Editor: Nick Rogers

DE
G
DE GRUYTER
OPEN

Published by De Gruyter Open Ltd, Warsaw/Berlin
Part of Walter de Gruyter GmbH, Berlin/Munich/Boston

ISBN 978-3-11-042603-8
e-ISBN 978-3-11-042604-5

Bibliographic information published by the Deutsche Nationalbibliothek
The Deutsche Nationalbibliothek lists this publication in the Deutsche Nationalbibliografie;
detailed bibliographic data are available in the Internet at http://dnb.dnb.de.

Managing Editor: Aleksandra Nowacka-Leverton
Associate Editor: Vicentiu Radulescu
Language Editor: Nick Rogers

www.degruyteropen.com

Contents

Preface

This book aims to provide a thorough introduction to smooth and nonsmooth (convex and nonconvex) optimization theory on finite dimensional normed vector spaces ($\mathbb{R}^p$ spaces with $p \in \mathbb{N} \setminus \{0\}$). We present several important achievements of nonlinear analysis that motivate optimization problems and offer deeper insights into the further developments. Many of the results in this book hold in more general normed vector spaces, but the fundamental ideas of the theory are similar in every setting. We chose the framework of finite dimensional vector spaces since it offers the possibility to simplify some of the proofs and permits an intuitive understanding of the main ideas.

This book is intended to support courses of optimization and/or nonlinear analysis for undergraduate students, but we hope that graduate students in pure and applied mathematics, researchers and engineers could also benefit from it. We base our hopes on the following five facts:

- This book is largely self-contained: the prerequisites are the main facts of the classical differential calculus for functions of several real variables and linear algebra. They are recalled in the first chapter, so that this book can be then used without other references.
- We give necessary (and sometimes sufficient) optimality conditions in several cases of regularity for each problems' data: in the case of smooth functions, convex nonsmooth functions, and locally Lipschitz nonconvex functions in order to cover a large part of optimization theory.
- We present many deep results of nonlinear analysis under natural assumptions and proofs.
- Basic theoretical algorithms and their effective implementation in Matlab, together with the results of these numerical simulations are presented, and this shows both the power and practical applicability of the theory.
- An extended chapter of problems and their solutions gives the reader the possibility to solidify theoretical facts and to have a better understanding on various aspects of the main results in this book.

It should be clearly stated that we do not claim any originality in this monograph, but the selection and the organization of the material reflects our point of view on optimization theory. Based on our scientific and teaching criteria, the material of this book is organized into seven chapters which we briefly describe here.

The first chapter is introductory and fixes the general framework, the notations and the prerequisites.

The second chapter contains several concepts and results of nonlinear analysis which are essential to the rest of the book. Convex sets and functions, cones, the Bouligand tangent cone to a set at a point are studied, and we give complete proofs of fundamental results, among which Farkas Lemma, Banach Principle of fixed point and Graves Theorem.

The third chapter is the first one fully dedicated to optimization problems; it presents in detail the main aspects of the theory for the case of smooth data. We present general necessary and sufficient optimality conditions of first and second-order for problems with differentiable cost functions and with geometrical or smooth functional constraints. We arrive at the famous Karush-Kuhn-Tucker optimality conditions and we investigate several qualification conditions needed in this celebrated result.

The fourth chapter concerns the case of convex nonsmooth optimization problems. We introduce here, in compensation for the missing differentiability, the concept of the subgradient and we deduce, in this setting, necessary optimality conditions in Fritz John and Karush-Kuhn-Tucker forms.

The fifth chapter generalizes the theory. We work with functions that are neither differentiable nor convex, but are locally Lipschitz. This is a good setting to present Clarke and Mordukhovich generalized differentiation calculus, which finally allows us to arrive, once again, at optimality conditions with similar formulations as in the previous two chapters.

The sixth chapter is dedicated to the presentation of some basic algorithms for smooth optimization problems. We show Matlab code that accurately approximate the solutions of some optimization problems or related nonlinear equations.

The seventh chapter contains more than one hundred exercises and problems which are organized according to main themes of the book: nonlinear analysis, smooth optimization, nonsmooth optimization.

In our presentation we used several important monographs as follows: for theoretical expositions we mainly used (Zălinescu 1998; Pachpatte 2005; Nocedal and Wright 2006; Niculescu and Persson 2006; Rădulescu et al. 2009; Mordukhovich 2006; Hiriart-Urruty 2008; Clarke 2013; Cârjă 2003;), while for examples, problems and exercises we used (Pedregal 2004; Nocedal and Write, 2006; Isaacson and Keller 1966; Hiriart-Urruty 2009; Hestenes 1975; Forsgren et al. 2002; Clarke, 1983) and (Bazaraa et al. 2006). Finally, for the Matlab numerical simulations we used (Quarteroni and Saleri 2006).

In the case of the Ekeland Variational Principle, which was obtained in 1974 in the framework of general metric spaces, and whose original proof was based on an iteration procedure, the simpler proof for the case of finite dimensional vector spaces we present here was obtained in 1983 by J.-B. Hiriart-Urruty in (Hiriart-Urruty 1983). The very simple and natural proof of Farkas Lemma that is given in this book is based on the paper of D. Bartl (Bartl 2012) and on a personal communication to the authors from C. Zălinescu. The Graves theorem is taken from (Cârjă 2003).

Of course, the main reference for convex analysis is the celebrated R. T. Rockafellar monograph (Rockafellar 1970), but we also used the books (Niculescu and Persson 2006; Zălinescu 1998) and (Zălinescu 2002). For the section concerning the fixed points for function of real variable, we used several problems presented in (Radulescu et al. 2009).

For the section dedicated to the generalized Clarke calculus, we used the monographs (Clarke 1983; Clarke 2013; Rockafellar and Wets 1998). Theorem 5.1.32 is taken from (Rockafellar 1985). For the second part of Chapter 5, dedicated to Mordukhovich calculus, we mainly used (Mordukhovich 2006). The calculus rules for the Fréchet subdifferential of difference of functions, as well as the chain rule for the Fréchet subdifferential were taken from (Mordukhovich et al. 2006).

Many of the optimization problems given as exercises are taken from (Hiriart-Urruty 2008) and (Pedregal 2004), but (Hiriart-Urruty 2008) was used as well for some other theoretical examples such as the second problem from Section 3.4 or the Kantorovich inequality. The rather complicated proof of the fact that the Mangasarian Fromovitz condition is a qualification condition is taken from (Nocedal and Write 2006) which is used as well for presenting the sufficient optimality conditions of second order in Section 3.3. The Hardy and Carleman inequalities correspond to material in (Pachpatte 2005).

Chapter 6 is dedicated to numerical algorithms. We used the monographs: (Isaacson and Keller 1966) (for the convergence of Picard iterations and the Aitken methods), (Nocedal and Write 2006) (for the Newton and the SQP methods). For the presentation of the barrier method, we used (Forsgren et al. 2002).

Acknowledgements: We would like to thank Professor Vicenţiu Rădulescu, who kindly showed us the opportunity to write this book. Then, our thanks are addressed to dr. Aleksandra Nowacka-Leverton, Managing Editor to De Gruyter Open, for her support during the preparation of the manuscript, and to the Technical Department of De Gruyter Open, for their professional contribution to the final form of the monograph. We also take this opportunity to thank our families for their endless patience during the many days (including weekends) of work on this book.

01.10.2014   Marius Durea
Iaşi, Romania   Radu Strugariu

1 Preliminaries

The aim of this chapter is to introduce the notations, notions and results which will be useful in subsequent chapters. The results will be given without proof as they refer to basic notions of mathematical analysis, differential calculus and linear algebra.

1.1 $\mathbb{R}^p$ Space

Let $\mathbb{N}$ be the set of natural numbers and $\mathbb{N}^* := \mathbb{N} \setminus \{0\}$. Take $p \in \mathbb{N}^*$. We denote by $\mathbb{R}$ the set of real numbers and we introduce the set

$$\mathbb{R}^p := \{(x^1, x^2, ..., x^p) \mid x^i \in \mathbb{R}, \ \forall i \in \overline{1, p}\}.$$

This set can be organized as a p dimensional real vector space, with respect to the standard operations defined as follows: for every $x = (x^1, x^2, ..., x^p)$, $y = (y^1, y^2, ..., y^p) \in \mathbb{R}^p$ and every $a \in \mathbb{R}$

$$x + y := (x^1 + y^1, x^2 + y^2, ..., x^p + y^p) \in \mathbb{R}^p,$$
$$ax := (ax^1, ax^2, ..., ax^p) \in \mathbb{R}^p.$$

Recall that the canonical base in $\mathbb{R}^p$ is the set $\{e_1, ..., e_p\}$, where for any $i \in \overline{1, p}$, $e_i := (0, ..., 1, ..., 0)$, and 1 is placed on the i-th coordinate. In some situations, when there is no risk of confusion, we will use the notation with subscript indices of the components. We will extend these operations also for sets: if $A, B \subset \mathbb{R}^p$ are nonempty, $\alpha \in \mathbb{R} \setminus \{0\}$ and $C \subset \mathbb{R}$ is nonempty, one defines $A + B := \{a + b \mid a \in A, b \in B\}$, $\alpha A := \{\alpha a \mid a \in A\}$, $CA := \{\alpha a \mid \alpha \in C, a \in A\}$, $A - B := A + (-1)B$.

One can consider an element x of $\mathbb{R}^p$ as a matrix of dimension $1 \times p$. The corresponding transposed matrix will be denoted by x^t. Also, one defines the usual scalar product of two vectors $x, y \in \mathbb{R}^p$ by

$$\langle x, y \rangle := \sum_{i=1}^{p} x^i y^i = xy^t.$$

Moreover, $\mathbb{R}^p$ can be seen as a normed vector space (in particular, as a metric space) endowed with the Euclidean norm $\|\cdot\| : \mathbb{R}^p \to \mathbb{R}_+$ given by

$$\|x\| := \sqrt{\langle x, x \rangle} = \sqrt{\sum_{i=1}^{p} (x^i)^2}.$$

It is easy to prove that for every $x, y \in \mathbb{R}^p$, the next relation (the parallelogram law) holds

$$\|x + y\|^2 + \|x - y\|^2 = 2 \|x\|^2 + 2 \|y\|^2.$$

The angle between two vectors $x, y \in \mathbb{R}^p \setminus \{0\}$ is the value $\theta \in [0, \pi]$ given by

$$\cos \theta := \frac{\langle x, y \rangle}{\|x\| \, \|y\|}.$$

The open (closed) ball and the sphere centered at $\overline{x} \in \mathbb{R}^p$ with radius $\varepsilon > 0$ are given, respectively, by:

$$B(\overline{x}, \varepsilon) := \{x \in \mathbb{R}^p \mid \|x - \overline{x}\| < \varepsilon\},$$

$$D(\overline{x}, \varepsilon) := \{x \in \mathbb{R}^p \mid \|x - \overline{x}\| \le \varepsilon\},$$

and

$$S(\overline{x}, \varepsilon) := \{x \in \mathbb{R}^p \mid \|x - \overline{x}\| = \varepsilon\}.$$

One says that a subset $A \subset \mathbb{R}^p$ is bounded if it is contained in an open ball centered in the origin i.e., if there exists $M > 0$ such that $A \subset B(0, M)$.

A neighborhood of an element $\overline{x} \in \mathbb{R}^p$ is a subset of $\mathbb{R}^p$ which contains an open ball centered in $\overline{x}$. We denote by $\mathcal{V}(\overline{x})$ the class of all neighborhoods of $\overline{x}$. Let us summarize some facts:

–   A subset of $\mathbb{R}^p$ is open if it is empty or it is neighborhood for all of its points.
–   A subset of $\mathbb{R}^p$ is closed if its complement with respect to $\mathbb{R}^p$ is open.
–   An element a is an interior point of the set $A \subset \mathbb{R}^p$ if A is a neighborhood of a. We denote by $\operatorname{int} A$ the interior of A (i.e., the set of all interior points of A).
–   An element a is an accumulation point (or a limit point) of $A \subset \mathbb{R}^p$ if every neighborhood of a has at least one element in common with the set A which is different from a. We denote by A' the set of all limit points of A. If $a \in A \setminus A'$, one says that a is an isolated point of A.
–   An element a is an adherent point (or a closure point) of $A \subset \mathbb{R}^p$ if every neighborhood of a has at least one element in common with the set A. We will use the notations $\operatorname{cl} A$ and $\overline{A}$ to denote the closure of A (i.e., the set of all the adherent points of A).
–   A subset of $\mathbb{R}^p$ is compact if it is bounded and closed.

We denote by $\operatorname{bd} A$ the set $\operatorname{cl} A \setminus \operatorname{int} A = \operatorname{cl} A \cap \operatorname{cl}(\mathbb{R}^p \setminus A)$ and we call it the boundary of A.

Proposition 1.1.1. *(i) A subset of $\mathbb{R}^p$ is open if and only if it coincides with its interior.*
(ii) A subset of $\mathbb{R}^p$ is closed if and only if it coincides with its closure.

Definition 1.1.2. *One says that a function $f : \mathbb{N} \to \mathbb{R}^p$ is a sequence of elements from $\mathbb{R}^p$.*

The value of the function f in $n \in \mathbb{N}$, $f(n)$, is denoted by x_n (or $y_n, z_n, \ldots$), and the sequence defined by f is denoted by (x_n) (respectively, by $(y_n), (z_n), \ldots$).

Definition 1.1.3. *A sequence is bounded if the set of its terms is bounded.*

Definition 1.1.4. *One says that (y_k) is a subsequence of (x_n) if for every $k \in \mathbb{N}$, one has $y_k = x_{n_k}$, where by (n_k) one denotes a strictly increasing sequence of natural numbers (i.e., $n_k < n_{k+1}$ for every $k \in \mathbb{N}$).*

Definition 1.1.5. *One says that a sequence $(x_n) \subset \mathbb{R}^p$ is convergent (or converges) if there exists $x \in \mathbb{R}^p$ such that*

$$\forall\, V \in \mathcal{V}(x),\ \exists n_V \in \mathbb{N},\ \forall n \geq n_V : x_n \in V.$$

The element x is called the limit of (x_n).

If it exists, the limit of a sequence is unique.

We will use the notations $x_n \to x$, $\lim\limits_{n\to\infty} x_n = x$ or, simplified, $\lim x_n = x$ to formalize the previous definition.

Proposition 1.1.6. *A sequence (x_n) is convergent to $x \in \mathbb{R}^p$ if and only if*

$$\forall \varepsilon > 0,\ \exists n_\varepsilon \in \mathbb{N}, \forall n \geq n_\varepsilon : \|x_n - x\| < \varepsilon.$$

Proposition 1.1.7. *The sequence $(x_n) \subset \mathbb{R}^p$ converges to $x \in \mathbb{R}^p$ if and only if the coordinate sequences (x_n^i) converge (in $\mathbb{R}$) to x^i for every $i \in \overline{1,p}$.*

Proposition 1.1.8. *A sequence is convergent to $x \in \mathbb{R}^p$ if and only if all of its subsequences are convergent to x.*

Proposition 1.1.9. *Every convergent sequence is bounded.*

Proposition 1.1.10 (Characterization of the closure points using sequences). *Consider $A \subset \mathbb{R}^p$. A point $x \in \mathbb{R}^p$ is a closure point of A if and only if there exists a sequence $(x_n) \subset A$ such that $x_n \to x$.*

Proposition 1.1.11. *The set $A \subset \mathbb{R}^p$ is closed if and only if every convergent sequence from A has its limit in A.*

Proposition 1.1.12. *The set $A \subset \mathbb{R}^p$ is compact if and only if every sequence from A has a subsequence which converges to a point of A.*

Theorem 1.1.13 (Cesàro Lemma). *Every bounded sequence contains a convergent subsequence.*

Definition 1.1.14. *One says that $(x_n) \subset \mathbb{R}^p$ is a Cauchy sequence or a fundamental sequence if*

$$\forall \varepsilon > 0,\ \exists n_\varepsilon \in \mathbb{N},\ \forall n, m \geq n_\varepsilon : \|x_n - x_m\| < \varepsilon.$$

The above definition can be reformulated as follows: (x_n) is a Cauchy sequence if:

$$\forall \varepsilon > 0, \ \exists n_\varepsilon \in \mathbb{N}, \ \forall n \geq n_\varepsilon, \ \forall p \in \mathbb{N} : \|x_{n+p} - x_n\| < \varepsilon.$$

Theorem 1.1.15 (Cauchy). *The space $\mathbb{R}^p$ is complete, i.e., a sequence from $\mathbb{R}^p$ is convergent if and only if it is a Cauchy sequence.*

The next results are specific to the case of real sequences.

Definition 1.1.16. *One says that a sequence (x_n) of real numbers is increasing (strictly increasing, decreasing, strictly decreasing) if for every $n \in \mathbb{N}$, $x_{n+1} \geq x_n$ ($x_{n+1} > x_n$, $x_{n+1} \leq x_n$, $x_{n+1} < x_n$). If (x_n) is either increasing or decreasing, then it is called monotone.*

Let $\overline{\mathbb{R}} := \mathbb{R} \cup \{-\infty, +\infty\}$ be the set of extended real numbers. A neighborhood of $+\infty$ is a subset of $\overline{\mathbb{R}}$ which contains an interval of the form $(x, +\infty]$, where $x \in \mathbb{R}$. The neighborhoods of $-\infty$ are defined in a similar manner.

Definition 1.1.17. *(i) One says that the sequence $(x_n) \subset \mathbb{R}$ has the limit equal to $+\infty$ if*

$$\forall V \in \mathcal{V}(+\infty), \ \exists n_V \in \mathbb{N}, \ \forall n \geq n_V : x_n \in V.$$

(ii) One says that the sequence $(x_n) \subset \mathbb{R}$ has the limit equal to $-\infty$ if

$$\forall V \in \mathcal{V}(-\infty), \ \exists n_V \in \mathbb{N}, \ \forall n \geq n_V : x_n \in V.$$

Proposition 1.1.18. *(i) A sequence $(x_n) \subset \mathbb{R}$ has the limit equal to $+\infty$ if and only if*

$$\forall A > 0, \ \exists n_A \in \mathbb{N}, \ \forall n \geq n_A : x_n > A.$$

(ii) A sequence $(x_n) \subset \mathbb{R}$ has the limit equal to $-\infty$ if and only if

$$\forall A > 0, \ \exists n_A \in \mathbb{N}, \ \forall n \geq n_A : x_n < -A.$$

Proposition 1.1.19. *Let (x_n), (y_n), (z_n) be sequences of real numbers, $x, y \in \mathbb{R}$ and $n_0 \in \mathbb{N}$. Then:*
(i) (Passing to the limit in inequalities) if $x_n \to x$, $y_n \to y$ and $x_n \leq y_n$ for every $n \geq n_0$, then $x \leq y$;
(ii) (The boundedness criterion) if $|x_n - x| \leq y_n$ for every $n \geq n_0$, and $y_n \to 0$, then $x_n \to x$;
(iii) if $x_n \geq y_n$ for every $n \geq n_0$, and $y_n \to +\infty$, then $x_n \to +\infty$;
(iv) if $x_n \geq y_n$ for every $n \geq n_0$, and $x_n \to -\infty$, then $y_n \to -\infty$;
(v) if (x_n) is bounded and $y_n \to 0$, then $x_n y_n \to 0$;
(vi) if $x_n \leq y_n \leq z_n$ for every $n \geq n_0$, and $x_n \to x$, $z_n \to x$, then $y_n \to x$;
(vii) $x_n \to 0 \Leftrightarrow |x_n| \to 0 \Leftrightarrow x_n^2 \to 0$.

We now present some fundamental results in the theory of real sequences.

Theorem 1.1.20. *Every monotone real sequence has its limit in* $\overline{\mathbb{R}}$*. Moreover, if the sequence is bounded, then it is convergent, as follows: if it is increasing, then its limit is the supremum of the set of its terms, and if it is decreasing, the limit is the infimum of the set of its terms. If it is unbounded, then its limit is either* $+\infty$ *if the sequence is increasing, or* $-\infty$ *if the sequence is decreasing.*

Theorem 1.1.21 (Weierstrass theorem for sequences). *If* (x_n) *is a bounded and monotone sequence of real numbers, then* (x_n) *is convergent.*

Definition 1.1.22. *Let* $(x_n)_{n\geq 0}$ *be a sequence of real numbers. An element* $x \in \overline{\mathbb{R}}$ *is called a limit point of* (x_n) *if there exists a subsequence* (x_{n_k}) *of* (x_n) *such that* $x = \lim\limits_{k\to\infty} x_{n_k}$*.*

We finalize this section with two useful convergence criteria.

Proposition 1.1.23. *Let* (x_n) *be a sequence of strictly positive real numbers such that there exists* $\lim \dfrac{x_{n+1}}{x_n} = x$*. If* $x < 1$*, then* $x_n \to 0$*, and if* $x > 1$*, then* $x_n \to +\infty$*.*

Proposition 1.1.24 (Stolz-Cesàro Criterion). *Let* (x_n) *and* (y_n) *be real sequences such that* (y_n) *is strictly increasing and its limit is equal to* $+\infty$*. If there exists* $\lim \dfrac{x_{n+1} - x_n}{y_{n+1} - y_n} = x \in \overline{\mathbb{R}}$*, then* $\lim \dfrac{x_n}{y_n}$ *exists and is equal to* x*.*

1.2 Limits of Functions and Continuity

In this section, we expand on some issues related to the concepts of limit and continuity for functions. Let $p, q \in \mathbb{N}^*$.

Definition 1.2.1. *Let* $f : A \to \mathbb{R}^q$*,* $A \subset \mathbb{R}^p$ *and* $a \in A'$*. One says that the element* $l \in \mathbb{R}^q$ *is the limit of the function* f *at* a*, if for every* $V \in \mathcal{V}(l)$*, there exists* $U \in \mathcal{V}(a)$ *such that if* $x \in U \cap A$*,* $x \neq a$*, then* $f(x) \in V$*. We will denote this situation by* $\lim\limits_{x\to a} f(x) = l$*.*

Theorem 1.2.2. *Let* $f : A \to \mathbb{R}^q$*,* $A \subset \mathbb{R}^p$ *and* $a \in A'$*. The next assertions are equivalent:*

(i) $\lim\limits_{x\to a} f(x) = l$*;*

(ii) for every $B(l, \varepsilon) \subset \mathbb{R}^q$*, there exists* $B(a, \delta) \subset \mathbb{R}^p$ *such that if* $x \in B(a, \delta) \cap A$*,* $x \neq a$*, then* $f(x) \in B(l, \varepsilon)$*;*

(iii) for every $\varepsilon > 0$*, there exists* $\delta > 0$*, such that if* $\|x - a\| < \delta$*,* $x \in A$*,* $x \neq a$*, then* $\|f(x) - l\| < \varepsilon$*;*

(iv) for every $\varepsilon > 0$*, there exists* $\delta > 0$*, such that if* $|x_i - a_i| < \delta$ *for every* $i \in \overline{1, p}$*, where* $x = (x_1, x_2, .., x_p) \in A$*,* $a = (a_1, a_2, .., a_p)$*,* $x \neq a$*, then* $\|f(x) - l\| < \varepsilon$*;*

(v) for every sequence $(x_n) \subset A \setminus \{a\}$*,* $x_n \to a$ *implies that* $f(x_n) \to l$*.*

Theorem 1.2.3. *Let $f : A \to \mathbb{R}^q$, $A \subset \mathbb{R}^p$, $l \in \mathbb{R}^q$ and $a \in A'$. If the function f has the limit l at a, then this limit is unique.*

Remark 1.2.4. *If there exist two sequences (x'_n), $(x''_n) \subset A \setminus \{a\}$, $x'_n \to a$, $x''_n \to a$ such that $f(x'_n) \to l'$, $f(x''_n) \to l''$ and $l' \neq l''$, then the limit of the function f at $a \in A'$ does not exist.*

Theorem 1.2.5. *Let $f : A \to \mathbb{R}^q$, $A \subset \mathbb{R}^p$, $f = (f_1, f_2, ..., f_q)$ and $a \in A'$. Then f has the limit $l = (l_1, l_2, ..., l_q) \in \mathbb{R}^q$ at a if and only if there exists $\lim_{x \to a} f_i(x) = l_i$, for every $i \in \overline{1, q}$.*

Definition 1.2.6. *Let $a \in \mathbb{R}$, $A \subset \mathbb{R}$ and denote $A_s = A \cap (-\infty, a]$, $A_d = A \cap [a, \infty)$. One says that the element a is a left (right) accumulation point for A, if it is an accumulation point for A_s (A_d, respectively). We will denote the set of left (right) accumulation points of A by A'_s (A'_d, respectively).*

Definition 1.2.7. *Let $f : A \to \mathbb{R}^q$, $A \subset \mathbb{R}$ and a be a left (right) accumulation point of A. One says that the element $l \in \mathbb{R}^q$ is the left-hand (right-hand) limit of the function f in a if for every neighborhood $V \in \mathcal{V}(l)$ there exists $U \in \mathcal{V}(a)$, such that if $x \in U \cap A_s$ ($x \in U \cap A_d$, respectively), $x \neq a$, then $f(x) \in V$. In this case we will write $\lim_{x \to a, x < a} f(x) = l$, or $\lim_{x \to a^-} f(x) = l$, or $\lim_{x \uparrow a} f(x) = l$ ($\lim_{x \to a, x > a} f(x) = l$, or $\lim_{x \to a^+} f(x) = l$, or $\lim_{x \downarrow a} f(x) = l$, respectively).*

Theorem 1.2.8. *Let $I \subset \mathbb{R}$ be an open interval, $f : I \to \mathbb{R}^q$, and $a \in I$. Then there exists $\lim_{x \to a} f(x) = l$ if and only if the left-hand and the right-hand limits of f at a exist and they are equal. In this case, all three limits are equal:*

$$\lim_{x \to a^-} f(x) = \lim_{x \to a^+} f(x) = l.$$

A well-known result says that the monotone real functions admit lateral limits at every accumulation point of their domains.

Theorem 1.2.9 (The boundedness criterion). *Let $f : A \to \mathbb{R}^q$, $g : A \to \mathbb{R}$, $A \subset \mathbb{R}^p$ and $a \in A'$. If there exist $l \in \mathbb{R}^q$ and $U \in \mathcal{V}(a)$ such that $\|f(x) - l\| \leq |g(x)|$ for every $x \in U \setminus \{a\}$, and $\lim_{x \to a} g(x) = 0$, then there exists $\lim_{x \to a} f(x) = l$.*

Theorem 1.2.10. *Let $f, g : A \subset \mathbb{R}^p \to \mathbb{R}^q$, and $a \in A'$. If $\lim_{x \to a} f(x) = 0$ and there exists $U \in \mathcal{V}(a)$ such that g is bounded on U, then there exists the limit $\lim_{x \to a} f(x) g(x) = 0$.*

Theorem 1.2.11. *Let $f : A \subset \mathbb{R}^p \to \mathbb{R}^q$, and $a \in A'$. If there exists $\lim_{x \to a} f(x) = l$, $l > 0$ ($l < 0$), then there exists $U \in \mathcal{V}(a)$ such that for every $x \in U \cap A$, $x \neq a$, one has $f(x) > 0$ (respectively, $f(x) < 0$).*

Theorem 1.2.12. *Let $f : A \subset \mathbb{R}^p \to \mathbb{R}^q$, and $a \in A'$. If there exists $\lim\limits_{x \to a} f(x) = l$, then there exists $U \in \mathcal{V}(a)$ such that f is bounded on U (i.e., there exists $M > 0$ such that for every $x \in U$, one has $\|f(x)\| \leq M$).*

Definition 1.2.13. *Let $f : A \subset \mathbb{R}^p \to \mathbb{R}$ and $a \in A'$. One says that the function f has the limit equal to $+\infty$ (respectively, $-\infty$) at a, if for every $V \in \mathcal{V}(+\infty)$ (respectively, $V \in \mathcal{V}(-\infty)$), there exists $U \in \mathcal{V}(a)$ such that for every $x \in U \cap A$, $x \neq a$, one has $f(x) \in V$. In this case, we will write $\lim\limits_{x \to a} f(x) = +\infty$ (respectively, $\lim\limits_{x \to a} f(x) = -\infty$).*

Theorem 1.2.14. *Let $f : A \subset \mathbb{R}^p \to \mathbb{R}$ and $a \in A'$. Then there exists $\lim\limits_{x \to a} f(x) = +\infty$ (respectively, $\lim\limits_{x \to a} f(x) = -\infty$) if and only if for every $\varepsilon > 0$, there exists $\delta > 0$, such that if $\|x - a\| < \delta$, $x \in A$, $x \neq a$, one has $f(x) > \varepsilon$ (respectively, $f(x) < -\varepsilon$).*

Definition 1.2.15. *Let $f : A \subset \mathbb{R} \to \mathbb{R}^q$, such that $+\infty$ (respectively, $-\infty$) is an accumulation point of A. One says that the element $l \in \mathbb{R}^q$ is the limit of f at $+\infty$ (respectively, $-\infty$), if for every $V \in \mathcal{V}(l)$, there exists $U \in \mathcal{V}(+\infty)$ (respectively, $U \in \mathcal{V}(-\infty)$) such that for every $x \in U \cap A$, one has $f(x) \in V$. In this case, we will write $\lim\limits_{x \to +\infty} f(x) = l$ (respectively, $\lim\limits_{x \to -\infty} f(x) = l$).*

Theorem 1.2.16. *Let $f : A \subset \mathbb{R} \to \mathbb{R}^q$, such that $+\infty$ (respectively, $-\infty$) is an accumulation point of A. Then there exists $\lim\limits_{x \to +\infty} f(x) = l$ (respectively, $\lim\limits_{x \to -\infty} f(x) = l$) if and only if for every $\varepsilon > 0$, there exists $\delta > 0$, such that if $x > \delta$ (respectively, $x < -\delta$), $x \in A$, one has $\|f(x) - l\| < \varepsilon$.*

Definition 1.2.17. *Let $f : A \subset \mathbb{R}^p \to \mathbb{R}^q$, and $a \in A$. One says that the function f is continuous at a if for every $V \in \mathcal{V}(f(a))$, there exists $U \in \mathcal{V}(a)$ such that for every $x \in U \cap A$, one has $f(x) \in V$.*

If the function f is not continuous at a, one says that f is discontinuous at a, or that a is a discontinuity point of the function f.

Theorem 1.2.18. *Let $f : A \subset \mathbb{R}^p \to \mathbb{R}^q$, and $a \in A' \cap A$. The function f is continuous at a if and only if $\lim\limits_{x \to a} f(x) = f(a)$. If a is an isolated point of A, then f is continuous at a.*

Theorem 1.2.19. *Let $f : A \subset \mathbb{R}^p \to \mathbb{R}^q$, and $a \in A$. The next assertions are equivalent:*
    (i) f is continuous at a;
    (ii) ($\varepsilon - \delta$ characterization) for every $\varepsilon > 0$, there exists $\delta > 0$, such that if $\|x - a\| < \delta$, $x \in A$, then $\left\|f(x) - f(a)\right\| < \varepsilon$;
    (iii) (sequential characterization) for every $(x_n) \subset A$, $x_n \to a$, one has $f(x_n) \to f(a)$.

Theorem 1.2.20. *The image of a compact set through a continuous function is a compact set.*

Theorem 1.2.21 (Weierstrass Theorem). *Let K be a compact subset of $\mathbb{R}^p$. If $f : K \to \mathbb{R}$ is a continuous function, then f is bounded and it attains its extreme values on the set K (i.e., there exist $a, b \in K$, such that $\sup_{x \in K} f(x) = f(a)$ and $\inf_{x \in K} f(x) = f(b)$).*

Definition 1.2.22. *Let $f : D \subset \mathbb{R}^p \to \mathbb{R}^q$. One says that the function f is uniformly continuous on the set D if for every $\varepsilon > 0$, there exists $\delta > 0$, such that for every $x', x'' \in D$ with $\|x' - x''\| < \delta$, one has $\|f(x') - f(x'')\| < \varepsilon$.*

Remark 1.2.23. *Every function which is uniformly continuous on D is continuous on D, i.e., it is continuous at every point of D.*

Theorem 1.2.24 (Cantor Theorem). *Every function which is continuous on a compact set $K \subset \mathbb{R}^p$ and takes values in $\mathbb{R}^q$ is uniformly continuous on K.*

Definition 1.2.25. *Let $L \geq 0$ be a real number. One says that a function $f : A \subset \mathbb{R}^p \to \mathbb{R}^q$ is Lipschitz on A with modulus L, or L–Lipschitz on A, if $\|f(x) - f(y)\| \leq L \|x - y\|$, for every $x, y \in A$.*

Proposition 1.2.26. *Every Lipschitz function on $A \subset \mathbb{R}^p$ is uniformly continuous on A.*

Theorem 1.2.27. *Let $I \subset \mathbb{R}$ be an interval. If $f : I \to \mathbb{R}$ is injective and continuous, then f is strictly monotone on I.*

Definition 1.2.28. *Let $I \subset \mathbb{R}$ be an interval. One says that the function $f : I \to \mathbb{R}$ has the Darboux property if for every $a, b \in I$, $a < b$ and every $\lambda \in (f(a), f(b))$ or $\lambda \in (f(b), f(a))$, there exists $c_\lambda \in (a, b)$ such that $f(c_\lambda) = \lambda$.*

Theorem 1.2.29. *Let $I \subset \mathbb{R}$ be an interval. If the function $f : I \to \mathbb{R}$ has the Darboux property and there exist $a, b \in I$, $a < b$, such that $f(a)f(b) < 0$, then the equation $f(x) = 0$ has at least one solution in (a, b).*

Theorem 1.2.30. *Let $I \subset \mathbb{R}$ be an interval. The function $f : I \to \mathbb{R}$ has the Darboux property if and only if for every interval $J \subset I$, $f(J)$ is an interval.*

Theorem 1.2.31. *Let $I \subset \mathbb{R}$ be an interval. If $f : I \to \mathbb{R}$ is continuous, then f has the Darboux property.*

Recall that every linear operator $T : \mathbb{R}^p \to \mathbb{R}^q$ is continuous. For such a map, one uses the constant

$$\|T\| := \inf\{M > 0 \mid \|Tx\| \le M\,\|x\|\,,\ \forall x \in \mathbb{R}^p\}$$
$$= \sup\left\{\|Tx\| \mid x \in D(0,1)\right\}.$$

The mapping $T \mapsto \|T\|$ satisfies the axioms of a norm, therefore it is called the norm of the operator T. Consequently, the set of linear operators from $\mathbb{R}^p$ to $\mathbb{R}^q$ is a real normed vector space, with respect to the usual algebraic operations and to the norm previously defined. This space is denoted by $L(\mathbb{R}^p, \mathbb{R}^q)$ and can be isomorphically identified with $\mathbb{R}^{pq}$. Every operator $T \in L(\mathbb{R}^p, \mathbb{R}^q)$ can be naturally associated with a $q \times p$ matrix, denoted by $A_T = (a_{ji})_{j\in\overline{1,q},i\in\overline{1,p}}$, as follows: if $(e_i)_{i\in\overline{1,p}}$ and $(e_i')_{i\in\overline{1,q}}$ are the canonical bases of the spaces $\mathbb{R}^p$ and $\mathbb{R}^q$, respectively, then $(a_{ji})_{j\in\overline{1,q},i\in\overline{1,p}}$ are the coordinates of the expressions of the images of the elements $(e_i)_{i\in\overline{1,p}}$ through T with respect to the basis $(e_i')_{i\in\overline{1,q}}$, i.e.,

$$T(e_i) = \sum_{j=1}^{q} a_{ji}e_j', \ \forall i \in \overline{1,p}.$$

Consequently, $T \mapsto A_T$ is an isomorphism of linear spaces between $L(\mathbb{R}^p, \mathbb{R}^q)$ and the space of real $q \times p$ matrices. Also, for every $x \in \mathbb{R}^p$:

$$T(x) = (A_T x^t)^t.$$

Moreover, for every $x \in \mathbb{R}^p$ and $y \in \mathbb{R}^q$, one has that

$$\left\langle (A_T x^t)^t, y \right\rangle = \left\langle x, (A_T^t y^t)^t \right\rangle.$$

If A is a $q \times p$ matrix, then the linear operator associated with A is surjective if and only if the map associated with A^t is injective.

Recall also that if $T : \mathbb{R}^p \to \mathbb{R}^q$ is a linear operator, then its kernel,

$$\text{Ker}(T) := \{x \in \mathbb{R}^p \mid T(x) = 0\},$$

is a linear subspace of $\mathbb{R}^p$, and its image,

$$\text{Im}(T) := \{T(x) \mid x \in \mathbb{R}^p\},$$

is a linear subspace of $\mathbb{R}^q$. Moreover,

$$p = \dim(\text{Ker}(T)) + \dim(\text{Im}(T)),$$

where by dim we denote the algebraic dimension.

From the theory of linear algebra, one knows that if A is a symmetric square matrix of order p, then its eigenvalues are real and, moreover, there exists an orthogonal matrix B (i.e., $BB^t = B^t B = I$) such that $B^t AB$ is the diagonal matrix having the eigenvalues on its main diagonal. Recall that, as usual, I denotes the identity matrix.

One says that a matrix A as above is positive semidefinite if $\langle (Ax^t)^t, x \rangle \geq 0$ for every $x \in \mathbb{R}^p$, and positive definite if $\langle (Ax^t)^t, x \rangle > 0$ for every $x \in \mathbb{R}^p \setminus \{0\}$. Actually, A is positive definite if and only if it is positive semidefinite and invertible.

We end this section by mentioning the celebrated result of Hahn-Banach. Recall that a function $f : \mathbb{R}^p \to \mathbb{R}$ is called sublinear if it is positive homogeneous (i.e., $f(\alpha x) = \alpha f(x)$ for all $\alpha \geq 0$ and $x \in \mathbb{R}^p$) and subadditive (i.e., $f(x + y) \leq f(x) + f(y)$ for all $x, y \in \mathbb{R}^p$).

Theorem 1.2.32 (Hahn-Banach). *Let X be a linear subspace of $\mathbb{R}^p$, $\chi : \mathbb{R}^p \to \mathbb{R}$ be a sublinear function, and $\varphi_0 : X \to \mathbb{R}$ be a linear function. If $\varphi_0(x) \leq \chi(x)$ for every $x \in X$, then there exists a linear function $\varphi : \mathbb{R}^p \to \mathbb{R}$ such that $\varphi|_X = \varphi_0$ and $\varphi(x) \leq \chi(x)$ for every $x \in \mathbb{R}^p$.*

1.3 Differentiability

Definition 1.3.1. *Let $f : D \subset \mathbb{R}^p \to \mathbb{R}^q$ and $a \in \text{int } D$. One says that f is Fréchet differentiable (or, simply, differentiable) at a if there exists a linear operator denoted by $\nabla f(a) : \mathbb{R}^p \to \mathbb{R}^q$ such that*

$$\lim_{h \to 0} \frac{f(a + h) - f(a) - \nabla f(a)(h)}{\|h\|} = \lim_{x \to a} \frac{f(x) - f(a) - \nabla f(a)(x - a)}{\|x - a\|} = 0.$$

The map $\nabla f(a)$ is called the Fréchet differential of the function f at a.

The previous relation is equivalent to the following conditions:

$$\forall \varepsilon > 0, \exists \delta > 0, \forall x \in B(a, \delta) : \|f(x) - f(a) - \nabla f(a)(x - a)\| \leq \varepsilon \|x - a\| ;$$

$$\exists \alpha : D - \{a\} \to \mathbb{R}^q : \lim_{h \to 0} \alpha(h) = \alpha(0) = 0,$$

$$f(a + h) = f(a) + \nabla f(a)(h) + \|h\| \alpha(h), \ \forall h \in D - \{a\}.$$

One says that $f : D \subset \mathbb{R}^p \to \mathbb{R}^q$ is of class C^1 on the open set D if f is Fréchet differentiable on D and ∇f is continuous on D. Obviously, f can be written as $f = (f_1, f_2, \ldots, f_q)$, where $f_i : \mathbb{R}^p \to \mathbb{R}$, $i \in \overline{1, q}$ and, in general, the map $\nabla f(a) \in L(\mathbb{R}^p, \mathbb{R}^q)$ will be identified to the $q \times p$ matrix

$$\begin{pmatrix} \dfrac{\partial f_1}{\partial x^1}(a) & \dfrac{\partial f_1}{\partial x^2}(a) & \cdots & \dfrac{\partial f_1}{\partial x^p}(a) \\ \dfrac{\partial f_2}{\partial x^1}(a) & \dfrac{\partial f_2}{\partial x^2}(a) & \cdots & \dfrac{\partial f_2}{\partial x^p}(a) \\ \vdots & \vdots & \ddots & \vdots \\ \dfrac{\partial f_q}{\partial x^1}(a) & \dfrac{\partial f_q}{\partial x^2}(a) & \cdots & \dfrac{\partial f_q}{\partial x^p}(a) \end{pmatrix},$$

called the Jacobian matrix of f at the point a, where $\dfrac{\partial f_i}{\partial x^j}(a)$ is the partial derivative of the function f_i with respect to the variable x^j at a.

We will subsequently refer several times to the Jacobian matrix instead of the differential. Based on a general result, if $f : D \subset \mathbb{R}^p \to \mathbb{R}^p$ and $a \in \mathrm{int}\, D$, $\nabla f(a)$ is an isomorphism of $\mathbb{R}^p$ if and only if the Jacobian matrix of f at a is invertible.

The next calculus rules hold.

–   Let $f : \mathbb{R}^p \to \mathbb{R}^q$ be an affine function, i.e., it takes the form $f(x) := g(x) + u$ for every $x \in \mathbb{R}^p$, where $g : \mathbb{R}^p \to \mathbb{R}^q$ is linear, and $u \in \mathbb{R}^q$. Then for every $x \in \mathbb{R}^p$, $\nabla f(x) = g$.

–   Let $f : \mathbb{R}^p \to \mathbb{R}$ be of the form $f(x) = \dfrac{1}{2}\left\langle (Ax^t)^t, x\right\rangle + \langle b, x\rangle$, where A is a symmetric square matrix of order p, and $b \in \mathbb{R}^p$. Then for every $x \in \mathbb{R}^p$, $\nabla f(x) = (Ax^t)^t + b$.

–   Let $D \subset \mathbb{R}^p$, $E \subset \mathbb{R}^q$, $\overline{x} \in \mathrm{int}\, D$, $\overline{y} \in \mathrm{int}\, E$ and $f, g : D \to \mathbb{R}^q$, $\varphi : D \to \mathbb{R}$, $h : E \to \mathbb{R}^k$.

    –   If f, g are differentiable at $\overline{x}$, and $\alpha, \beta \in \mathbb{R}$, then the function $\alpha f + \beta g$ is differentiable at $\overline{x}$ and

    $$\nabla(\alpha f + \beta g)(\overline{x}) = \alpha \nabla f(\overline{x}) + \beta \nabla g(\overline{x}).$$

    –   If f, φ are differentiable at $\overline{x}$, then φf is differentiable at $\overline{x}$ at

    $$\nabla(\varphi f)(\overline{x}) = \varphi(\overline{x})\nabla f(\overline{x}) + f(\overline{x})\nabla \varphi(\overline{x}),$$

    where $\left(f(\overline{x})\nabla\varphi(\overline{x})\right)(x) := \nabla\varphi(\overline{x})(x) \cdot f(\overline{x})$.

    –   (Chain rule) If $f(D) \subset E$, $\overline{y} = f(\overline{x})$, f is differentiable at $\overline{x}$ and h is differentiable at $\overline{y}$, then $h \circ f$ is differentiable at $\overline{x}$ and

    $$\nabla(h \circ f)(\overline{x}) = \nabla h(\overline{y}) \circ \nabla f(\overline{x}).$$

A case which deserves special attention is $p = 1$. In this case one says that f is derivable at a if there exists

$$\lim_{h \to 0} \frac{f(a + h) - f(a)}{h} \in \mathbb{R}^q. \tag{1.3.1}$$

One denotes this limit by $f'(a)$ and it is called the derivative of f at a.

Proposition 1.3.2. *Let $f : D \subset \mathbb{R} \to \mathbb{R}^q$ and $a \in \mathrm{int}\, D$. The next assertions are equivalent:*

(i) f is derivable at a;

(ii) f is Fréchet differentiable at a.

In every one of these cases, $\nabla f(a)(x) = xf'(a)$ for every $x \in \mathbb{R}$.

Let $r \in \mathbb{N}^*$. If $f : D \subset \mathbb{R}^p \times \mathbb{R}^q \to \mathbb{R}^r$, and $(a, b) \in \mathrm{int}\, D$ is fixed, one defines $D_1 := \{x \in \mathbb{R}^p \mid (x, b) \in D\}$ and $f_1 : D_1 \to \mathbb{R}^r$, $f_1(x) := f(x, b)$. One says that f is Fréchet differentiable with respect to x at a if f_1 is Fréchet differentiable at a, and in this case the differential is denoted by $\nabla_x f(a, b)$. If f is differentiable at (a, b), then f is differentiable with respect to x and y at a and b, respectively, and

$$\nabla_x f(a, b) = \nabla f(a, b)(\cdot, 0), \quad \nabla_y f(a, b) = \nabla f(a, b)(0, \cdot).$$

In the general case, one says that $f : D \subset \mathbb{R}^p \to \mathbb{R}^q$ is twice Fréchet differentiable at $a \in \operatorname{int} D$ if f is Fréchet differentiable on a neighborhood $V \subset D$ of a and $\nabla f : V \to L(\mathbb{R}^p, \mathbb{R}^q)$ is Fréchet differentiable at a, i.e., there exists a functional denoted by $\nabla^2 f(a)$, from the space $L^2(\mathbb{R}^p, \mathbb{R}^q) := L(\mathbb{R}^p, L(\mathbb{R}^p, \mathbb{R}^q))$, and $\alpha : D - \{a\} \to L(\mathbb{R}^p, \mathbb{R}^q)$, such that $\lim_{h \to 0} \alpha(h) = \alpha(0) = 0$ and for every $h \in D - \{a\}$, one has

$$\nabla f(a + h) = \nabla f(a) + \nabla^2 f(a)(h, \cdot) + \|h\| \, \alpha(h).$$

Recall that the space $L^2(\mathbb{R}^p, \mathbb{R}^q)$ mentioned above can be identified with the space of bilinear maps from $\mathbb{R}^p \times \mathbb{R}^p$ to $\mathbb{R}^q$.

One says that f is of class C^2 on the open set D if it is twice Fréchet differentiable on D and $\nabla^2 f : D \to L^2(\mathbb{R}^p, \mathbb{R}^q)$ is continuous.

Theorem 1.3.3. *Let $f : D \subset \mathbb{R}^p \to \mathbb{R}^q$ and $a \in \operatorname{int} D$. If f is twice Fréchet differentiable at a, then $\nabla^2 f(a)$ is a symmetric bilinear map.*

In the case when $q = 1$, the map $\nabla^2 f(a)$ is defined by the symmetric square matrix $H(a) = \left(\dfrac{\partial^2 f}{\partial x^i \partial x^j}(a) \right)_{i,j \in \overline{1,p}}$, which is called the Hessian matrix of f at a. Moreover, $\left\langle \left(H(a)u^t \right)^t, v \right\rangle = \nabla^2 f(a)(u, v)$ for every $u, v \in \mathbb{R}^p$, i.e.,

$$\nabla^2 f(a)(u, v) = \sum_{i,j=1}^{n} \frac{\partial^2 f}{\partial x^i \partial x^j}(a) u_i v_j.$$

If $a, b \in \mathbb{R}^p$, one defines the closed and the open line segments between a and b as follows:

$$[a, b] := \{\alpha a + (1 - \alpha)b \mid \alpha \in [0, 1]\},$$
$$(a, b) := \{\alpha a + (1 - \alpha)b \mid \alpha \in (0, 1)\}.$$

Theorem 1.3.4 (Lagrange and Taylor Theorems). *Let $U \subset \mathbb{R}^p$ be an open set, $f : U \to \mathbb{R}$ and $a, b \in U$ with $[a, b] \subset U$. If f is of class C^1 on U, then there exists $c \in (a, b)$ such that*

$$f(b) = f(a) + \nabla f(c)(b - a).$$

If f is of class C^2 on U, then there exists $c \in (a, b)$ such that

$$f(b) = f(a) + \nabla f(a)(b - a) + \frac{1}{2} \nabla^2 f(c)(b - a, b - a).$$

Theorem 1.3.5 (Implicit Function Theorem). *Let $D \subset \mathbb{R}^p \times \mathbb{R}^q$ be an open set, $h : D \rightarrow \mathbb{R}^q$ be a function and $\overline{x} \in \mathbb{R}^p$, $\overline{y} \in \mathbb{R}^q$ be such that:*
  (i) $h(\overline{x}, \overline{y}) = 0$;
  (ii) the function h is of class C^1 on D;
  (iii) $\nabla_y h(\overline{x}, \overline{y})$ is invertible.
  Then there exist two neighborhoods U and V of $\overline{x}$ and $\overline{y}$, respectively, and a unique continuous function $\varphi : U \rightarrow V$ such that:
  (a) $h(x, \varphi(x)) = 0$ for every $x \in U$;
  (b) if $(x, y) \in U \times V$ and $h(x, y) = 0$, then $y = \varphi(x)$;
  (c) φ is differentiable on U and

$$\nabla\varphi(x) = -[\nabla_y h(x, \varphi(x))]^{-1} \nabla_x h(x, \varphi(x)), \ \forall x \in U.$$

Some fundamental results from the theory of differentiability of the real functions are briefly given at the end of this section.

In the case $p = q = 1$ one can apply Proposition 1.3.2. It is also sensible to speak of the existence of the derivative at points of the domain which are accumulation points of it: consider the limit from the relation (1.3.1) at accumulation points. Moreover, as in the case of the lateral limits, one can speak about the left and right-hand derivatives, by considering the lateral limits in the expression from relation (1.3.1). When they exist, we will call these limits the left, and the right-hand derivatives of the function f at a and we will denote them by $f'_-(a)$ and $f'_+(a)$, respectively.

Definition 1.3.6. *Let $A \subset \mathbb{R}$ and $f : A \rightarrow \mathbb{R}$. One says that $a \in A$ is a local minimum (maximum) point for f if there exists a neighborhood V of a such that $f(a) \leq f(x)$ (respectively, $f(a) \geq f(x)$), for every $x \in A \cap V$. One says that a point is a local extremum if it is a local minimum or a local maximum.*

Theorem 1.3.7 (Fermat Theorem). *Let $I \subset \mathbb{R}$ be an interval and $a \in \operatorname{int} I$. If $f : I \rightarrow \mathbb{R}$ is derivable at a, and a is a local extremum point for f, then $f'(a) = 0$.*

Theorem 1.3.8 (Rolle Theorem). *Let $a, b \in \mathbb{R}$, $a < b$, and $f : [a, b] \rightarrow \mathbb{R}$ be a function which is continuous on $[a, b]$, derivable on (a, b), and satisfies $f(a) = f(b)$. Then there exists $c \in (a, b)$ such that $f'(c) = 0$.*

Theorem 1.3.9 (Lagrange Theorem). *Let $a, b \in \mathbb{R}$, $a < b$, and $f : [a, b] \rightarrow \mathbb{R}$ be a function which is continuous on $[a, b]$, derivable on (a, b). Then there exists $c \in (a, b)$ such that $f(b) - f(a) = f'(c)(b - a)$.*

Proposition 1.3.10. *Let $I \subset \mathbb{R}$ be an interval and $f : I \rightarrow \mathbb{R}$ be derivable on I.*
  (i) If $f'(x) = 0$ for every $x \in I$, then f is constant on I.

(ii) If $f'(x) > 0$ (respectively, if $f'(x) \geq 0$) for every $x \in I$, then f is strictly increasing (respectively, it is increasing) on I.

(iii) If $f'(x) < 0$ (respectively, if $f'(x) \leq 0$), for every $x \in I$, then f is strictly decreasing (respectively, it is decreasing) on I.

Theorem 1.3.11 (Rolle Sequence). *Let $I \subset \mathbb{R}$ be an interval and $f : I \to \mathbb{R}$ be a derivable function. If $x_1, x_2 \in I$, $x_1 < x_2$ are consecutive roots of the derivative f' (i.e., $f'(x_1) = 0$, $f'(x_2) = 0$ and $f'(x) \neq 0$ for any $x \in (x_1, x_2)$) then:*

(i) if $f(x_1)f(x_2) < 0$, the equation $f(x) = 0$ has exactly one root in the interval (x_1, x_2);

(ii) if $f(x_1)f(x_2) > 0$, the equation $f(x) = 0$ has no roots in the interval (x_1, x_2);

(iii) if $f(x_1) = 0$ or $f(x_2) = 0$, then x_1 or x_2 is a multiple root of the equation $f(x) = 0$ and this equation has no other root in the interval (x_1, x_2).

Theorem 1.3.12 (Cauchy Rule). *Let $I \subset \mathbb{R}$ be an interval and $f, g : I \to \mathbb{R}$, $a \in I$, which satisfy:*

(i) $f(a) = g(a) = 0$;

(ii) f, g are derivable at a;

(iii) $g'(a) \neq 0$.

Then there exists $V \in \mathcal{V}(a)$ such that $g(x) \neq 0$, for any $x \in V \setminus \{a\}$ and

$$\lim_{x \to a} \frac{f(x)}{g(x)} = \frac{f'(a)}{g'(a)}.$$

Theorem 1.3.13 (L'Hôpital Rule). *Let $f, g : (a, b) \to \mathbb{R}$, where $-\infty \leq a < b \leq \infty$. If:*

(i) f, g are derivable on (a, b) with $g' \neq 0$ on (a, b);

(ii) there exists $\lim_{\substack{x \to a \\ x > a}} \dfrac{f'(x)}{g'(x)} = L \in \overline{\mathbb{R}}$;

(iii) $\lim_{\substack{x \to a \\ x > a}} f(x) = \lim_{\substack{x \to a \\ x > a}} g(x) = 0$ or

(iii)' $\lim_{\substack{x \to a \\ x > a}} g(x) = \infty$,

then there exists $\lim_{x \to a} \dfrac{f(x)}{g(x)} = L$.

Theorem 1.3.14. *Let $I \subset \mathbb{R}$ be an open interval, $f : I \to \mathbb{R}$ be a n–times derivable function at $a \in I$, $(n \in \mathbb{N}, n \geq 2)$, such that*

$$f'(a) = 0, \ f''(a) = 0, \ \ldots, f^{(n-1)}(a) = 0, \ f^{(n)}(a) \neq 0.$$

(i) If n is even, then a is an extremum point, more precisely: a local maximum if $f^{(n)}(a) < 0$, and a local minimum if $f^{(n)}(a) > 0$.

(ii) If n is odd, then a is not an extremum point.

1.4 The Riemann Integral

At the end of this chapter we discuss the main aspects concerning the Riemann integral. Let $a, b \in \mathbb{R}$, $a < b$.

Definition 1.4.1. *(i) A partition of the interval $[a, b]$ is a finite set of real numbers $x_0, x_1, ..., x_n$ ($n \in \mathbb{N}^*$), denoted by Δ, such that*

$$a = x_0 < x_1 < ... < x_{n-1} < x_n = b.$$

(ii) The norm of the partition Δ is the number

$$\|\Delta\| := \max\{x_i - x_{i-1} \mid i \in \overline{1, n}\}.$$

(iii) A tagged partition of the interval $[a, b]$ is a partition Δ, together with a finite set of real numbers $\Xi := \{\xi_i \mid i \in \overline{1, n}\}$, such that $\xi_i \in [x_{i-1}, x_i]$ for any $i \in \overline{1, n}$. The set Ξ is called the intermediate points system associated to Δ.

(iv) Let $f : [a, b] \to \mathbb{R}$ be a function. The Riemann sum associated to a tagged partition of the interval $[a, b]$ is

$$S(f, \Delta, \Xi) := \sum_{i=1}^{n} f(\xi_i)(x_i - x_{i-1}).$$

Definition 1.4.2. *Let $f : [a, b] \to \mathbb{R}$ be a function. One says that f is Riemann integrable on $[a, b]$ if there exists $I \in \mathbb{R}$ such that for every $\varepsilon > 0$, there exists $\delta > 0$ such that for any partition Δ of the interval $[a, b]$ with the property $\|\Delta\| < \delta$, and for any intermediate points system Ξ associated to Δ, the next inequality holds:*

$$\left| S(f, \Delta, \Xi) - I \right| < \varepsilon.$$

The real number I from the previous definition, which is unique, is called the Riemann integral of f an $[a, b]$ and is denoted by

$$\int_a^b f(x)dx.$$

Theorem 1.4.3. *Any function which is Riemann integrable on $[a, b]$ is bounded on $[a, b]$.*

Definition 1.4.4. *Let $f : [a, b] \to \mathbb{R}$ be a function. One says that a function $F : [a, b] \to \mathbb{R}$ is an antiderivative (or, equivalently, a primitive integral, or an indefinite integral) of f on $[a, b]$ if F is derivable on $[a, b]$ and $F'(x) = f(x)$ for any $x \in [a, b]$.*

If an antiderivative exists for a given function, then infinitely many antiderivatives exist for that function and the difference of any two such antiderivatives is a constant.

The next result is sometimes called the fundamental theorem of calculus.

Theorem 1.4.5 (Leibniz-Newton). *If $f : [a, b] \to \mathbb{R}$ is Riemann integrable on the interval $[a, b]$ and it admits an antiderivative F on $[a, b]$, then*

$$\int_a^b f(x)dx = F(b) - F(a).$$

Continuous functions satisfy both hypotheses of the preceding theorem.

Theorem 1.4.6. *If $f : [a, b] \to \mathbb{R}$ is continuous on $[a, b]$, then f is Riemann integrable on $[a, b]$ and it admits antiderivatives on $[a, b]$.*

Theorem 1.4.7. *If $f : [a, b] \to \mathbb{R}$ is bounded and has a finite set of discontinuity points, then f is Riemann integrable on $[a, b]$. Every function which is monotone on $[a, b]$ is Riemann integrable on $[a, b]$.*

We present now the main properties of the Riemann integral.

Theorem 1.4.8. *(i) If $f, g : [a, b] \to \mathbb{R}$ are Riemann integrable on $[a, b]$, and $\alpha, \beta \in \mathbb{R}$, then $\alpha f + \beta g$ is Riemann integrable on $[a, b]$ and*

$$\int_a^b (\alpha f(x) + \beta g(x))dx = \alpha \int_a^b f(x)dx + \beta \int_a^b g(x)dx.$$

(ii) If $f : [a, b] \to \mathbb{R}$ is Riemann integrable on $[a, b]$, and $m \le f(x) \le M$ for every $x \in [a, b]$ $(m, M \in \mathbb{R})$, then

$$m(b - a) \le \int_a^b f(x)dx \le M(b - a).$$

In particular, if $f(x) \ge 0$ for every $x \in [a, b]$, then

$$\int_a^b f(x)dx \ge 0,$$

and if $f, g : [a, b] \to \mathbb{R}$ are Riemann integrable and $f(x) \le g(x)$ for every $x \in [a, b]$, then

$$\int_a^b f(x)dx \le \int_a^b g(x)dx.$$

(iii) If $f : [a, b] \to \mathbb{R}$ is Riemann integrable on $[a, b]$, then $|f|$ is Riemann integrable on $[a, b]$.

(iv) If $f, g : [a, b] \to \mathbb{R}$ are Riemann integrable $[a, b]$, then $f \cdot g$ is Riemann integrable on $[a, b]$.

Theorem 1.4.9. *(i) If $f : [a, b] \to \mathbb{R}$ is Riemann integrable on $[a, b]$, then f is Riemann integrable on every subinterval of $[a, b]$.*

(ii) If $c \in (a, b)$ and f is Riemann integrable on $[a, c]$ and on $[c, b]$, then f is Riemann integrable on $[a, b]$ and

$$\int_a^b f(x)dx = \int_a^c f(x)dx + \int_c^b f(x)dx.$$

Theorem 1.4.10. *Let $f : [a, b] \to \mathbb{R}$ be a function, and $f^* : [a, b] \to \mathbb{R}$ be another function which coincides with f on $[a, b]$, except on a finite set of points. If f^* is Riemann integrable on $[a, b]$, then f is Riemann integrable on $[a, b]$ and*

$$\int_a^b f(x)dx = \int_a^b f^*(x)dx.$$

Theorem 1.4.11 (integration by parts). *If $f, g : [a, b] \to \mathbb{R}$ are C^1 functions, then*

$$\int_a^b f(x)g'(x)\, dx = f(x)g(x)\big|_a^b - \int_a^b f'(x)g(x)\, dx.$$

Theorem 1.4.12 (change of variable). *Let $\varphi : [a, b] \to [c, d]$ be a C^1 function, and let $f : [c, d] \to \mathbb{R}$ be a continuous function. Then*

$$\int_a^b f(\varphi(t)) \cdot \varphi'(t)\, dt = \int_{\varphi(a)}^{\varphi(b)} f(x)\, dx. \tag{1.4.1}$$

We end this section by the next multidimensional variant of Taylor Theorem. In what follows, the equality is understood on components (i.e., for every function $f_i : \mathbb{R}^p \to \mathbb{R}$, $i = \overline{1, p}$, where $f = (f_1, ..., f_p)$).

Theorem 1.4.13. *Suppose $f : \mathbb{R}^p \to \mathbb{R}^p$ is continuously differentiable on some convex open set D and that $x, x + y \in D$. Then there is $t \in (0, 1)$ such that*

$$f(x + y) = f(x) + \int_0^1 \nabla f(x + ty)(y)\, dt.$$

2 Nonlinear Analysis Fundamentals

In this chapter we study convex sets and functions, convex cones, the Bouligand tangent cone to a set at a point, and semicontinuous functions. We prove fundamental results, which will be the main investigation tools in subsequent chapters. The Farkas Lemma and its consequences will be decisive for establishing optimality conditions in Karush-Kuhn-Tucker form, while the Banach fixed point Principle will be used for discussing some optimization algorithms. We also study some important inequalities related to the study of convex functions.

2.1 Convex Sets and Cones

Definition 2.1.1. *One says that a nonempty set $D \subset \mathbb{R}^p$ is convex if for every $x, y \in D$,*

$$[x, y] = \{\alpha x + (1 - \alpha)y \mid \alpha \in [0, 1]\} \subset D.$$

In other words, D is convex if and only if together with two points a_1, a_2, it contains the whole segment $[a_1, a_2]$. It is sufficient to take $\alpha \in (0, 1)$. By mathematical induction one can show that if D is convex, then for every $n \in \mathbb{N}^*$, $x_1, x_2, ..., x_n \in D$, $\alpha_1, \alpha_2, ..., \alpha_n \in [0, 1]$ with $\sum_{i=1}^n \alpha_i = 1$:

$$\sum_{i=1}^n \alpha_i x_i \in D.$$

A sum like the one above is called a convex combination of the elements (x_i). In $\mathbb{R}$, the convex sets are intervals.

One of the main objects for our study is defined next.

Definition 2.1.2. *A nonempty subset $K \subset \mathbb{R}^p$ is a cone if the next relation holds:*

$$\forall y \in K, \ \forall \lambda \in \mathbb{R}_+ := [0, \infty) : \lambda y \in K.$$

According to the definition, every cone contains the origin.

Proposition 2.1.3. *A cone C is convex if and only if $C + C = C$.*

Proof Suppose first that C is a convex cone. As $0 \in C$, it is clear that $C \subset C + C$. Let $u \in C + C$. Then there exist $c_1, c_2 \in C$ such that $c_1 + c_2 = u$. But, using the properties of C and the obvious relation

$$u = 2 \left(2^{-1}c_1 + 2^{-1}c_2 \right),$$

one can deduce that $u \in C$. For the converse implication, suppose that C is a cone which satisfies $C + C = C$. Fix $\alpha \in (0, 1)$ and $c_1, c_2 \in C$. Then, from the cone property of C, one knows that $\alpha c_1, (1 - \alpha)c_2 \in C$, therefore $\alpha c_1 + (1 - \alpha)c_2 \in C + C = C$, which ends the proof. $\qquad\square$

Definition 2.1.4. *Let $A \subset \mathbb{R}^p$ be a nonempty set and $x \in \mathbb{R}^p$. One defines the distance from x to A by the relation:*

$$d(x, A) := \inf\{\|x - a\| \mid a \in A\}.$$

We also consider the function $d_A : \mathbb{R}^p \to \mathbb{R}$ given by

$$d_A(x) := d(x, A).$$

We now introduce some basic properties of the distance from a point to a (nonempty) set.

Theorem 2.1.5. *Let $A \subset \mathbb{R}^p$, $A \neq \emptyset$. Then:*
(i) $d(x, A) = 0$ if and only if $x \in \mathrm{cl}\, A$.
(ii) The function d_A is 1–Lipschitz.
(iii) If A is closed, then for every $x \in \mathbb{R}^p$, there exists $a_x \in A$ such that $d(x, A) = \|x - a_x\|$. If, moreover, A is convex, then a_x having the previous property is unique and it is characterized by the relations

$$\begin{cases} a_x \in A \\ \langle x - a_x, u - a_x \rangle \leq 0, \ \forall u \in A. \end{cases}$$

Proof (i) The following equivalences hold

$$d(x, A) = 0 \Leftrightarrow \inf_{a \in A} \|x - a\| = 0$$

$$\Leftrightarrow \exists (a_n) \subset A \text{ with } \lim_{n \to \infty} \|x - a_n\| = 0 \Leftrightarrow x \in \overline{A}.$$

(ii) For every $x, y \in \mathbb{R}^p$ and every $a \in A$, these relations hold:

$$d(x, A) \leq \|x - a\| \leq \|x - y\| + \|y - a\|.$$

As a is taken arbitrary from A, one deduces

$$d(x, A) \leq \|x - y\| + d(y, A),$$

i.e.,

$$d(x, A) - d(y, A) \leq \|x - y\|.$$

By reversing the roles of x and y, one has:

$$|d(x, A) - d(y, A)| \leq \|x - y\|,$$

which is the desired conclusion.

(iii) If $x \in A$, then $a_x := x$ is the unique element having the previously introduced property. Let $x \notin A$. As $d(x, A)$ is a real number, there exists $r > 0$ such that $A_1 := A \cap D(x, r) \neq \emptyset$. Since A_1 is a compact set and the function $g : A_1 \to \mathbb{R}$, $g(y) = d(x, y)$ is continuous, according to Weierstrass Theorem, g attains its minimum on A_1, i.e., there exists $a_x \in A_1$ with $g(a_x) = \inf_{y \in A_1} g(y) = d(x, A_1)$. Now, one can check that $d(x, A_1) = d(x, A)$ and the first conclusion follows. Suppose that, moreover, A is convex. If $x \in A$, there is nothing to prove. Take $x \notin A$. Consider $a_1, a_2 \in A$ with $d(x, A) = \|x - a_1\| = \|x - a_2\|$. Using the parallelogram law we know:

$$\left\|(x - a_1) + (x - a_2)\right\|^2 + \left\|(x - a_1) - (x - a_2)\right\|^2 = 2\|x - a_1\|^2 + 2\|x - a_2\|^2,$$

i.e.,

$$\|2x - a_1 - a_2\|^2 + \|a_2 - a_1\|^2 = 4d^2(x, A),$$

and dividing by 4 one gets

$$\left\|x - \frac{a_1 + a_2}{2}\right\|^2 + 4^{-1}\|a_2 - a_1\|^2 = d^2(x, A).$$

Since A is convex, $2^{-1}(a_1 + a_2) \in A$, hence $\left\|x - \dfrac{a_1 + a_2}{2}\right\|^2 \geq d^2(x, A)$. This relation and the previous equality show that $\|a_2 - a_1\| = 0$, hence $a_1 = a_2$. The proof of uniqueness is now complete. Let us prove now that a_x verifies the relation $\langle x - a_x, u - a_x \rangle \leq 0$ for any $u \in A$. For this, take $u \in A$. Then for every $\alpha \in (0, 1]$, one has

$$v = \alpha u + (1 - \alpha)a_x \in A.$$

Hence,

$$\|x - a_x\| \leq \|x - \alpha u - (1 - \alpha)a_x\| = \|x - a_x - \alpha(u - a_x)\|,$$

and, consequently,

$$\|x - a_x\|^2 \leq \|x - a_x\|^2 - 2\alpha \langle x - a_x, u - a_x \rangle + \alpha^2 \|u - a_x\|^2.$$

After reducing terms and dividing by $\alpha > 0$, we can see that

$$0 \leq -2 \langle x - a_x, u - a_x \rangle + \alpha \|u - a_x\|^2.$$

If we let $\alpha \to 0$, the desired inequality follows. For the converse, if an element $a \in A$ satisfies $\langle x - a, u - a \rangle \leq 0$ for any $u \in A$, then for every $v \in A$ one has

$$\|x - a\|^2 - \|x - v\|^2 = 2 \langle x - a, v - a \rangle - \|a - v\|^2 \leq 0,$$

hence a coincides with a_x. The proof is now complete. $\qquad\square$

In the case when A is closed, then for $x \in \mathbb{R}^p$ one denotes the projection set of x on A by

$$\mathrm{pr}_A x := \{a \in A \mid d(x, A) = \|x - a\|\}.$$

If, moreover, A is convex, then, according to the above theorem, this set consists of only one element, which we still denote by $\mathrm{pr}_A\, x$, and we call this the projection of x on A.

Let $S \subset \mathbb{R}^p$ be a nonempty set. The polar of S is the set

$$S^- := \{u \in \mathbb{R}^p \mid \langle u, x \rangle \le 0, \ \forall x \in S\}.$$

It is easy to observe that S^- is a closed convex cone and that, in general, $S \subset (S^-)^-$. If we consider the reverse inclusion, the next result follows.

Theorem 2.1.6. *Let $C \subset \mathbb{R}^p$ be a closed convex cone. Then $C = (C^-)^-$.*

Proof Consider $z \in (C^-)^-$ and $\bar{z} = \mathrm{pr}_C\, z$. We will prove that $z = \bar{z}$. From the last part of Theorem 2.1.5, for any $c \in C$, one has

$$\langle z - \bar{z}, c - \bar{z} \rangle \le 0.$$

As $0 \in C$ and $2\bar{z} \in C$, we deduce

$$- \langle \bar{z}, z - \bar{z} \rangle \le 0, \quad \langle \bar{z}, z - \bar{z} \rangle \le 0$$

hence

$$\langle z - \bar{z}, c \rangle \le 0$$

for every $c \in C$, i.e., $z - \bar{z} \in C^-$. As $z \in (C^-)^-$, one gets

$$\langle z, z - \bar{z} \rangle \le 0.$$

But

$$\|z - \bar{z}\|^2 = \langle z - \bar{z}, z \rangle - \langle z - \bar{z}, \bar{z} \rangle \le 0$$

which means that $z = \bar{z}$, so $z \in C$. This establishes the theorem. $\qquad\square$

Example 2.1.7. *1. Consider $S = \{(x, 0) \in \mathbb{R}^2 \mid x \ge 0\}$. One can observe that $S^- = \{(x, y) \in \mathbb{R}^2 \mid x \le 0\}$. Obviously, $(S^-)^- = S^-$.*

2. The polar of $\mathbb{R}_+^2 := \{(x, y) \in \mathbb{R}^2 \mid x, y \ge 0\}$ is $\mathbb{R}_-^2 := \{(x, y) \in \mathbb{R}^2 \mid x, y \le 0\}$. The polar of the set $S = \{(x, 0) \in \mathbb{R}^2 \mid x \ge 0\} \cup \{(0, y) \in \mathbb{R}^2 \mid y \ge 0\}$ is also $\mathbb{R}_-^2$. From this example one can see that, in general, $S_1^- = S_2^-$ does not imply $S_1 = S_2$.

The next result, which has an algebraic character, it was obtained by the Hungarian mathematician Julius Farkas in 1902.

Theorem 2.1.8 (Farkas' Lemma). *Let $n \in \mathbb{N}^*$, $(\varphi_i)_{i \in \overline{1,n}} \subset L(\mathbb{R}^p, \mathbb{R})$ and $\varphi \in L(\mathbb{R}^p, \mathbb{R})$. Then*

$$\forall x \in \mathbb{R}^p : [\varphi_1(x) \le 0, \dots, \varphi_n(x) \le 0] \Rightarrow \varphi(x) \le 0 \qquad (2.1.1)$$

if and only if there exists $(\alpha_i)_{i \in \overline{1,n}} \subset [0, \infty)$ such that $\varphi = \sum_{i=1}^{n} \alpha_i \varphi_i$.

Proof The converse implication is obvious. We prove the other one by induction for $n \geq 1$. Define the proposition $P(n)$ which says that for every $\varphi, \varphi_1, \ldots, \varphi_n \in L(\mathbb{R}^p, \mathbb{R})$ satisfying (2.1.1), there exist $(\alpha_i)_{i \in \overline{1,n}} \subset [0, \infty)$ such that $\varphi = \sum_{i=1}^n \alpha_i \varphi_i$.

Let us prove that $P(1)$ is true. Indeed, let $\varphi, \varphi_1 \in L(\mathbb{R}^p, \mathbb{R})$ such that

$$\varphi_1(x) \leq 0 \Rightarrow \varphi(x) \leq 0.$$

If $\varphi = 0$ then, obviously, $\varphi = 0\varphi_1$. Suppose $\varphi \neq 0$. Then, by the assumption:

$$\varphi_1(x) = 0 \Leftrightarrow [\varphi_1(x) \leq 0, \varphi_1(-x) \leq 0] \Rightarrow [\varphi(x) \leq 0, \varphi(-x) \leq 0] \Leftrightarrow \varphi(x) = 0,$$

hence $\operatorname{Ker} \varphi_1 \subset \operatorname{Ker} \varphi$. Since $\varphi \neq 0$, one has $\varphi_1 \neq 0$, so there exists $x_1 \in \mathbb{R}^p$ with $\varphi_1(\overline{x}) = -1$. Also by the assumption, $\varphi(\overline{x}) \leq 0$. Take $x \in \mathbb{R}^p$ arbitrarily. Then it is easy to verify that

$$x + \varphi_1(x)\overline{x} \in \operatorname{Ker} \varphi_1,$$

hence

$$x + \varphi_1(x)\overline{x} \in \operatorname{Ker} \varphi,$$

i.e.,

$$\varphi\left(x + \varphi_1(x)\overline{x}\right) = 0,$$

which proves that

$$\varphi(x) = -\varphi(\overline{x})\varphi_1(x).$$

Notice that x was arbitrarily chosen, so the desired relation is proved for $\alpha_1 := -\varphi(\overline{x}) \geq 0$.

Suppose now that $P(n)$ is true for a fixed $n \geq 1$ and we will try to prove that $P(n+1)$ is true.

Take $\varphi, \varphi_1, \ldots, \varphi_n, \varphi_{n+1} \in L(\mathbb{R}^p, \mathbb{R})$ such that

$$\forall x \in \mathbb{R}^p : [\varphi_1(x) \leq 0, \ldots, \varphi_n(x) \leq 0, \varphi_{n+1}(x) \leq 0] \Rightarrow \varphi(x) \leq 0. \tag{2.1.2}$$

If

$$\forall x \in \mathbb{R}^p : [\varphi_1(x) \leq 0, \ldots, \varphi_n(x) \leq 0] \Rightarrow \varphi(x) \leq 0, \tag{2.1.3}$$

then, from $P(n)$, there exist $(\alpha_i)_{i \in \overline{1,n}} \subset [0, \infty)$ such that $\varphi = \sum_{i=1}^n \alpha_i \varphi_i$; take $\alpha_{n+1} := 0$ and the conclusion follows.

Suppose relation (2.1.3) is not satisfied. Then there exists $\overline{x} \in \mathbb{R}^p$ such that $\varphi(\overline{x}) > 0$ and $\varphi_i(\overline{x}) \leq 0$ for any $i \in \overline{1, n}$. As (2.1.2) holds, $\varphi_{n+1}(\overline{x}) > 0$; we may suppose (by multiplying by the appropriate positive scalar) that $\varphi_{n+1}(\overline{x}) = 1$. But

$$\varphi_{n+1}(x - \varphi_{n+1}(x)\overline{x}) = 0, \ \forall x \in \mathbb{R}^p,$$

and from (2.1.2) we deduce

$$\forall x \in \mathbb{R}^p : [\varphi_1(x - \varphi_{n+1}(x)\overline{x}) \leq 0, \ldots, \varphi_n(x - \varphi_{n+1}(x)\overline{x}) \leq 0] \Rightarrow \varphi(x - \varphi_{n+1}(x)\overline{x}) \leq 0. \tag{2.1.4}$$

Take $\varphi'_i := \varphi_i - \varphi_i(\overline{x})\varphi_{n+1}$ for $i \in \overline{1, n}$ and $\varphi' := \varphi - \varphi(\overline{x})\varphi_{n+1}$, and then relation (2.1.4) becomes

$$\forall x \in X : [\varphi'_1(x) \le 0, \ldots, \varphi'_n(x) \le 0] \Rightarrow \varphi'(x) \le 0.$$

As $P(n)$ is true, there exist $(\alpha_i)_{i \in \overline{1,n}} \subset [0, \infty)$ such that $\varphi' = \sum_{i=1}^n \alpha_i \varphi'_i$. We deduce that

$$\varphi - \varphi(\overline{x})\varphi_{n+1} = \sum_{i=1}^n \alpha_i \left[\varphi_i - \varphi_i(\overline{x})\varphi_{n+1}\right],$$

hence $\varphi = \sum_{i=1}^{n+1} \alpha_i \varphi_i$, where $\alpha_{n+1} = \varphi(\overline{x}) - \sum_{i=1}^n \alpha_i \varphi_i(\overline{x}) \ge 0$ from the choice of $\overline{x}$ and from the fact that $\alpha_i \ge 0$ for any $i \in \overline{1, n}$). The proof is now complete. □

Throughout this book we shall use several different concepts of tangent vectors to a set at a point. We introduce now one of these concepts.

Definition 2.1.9. *Let $M \subset \mathbb{R}^p$ be a nonempty set and $\overline{x} \in \mathrm{cl}\, M$. One says that a vector $u \in \mathbb{R}^p$ is tangent in the sense of Bouligand to the set M at $\overline{x}$ if there exist $(t_n) \subset (0, \infty)$, $t_n \to 0$ and $(u_n) \to u$ such that for any $n \in \mathbb{N}$, one has*

$$\overline{x} + t_n u_n \in M.$$

It is sufficient that the above inclusion holds for every $n \in \mathbb{N}$ sufficiently large.

Theorem 2.1.10. *The set, denoted by $T_B(M, \overline{x})$, which contains all the tangent vectors to the set M at $\overline{x}$ is a closed cone, which we call the Bouligand tangent cone (or the contingent cone) to the set M at the point $\overline{x}$.*

Proof Let us prove first that $0 \in T_B(M, \overline{x})$. If $\overline{x} \in M$, then the assertion trivially follows, because it is sufficient to take (u_n) constantly equal to 0. If $\overline{x} \notin M$, then there exists $(x_n)_{n \in \mathbb{N}} \subset M$ such that $x_n \to \overline{x}$, and we consider $t_n := \sqrt{\|x_n - \overline{x}\|}$ and $u_n := \left(\sqrt{\|x_n - \overline{x}\|}\right)^{-1} (x_n - \overline{x})$ for every $n \in \mathbb{N}$. Since $t_n \to 0$ and $u_n \to 0$, one obtains the conclusion.

Consider now $u \in T_B(M, \overline{x})$ and $\lambda > 0$. According to the definition, there exist $(t_n) \subset (0, \infty)$, $t_n \to 0$ and $(u_n) \to u$ such that for every $n \in \mathbb{N}$,

$$\overline{x} + t_n u_n \in M.$$

This is equivalent to

$$\overline{x} + \frac{t_n}{\lambda}(\lambda u_n) \in M.$$

As $\left(\dfrac{t_n}{\lambda}\right) \to 0$ and $(\lambda u_n) \to \lambda u$, one deduces that $\lambda u \in T_B(M, \overline{x})$, hence $T_B(M, \overline{x})$ is a cone. We prove that the closure of $T_B(M, \overline{x})$ is contained in $T_B(M, \overline{x})$. Take $(u_n) \subset$

$T_B(M, \overline{x})$ and $(u_n) \to u$. One must prove that $u \in T_B(M, \overline{x})$. For every $n \in \mathbb{N}$, there exist $(t_n^k)_k \subset (0, \infty)$, $t_n^k \overset{k \to \infty}{\to} 0$ and $(u_n^k) \overset{k \to \infty}{\to} u_n$ such that for every $k \in \mathbb{N}$,

$$\overline{x} + t_n^k u_n^k \in M.$$

By using a diagonalization procedure, for every $n \in \mathbb{N}^*$, there exists $k_n \in \mathbb{N}$ such that the next relations hold:

$$t_n^{k_n} < \frac{1}{n}$$

$$\left\| u_n^{k_n} - u_n \right\| \le \frac{1}{n}.$$

It is easy to observe that the positive sequence $(t_n^{k_n})_n$ converges to 0, and using the inequality

$$\left\| u_n^{k_n} - u \right\| \le \left\| u_n^{k_n} - u_n \right\| + \left\| u_n - u \right\|,$$

one can deduce that $(u_n^{k_n}) \to u$. Moreover, for every $n \in \mathbb{N}$,

$$\overline{x} + t_n^{k_n} u_n^{k_n} \in M,$$

hence $u \in T_B(M, \overline{x})$ and the proof is complete. $\qquad\square$

Our first example is given next.

Example 2.1.11. *1. Consider the ball $M \subset \mathbb{R}^2$, $M := \{(x, y) \in \mathbb{R}^2 \mid (x - 1)^2 + y^2 \le 1\}$. Then $T_B(M, (0, 0)) = \{(x, y) \in \mathbb{R}^2 \mid x \ge 0\}$.*
2. One can easily observe that if $C \subset \mathbb{R}^p$ is a closed cone, then $T_B(C, 0) = C$.

Proposition 2.1.12. *If $\emptyset \ne M \subset \mathbb{R}^p$ and $\overline{x} \in \mathrm{cl}\, M$, then $T_B(M, \overline{x}) = T_B(\mathrm{cl}\, M, \overline{x})$. If $\overline{x} \in \mathrm{int}\, M$, then $T_B(M, \overline{x}) = \mathbb{R}^p$.*

Proof The inclusion $T_B(M, \overline{x}) \subset T_B(\mathrm{cl}\, M, \overline{x})$ is obvious by the use of $M \subset \mathrm{cl}\, M$. Take $u \in T_B(\mathrm{cl}\, M, \overline{x})$. There exist $(t_n) \subset (0, \infty)$, $t_n \to 0$ and $(u_n) \to u$ such that for every $n \in \mathbb{N}$,

$$\overline{x} + t_n u_n \in \mathrm{cl}\, M.$$

Using the sequence characterization of the closure, for any fixed n, there exist $(v_n^k)_k \subset M$ such that

$$v_n^k \overset{k}{\to} \overline{x} + t_n u_n.$$

As above, for any fixed n, there exists $k_n \in \mathbb{N}$ such that

$$\left\| v_n^{k_n} - (\overline{x} + t_n u_n) \right\| \le t_n^2.$$

Then, one can write

$$\left\| \frac{v_n^{k_n} - \overline{x}}{t_n} - u \right\| \le \left\| \frac{v_n^{k_n} - \overline{x}}{t_n} - u_n \right\| + \left\| u_n - u \right\| \le t_n + \left\| u_n - u \right\|,$$

hence $u'_n := \dfrac{v_n^{k_n} - \overline{x}}{t_n} \overset{n \to \infty}{\to} u$. But

$$\overline{x} + t_n u'_n = v_n^{k_n} \in M,$$

hence $u \in T_B(M, \overline{x})$. The second part of the conclusion easily follows: if $\overline{x} \in \operatorname{int} M$, then for every $u \in \mathbb{R}^p$ and every $(t_n) \subset (0, \infty)$, $t_n \to 0$, one has $\overline{x} + t_n u_n \in M$ for any n sufficiently large. This shows, in particular, that $u \in T_B(M, \overline{x})$, and the conclusion follows. □

In general, the Bouligand tangent cone is not convex and the relation $T_B(M, \overline{x}) = \mathbb{R}^p$ can be satisfied, even if $\overline{x} \notin \operatorname{int} M$.

Example 2.1.13. *1. Consider the set $M \subset \mathbb{R}^2$, $M = \{(x, y) \mid x \geq 0, y = 0\} \cup \{(x, y) \mid x = 0, y \geq 0\}$. Then $T_B(M, (0, 0)) = M$ is not a convex set.*

2. Let set M represent the plane domain bounded by the curve (the cardioid) which has the parametric representation

$$\begin{cases} x = -2\cos t + \cos 2t + 1 \\ y = 2\sin t - \sin 2t \end{cases}, \quad t \in [0, 2\pi].$$

Then $T_B(M, (0, 0)) = \mathbb{R}^2$, but $(0, 0) \notin \operatorname{int} M$.

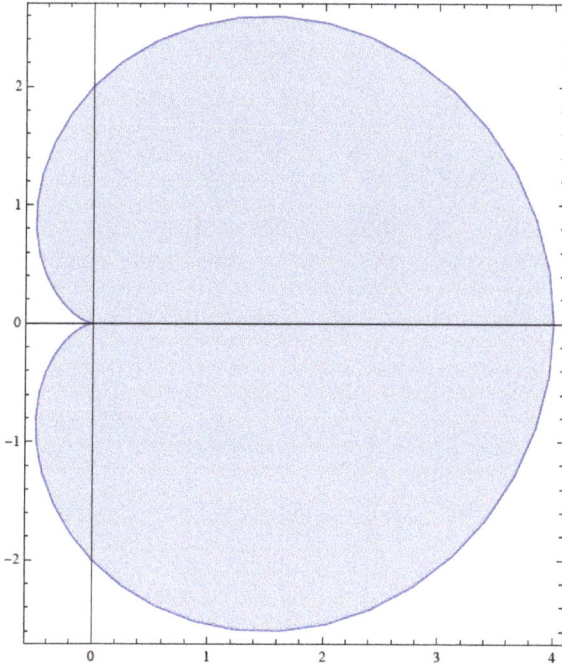

Figure 2.1: The cardioid.

Proposition 2.1.14. *Let $A_1, A_2 \subset \mathbb{R}^p$ be closed sets. Then the next relations hold:*
  (i) if $\overline{x} \in A_1 \cap A_2$, then $T_B(A_1 \cup A_2, \overline{x}) = T_B(A_1, \overline{x}) \cup T_B(A_2, \overline{x})$;
  (ii) if $\overline{x} \in A_1 \cap A_2$, then $T_B(A_1 \cap A_2, \overline{x}) \subset T_B(A_1, \overline{x}) \cap T_B(A_2, \overline{x})$;
  (iii) if $\overline{x} \in \mathrm{bd}\, A_1$, then $T_B(\mathrm{bd}\, A_1, \overline{x}) = T_B(A_1, \overline{x}) \cap T_B(\mathbb{R}^p \setminus A_1, \overline{x})$.

Proof The first two relations easily follow, as well as the inclusion $T_B(\mathrm{bd}\, A_1, \overline{x}) \subset T_B(A_1, \overline{x}) \cap T_B(\mathbb{R}^p \setminus A_1, \overline{x})$ from (*iii*), which can be proved by using (*ii*) and Proposition 2.1.12. Let us prove now the other inclusion from (*iii*). Take $u \in T_B(A_1, \overline{x}) \cap T_B(\mathbb{R}^p \setminus A_1, \overline{x})$. According to the definition, there exist $(t_n), (t'_n) \subset (0, \infty), t_n, t'_n \to 0$ and $(u_n), (u'_n) \to u$ such that for every $n \in \mathbb{N}$,

$$\overline{x} + t_n u_n \in A_1,$$

and

$$\overline{x} + t'_n u'_n \in \mathbb{R}^p \setminus A_1.$$

If an infinite number of terms from the first or the second relation are on the boundary of A_1, there is nothing to prove. Suppose next, without loss of generality, that for every $n \in \mathbb{N}$, there exists $\lambda_n \in (0, 1)$ such that

$$\lambda_n (\overline{x} + t_n u_n) + (1 - \lambda_n)(\overline{x} + t'_n u'_n) \in \mathrm{bd}\, A_1.$$

Consider the sequences

$$(t''_n) := (\lambda_n t_n + (1 - \lambda_n) t'_n) \subset (0, \infty)$$
$$(u''_n) := \frac{t_n \lambda_n}{t''_n} u_n + \frac{t'_n (1 - \lambda_n)}{t''_n} u'_n.$$

It is clear that $(t''_n) \to 0$. On the other hand,

$$\left\| u''_n - u \right\| \le \left\| u_n - u \right\| + \left\| u'_n - u \right\|,$$

hence $(u''_n) \to u$. Since

$$\overline{x} + t''_n u''_n \in \mathrm{bd}\, A_1,$$

one gets the desired conclusion. □

Denote by $N_B(M, \overline{x})$ the polar of $T_B(M, \overline{x})$ (i.e., $N_B(M, \overline{x}) := T_B(M, \overline{x})^-$), and we call this set the Bouligand normal cone to M at $\overline{x}$.

If the set M is convex, then the Bouligand tangent and normal cones have a special form.

Proposition 2.1.15. *Let $\emptyset \ne M \subset \mathbb{R}^p$ be a convex set and $\overline{x} \in M$. Then*

$$T_B(M, \overline{x}) = \mathrm{cl}\, \mathbb{R}_+(M - \overline{x}),$$

and

$$N_B(M, \overline{x}) = \{u \in \mathbb{R}^p \mid \langle u, c - \overline{x} \rangle \le 0, \forall c \in M\}.$$

Proof Take $c \in M$ and $d := c - \bar{x}$. Consider $(t_k)_k \to 0$. Then

$$\bar{x} + t_k d = (1 - t_k)\bar{x} + t_k c \in M,$$

hence $M - \bar{x} \subset T_B(M, \bar{x})$. Since $T_B(M, \bar{x})$ is a closed cone, one gets that $\mathrm{cl}\, \mathbb{R}_+(M - \bar{x}) \subset T_B(M, \bar{x})$. Take $u \in T_B(M, \bar{x})$. Then there exist $(t_k) \subset (0, \infty)$, $t_k \to 0$ and $(u_k) \to u$ such that for every $k \in \mathbb{N}$,

$$x_{k:} = \bar{x} + t_k u_k \in M.$$

Hence $u = \lim_k \dfrac{x_k - \bar{x}}{t_k}$. But $\left(\dfrac{x_k - \bar{x}}{t_k} \right)_k \subset \mathbb{R}_+(M - \bar{x})$. One can deduce that $T_B(M, \bar{x}) \subset \mathrm{cl}\, \mathbb{R}_+(M - \bar{x})$. Recall that, by definition,

$$N_B(M, \bar{x}) = T_B(M, \bar{x})^- = \{u \in \mathbb{R}^p \mid \langle u, v \rangle \le 0, \ \forall v \in T_B(M, \bar{x})\}.$$

Now, taking into account the particular form of $T_B(M, \bar{x})$, the conclusion follows. $\square$

For a reason we make clear later on, in the case of convex sets, we do not use the subscript B in the notation of these cones.

Example 2.1.16. *We want to compute the tangent and normal cones, at different points, to the set $M \subset \mathbb{R}^p$,*

$$M = \left\{ x = (x_1, x_2, \ldots, x_p) \in \mathbb{R}^p \mid x_i \ge 0, \ \forall i \in \overline{1, p}, \ \sum_{i=1}^{p} x_i = 1 \right\},$$

which is called the unit simplex. This set is convex and closed. According to the previous result, for every $\bar{x} \in M$,

$$T(M, \bar{x}) = \mathrm{cl}\, \mathbb{R}_+(M - \bar{x})$$
$$= \mathrm{cl}\, \left\{ u \in \mathbb{R}^p \mid \exists \alpha \ge 0, \ x \in M, \ u = \alpha(x - \bar{x}) \right\}.$$

Take u from the right-hand set. It is clear that, on one hand, $\sum_{i=1}^{p} u_i = 0$, and, on the other hand, if $\bar{x}_i = 0$, then $u_i \ge 0$. Denote by $I(\bar{x}) := \left\{ i \in \overline{1, p} \mid \bar{x}_i = 0 \right\}$. It follows that

$$T(M, \bar{x}) \subset \left\{ u \in \mathbb{R}^p \mid \sum_{i=1}^{p} u_i = 0 \text{ and } u_i \ge 0, \ \forall i \in I(\bar{x}) \right\}.$$

Let us now prove the reverse inclusion. It is easy to verify that the right-hand set is closed. Take u from this set. If $u = 0$, then, obviously, $u \in T(M, \bar{x})$. If $u \ne 0$, then we must prove that there exists $\alpha > 0$ such that $\bar{x} + \alpha u \in M$. On one hand, it is clear that $\sum_{i=1}^{p} (\bar{x}_i + \alpha u_i) = 1$ is satisfied for any α. If there is no i with $u_i < 0$, then it is also easy to observe that $\bar{x}_i + \alpha u_i \ge 0$, for any $i \in \overline{1, p}$, hence $u \in T(M, \bar{x})$. Suppose that the set J of indices for which $u_j < 0$ is nonempty. Then $J \subset \overline{1, p} \setminus I(\bar{x})$, hence $\bar{x}_j > 0$ for any $j \in J$. One can choose the positive α such that

$$\alpha < \min\{-u_j^{-1}\bar{x}_j \mid j \in J\},$$

and again one has $\bar{x}_i + \alpha u_i \geq 0$, for any $i \in \overline{1,p}$. Hence, $u \in T(M, \bar{x})$, and the double inclusion follows.

We prove next that

$$N(M, \bar{x}) = \left\{(a, a, \ldots, a) \in \mathbb{R}^p \mid a \in \mathbb{R}\right\} + \left\{v \in \mathbb{R}^p \mid v_i \leq 0, \ \forall i \in I(\bar{x}), \ v_i = 0, \ i \notin I(\bar{x})\right\}.$$

For this, consider the elements

$$a_0 = (1, 1, \ldots, 1), \ a_1 = -(1, 0, \ldots, 0), \ \ldots, \ a_n = -(0, 0, \ldots, 1)$$

and observe that $T(M, \bar{x})$ can be equivalently written as:

$$T(M, \bar{x}) = \left\{u \in \mathbb{R}^p \mid \langle a_0, u \rangle \leq 0, \ \langle -a_0, u \rangle \leq 0, \ \langle a_i, u \rangle \leq 0, \ \forall i \in I(\bar{x})\right\}.$$

The polar of this set is

$$N(M, \bar{x}) = \left\{\alpha a_0 - \beta a_0 + \sum_{i \in I(\bar{x})} \alpha_i a_i \mid \alpha, \beta, \alpha_i \geq 0, \ \forall i \in I(\bar{x})\right\}.$$

Indeed, the fact that the right-hand set is contained in the normal cone is obvious, and the reverse inclusion follows from Farkas' Lemma (Theorem 2.1.8). We now obtain the desired form of the normal cone.

At the end of this section, we discuss the concepts of convex hull and conic hull of a set. Let $A \subset \mathbb{R}^p$ be a nonempty set. The convex hull of A is the set

$$\text{conv } A = \left\{\sum_{i=1}^{n} \alpha_i x_i \mid n \in \mathbb{N}^*, \ (\alpha_i)_{i \in \overline{1,n}} \subset [0, \infty), \ \sum_{i=1}^{n} \alpha_i = 1, \ (x_i)_{i \in \overline{1,n}} \subset A\right\}.$$

It is not difficult to see that conv A is a convex set which contains A. One can easily verify that conv A is the smallest set (in the sense of inclusion) with these properties (see Problem 7.26).

The conic hull of the set A is

$$\text{cone } A := [0, \infty)A := \{\alpha x \mid \alpha \geq 0, \ x \in A\}.$$

In fact, cone A is the smallest cone which contains A.

We give next two results concerning these sets. The first one refers to the structure of the set conv A and it is called the Carathéodory Theorem, after the name of the Greek mathematician Constantin Carathéodory, who proved this result in 1911 for compact sets.

Theorem 2.1.17 (Carathéodory Theorem). *Let $A \subset \mathbb{R}^p$ be a nonempty set. Then*

$$\text{conv } A = \left\{\sum_{i=1}^{p+1} \alpha_i x_i \mid (\alpha_i)_{i \in \overline{1,p+1}} \subset [0, \infty), \ \sum_{i=1}^{p+1} \alpha_i = 1, \ (x_i)_{i \in \overline{1,p+1}} \subset A\right\}.$$

Proof We must prove that every element from conv A can be written as a combination of at most $p + 1$ elements from A. Consider $x \in \operatorname{conv} A$. According to the definition of conv A, x can be written as a convex combination of elements from A. Suppose, by means of contradiction, that the minimal number of elements from A which can form a convex combination equal to x is $n > p + 1$. So there exist $x_1, x_2, \ldots, x_n \in A$, $\alpha_1, \alpha_2, \ldots, \alpha_n \in (0, 1)$ with $\sum_{i=1}^{n} \alpha_i = 1$ such that $\sum_{i=1}^{n} \alpha_i x_i = x$. Then the elements $(x_i - x_n)_{i=\overline{1,n-1}}$ are linearly dependent (their number is greater than the dimension p of the space), so there exist $(\lambda_i)_{i=\overline{1,n-1}}$, not all equal to 0, such that

$$\sum_{i=1}^{n-1} \lambda_i(x_i - x_n) = 0,$$

which means

$$\sum_{i=1}^{n-1} \lambda_i x_i - \left(\sum_{i=1}^{n-1} \lambda_i\right) x_n = 0.$$

By denoting $-\left(\sum_{i=1}^{n-1} \lambda_i\right) = \lambda_n$, one has $\sum_{i=1}^{n} \lambda_i = 0$ and $\sum_{i=1}^{n} \lambda_i x_i = 0$. Then for every $t \in \mathbb{R}$,

$$x = \sum_{i=1}^{n} \alpha_i x_i + t \sum_{i=1}^{n} \lambda_i x_i = \sum_{i=1}^{n} (\alpha_i + t\lambda_i) x_i$$

and

$$\sum_{i=1}^{n} (\alpha_i + t\lambda_i) = 1.$$

As $\sum_{i=1}^{n} \lambda_i = 0$ and there is at least one nonzero element, there exists at least one negative value among the numbers $(\lambda_i)_{i \in \overline{1,n}}$. Denote $\bar{t} := \min\{-\alpha_i \lambda_i^{-1} \mid \lambda_i < 0\}$. Then all the values $(\alpha_i + \bar{t}\lambda_i)$ are in the interval $[0, \infty)$, and the corresponding value of the index which gives the minimum from above is zero, whence x is a convex combination of less than n elements from A, contradicting the minimality of n. Therefore, the assumption that we made was false, and the conclusion follows. $\qquad\square$

We now discuss the necessary conditions one needs in order that the conic hull of a set is closed. This does not happens automatically, as one can see from the example given by $A \subset \mathbb{R}^2$, $A := \{(x, y) \in \mathbb{R}^2 \mid (x - 1)^2 + y^2 = 1\}$, for which cone $A = \{(x, y) \in \mathbb{R}^2 \mid x > 0\} \cup \{(0, 0)\}$.

First, one defines for a nonempty set $A \subset X$ the asymptotic cone of A as

$$A^{\infty} = \{u \in X \mid \exists(t_n) \to 0, \ \exists(a_n) \subset A, \ t_n a_n \to u\}.$$

It is clear, by repeating the arguments from the case of the Bouligand tangent cone, that A^{∞} is a closed cone. If A is bounded, then $A^{\infty} = \{0\}$, and the converse also holds (if A would contain an unbounded sequence (a_n), then $\left(\dfrac{a_n}{\|a_n\|}\right)$ would also have a subsequence which converges to a nonzero element, which must be from A^∞). Let us observe also that if A is a cone, then $A^\infty = T_B(A, 0) = \text{cl } A$.

The next result concerns decomposition.

Theorem 2.1.18. *Let $A \subset \mathbb{R}^p$ be a nonempty closed set.*
(i) If $0 \notin A$, then $\text{cl cone } A = \text{cone } A \cup A^\infty$.
(ii) If $0 \in A$, then $\text{cl cone } A = \text{cone } A \cup A^\infty \cup T_B(A, 0)$.

Proof (i) Suppose that $0 \notin A$. It is clear that $\text{cone } A \subset \text{cl cone } A$. From the definition of A^∞, one also has that $A^\infty \subset \text{cl cone } A$. For the reverse inclusion, take $(u_n) \subset \text{cone } A, u_n \to u$. We must prove that $u \in \text{cone } A \cup A^\infty$. If $u = 0$, then the relation $u \in \text{cone } A$ is obvious. Suppose that $u \neq 0$, then for every $n \in \mathbb{N}$, there exist $t_n \geq 0$ and $a_n \in A$ such that $u_n = t_n a_n$. If (a_n) is unbounded, one can pass to a subsequence (a_{n_k}), where $\|a_{n_k}\| \to \infty$. Hence $t_{n_k} \to 0$, and $u \in A^\infty$. Suppose (a_n) is bounded. Since $0 \notin A$ and A is closed, there exists $y > 0$ such that $\|a_n\| \geq y$ for every n. One can deduce that (t_n) is bounded, so it converges (on a subsequence (t_{n_k}), eventually) to a number $t \geq 0$. If $t = 0$, then $u_{n_k} \to 0 = u$ (a situation which is excluded at this point of the proof). Accordingly, $t > 0$ and

$$\left\| a_{n_k} - t^{-1} u \right\| = t^{-1} \|t a_{n_k} - u\| = t^{-1} \left\| t_{n_k} a_{n_k} - u + (t - t_{n_k}) a_{n_k} \right\|$$

$$\leq t^{-1} \|t_{n_k} a_{n_k} - u\| + |t - t_{n_k}| t^{-1} \|a_{n_k}\| \to 0.$$

Hence $a_{n_k} \to t^{-1} u$ and since A is closed, $u \in \text{cone } A$. The proof of this part is complete.

(ii) The inclusion $\text{cone } A \cup A^\infty \cup T_B(A, 0) \subset \text{cl cone } A$ is obvious. Take $u \in \text{cl cone } A$. In the above it is possible as well that (a_n) converges to 0. Then $t_n \to \infty$, hence $u \in T_B(A, 0)$, which completes the proof. □

We finish with the following characterization result:

Corollary 2.1.19. *Let $A \subset \mathbb{R}^p$ be a nonempty closed set.*
(i) If $0 \notin A$, then $\text{cone } A$ is closed if and only if $A^\infty \subset \text{cone } A$.
(ii) If $0 \in A$, then $\text{cone } A$ is closed if and only if $A^\infty \cup T_B(A, 0) \subset \text{cone } A$.

2.2 Convex Functions

2.2.1 General Results

In this section we present the special class of convex functions. These functions are defined on convex sets.

Definition 2.2.1. *Let $D \subset \mathbb{R}^p$ be a convex set. One says that a function $f : D \to \mathbb{R}$ is convex if*

$$f(\lambda x + (1 - \lambda)y) \leq \lambda f(x) + (1 - \lambda)f(y), \ \forall x, y \in D, \ \forall \lambda \in [0, 1]. \tag{2.2.1}$$

It is clear that in the above definition it is sufficient to take $\lambda \in (0, 1)$.

As said before, in $\mathbb{R}$ the convex sets are exactly the intervals. In this framework, the convexity has the following geometric meaning: for every two points $x, y \in D$, $x < y$, the graph of the restriction of f to the $[x, y]$ interval lies below the line segment joining the points $(x, f(x))$ and $(y, f(y))$. This can be written as follows: for every $u \in [x, y]$,

$$f(u) \le f(x) + \frac{f(y) - f(x)}{y - x}(u - x), \qquad (2.2.2)$$

inequality which can be deduced from (2.2.1) by replacing λ with the value given by the relation $u = \lambda x + (1 - \lambda)y$. Therefore, (2.2.1) and (2.2.2) are equivalent (for functions defined on $\mathbb{R}$).

Definition 2.2.2. *Let $D \subset \mathbb{R}^p$ be a convex set. One says that a function $f : D \to \mathbb{R}$ is concave if $-f$ is convex.*

All of the properties of concave functions can be easily deduced from the similar properties of the convex functions, so in what follows we will consider only the later case.

We first deduce some general properties of the convex functions.

Proposition 2.2.3. *Let $D \subset \mathbb{R}^p$ be a convex set and $f : D \to \mathbb{R}$. The following relations are equivalent:*

(i) f is convex;
(ii) the epigraph of f,

$$\mathrm{epi}\, f := \{(x, t) \in D \times \mathbb{R} \mid f(x) \le t\},$$

is a convex subset of $\mathbb{R}^p \times \mathbb{R}$;
(iii) for any $x, y \in D$, define

$$I_{x,y} := \{t \in \mathbb{R} \mid tx + (1 - t)y \in D\};$$

then the function $\varphi_{x,y} : I_{x,y} \to \mathbb{R}$, $\varphi_{x,y}(t) = f(tx + (1 - t)y)$ is convex.

Proof We prove first the implication from (*i*) to (*ii*). Take $\lambda \in (0, 1)$ and $(x, t), (y, s) \in \mathrm{epi}\, f$. By the convexity of D, one knows that $\lambda x + (1 - \lambda)y \in D$, and using the convexity of f, one can say:

$$f(\lambda x + (1 - \lambda)y) \le \lambda f(x) + (1 - \lambda)f(y) \le \lambda t + (1 - \lambda)s,$$

i.e., $\big(\lambda x + (1 - \lambda)y, \lambda t + (1 - \lambda)s\big) \in \mathrm{epi}\, f$. Therefore, $\mathrm{epi}\, f$ is a convex set.

We prove now the converse implication. Take $x, y \in D$ and $\lambda \in [0, 1]$. Then $(x, f(x)), (y, f(y)) \in \mathrm{epi}\, f$ and by assumption, $\lambda(x, f(x)) + (1 - \lambda)(y, f(y)) \in \mathrm{epi}\, f$, hence

$$f(\lambda x + (1 - \lambda)y) \le \lambda f(x) + (1 - \lambda)f(y),$$

which shows that f is a convex function.

We prove now the equivalence between (*i*) and (*iii*). Observe first that $I_{x,y}$ is an interval which contains $[0, 1]$. Suppose that f is convex and take $u, v \in I_{x,y}$, $\lambda \in [0, 1]$. One knows that:

$$\begin{aligned}
\varphi_{x,y}(\lambda u + (1 - \lambda)v) &= f([\lambda u + (1 - \lambda)v]x + [1 - \lambda u - (1 - \lambda)v]y) \\
&= f(\lambda(ux + (1 - u)y) + (1 - \lambda)(vx + (1 - v)y)) \\
&\leq \lambda f(ux + (1 - u)y) + (1 - \lambda)f(vx + (1 - v)y) \\
&= \lambda \varphi_{x,y}(u) + (1 - \lambda)\varphi_{x,y}(v).
\end{aligned}$$

For the converse implication, take $x, y \in D$ and $t \in [0, 1]$. Then $\varphi_{x,y}$ is convex, hence for every $\lambda \in [0, 1]$, $u, v \in I_{x,y}$

$$\begin{aligned}
\varphi_{x,y}(\lambda u + (1 - \lambda)v) &\leq \lambda \varphi_{x,y}(u) + (1 - \lambda)\varphi_{x,y}(v) \\
&= \lambda f(ux + (1 - u)y) + (1 - \lambda)f(vx + (1 - v)y).
\end{aligned}$$

By taking $u = 1$, $v = 0$, $\lambda = t$ we deduce that

$$\varphi_{x,y}(t) \leq tf(x) + (1 - t)f(y),$$

hence f is convex. $\qquad\square$

Theorem 2.2.4. *Let $D \subset \mathbb{R}^p$ be a convex set and $f : D \to \mathbb{R}$ be a convex function. Then f is continuous at every interior point of D.*

Proof Take $\overline{x} \in \operatorname{int} D$. A translation permits us to consider the case $\overline{x} = 0$. We prove first that f is bounded on a neighborhood of 0. If we denote by $(e_i)_{i \in \overline{1,p}}$ the canonical base of $\mathbb{R}^p$, then there exists an $a > 0$ such that ae_i and $-ae_i$ are in D for any $i \in \overline{1, p}$. Under these conditions, the set

$$V := \left\{ x \in \mathbb{R}^p \mid x = \sum_{i=1}^{p} x_i e_i, \ |x_i| < \frac{a}{p}, \ \forall i \in \overline{1, p} \right\}$$

is a neighborhood of 0 contained in D. For $x \in V$, there exist $(x_i)_{i \in \overline{1,p}}$ with $|x_i| < \frac{a}{p}$, $i \in \overline{1, p}$ and $x = \sum_{i=1}^{p} x_i e_i$. Suppose first that $x_i \neq 0$ for any $i \in \overline{1, p}$. One has

$$\begin{aligned}
f(x) = f\left(\sum_{i=1}^{p} x_i e_i\right) &= f\left(\sum_{i=1}^{p} \frac{|x_i|}{a} a \frac{x_i}{|x_i|} e_i + \left(1 - \sum_{i=1}^{p} \frac{|x_i|}{a}\right) 0\right) \\
&\leq \sum_{i=1}^{p} \frac{|x_i|}{a} f\left(a \frac{x_i}{|x_i|} e_i\right) + \left(1 - \sum_{i=1}^{p} \frac{|x_i|}{a}\right) f(0) \\
&\leq \max\{f(ae_i), f(-ae_i) \mid i \in \overline{1, p}\} + |f(0)|.
\end{aligned}$$

We can now observe that if there are indices i for which $x_i = 0$, then these can be excluded from the above calculations, and the estimation holds.

Since the right-hand part is a constant (which we denote by M), the proof is finished. Take $\varepsilon \in (0, 1)$ and U a symmetric neighborhood of 0 such that $\varepsilon^{-1}U \subset V$. Then, for any $x \in U$,

$$f(x) = f\left(\varepsilon(\varepsilon^{-1}x) + (1 - \varepsilon)0\right) \le \varepsilon f(\varepsilon^{-1}x) + (1 - \varepsilon)f(0) \le \varepsilon M + (1 - \varepsilon)f(0),$$

i.e.,

$$f(x) - f(0) \le \varepsilon M - \varepsilon f(0).$$

From the fact that U is symmetric, one deduces that for every $x \in U$

$$f(-x) \le \varepsilon M + (1 - \varepsilon)f(0).$$

Moreover,

$$f(0) = f\left(\frac{1}{2}x + \frac{1}{2}(-x)\right) \le \frac{1}{2}f(x) + \frac{1}{2}f(-x) \le \frac{1}{2}f(x) + \frac{1}{2}\left(\varepsilon M + (1 - \varepsilon)f(0)\right),$$

hence

$$f(0) - f(x) \le \varepsilon M - \varepsilon f(0).$$

This relation can be combined with the similar one from above, and gives

$$\left|f(x) - f(0)\right| \le \varepsilon M - \varepsilon f(0).$$

This inequality proves the continuity of f at 0. □

We want to emphasize now some characterizations of differentiable convex functions. Some preliminary results on convex functions defined on a real intervals are necessary.

Proposition 2.2.5. *Let $I \subset \mathbb{R}$ be an interval and $f : I \to \mathbb{R}$ be a function. The next relations are equivalent:*

(i) f is convex;

(ii) for every $x_1, x_2, x_3 \in I$ satisfying the relation $x_1 < x_2 < x_3$ one has

$$\frac{f(x_2) - f(x_1)}{x_2 - x_1} \le \frac{f(x_3) - f(x_1)}{x_3 - x_1} \le \frac{f(x_3) - f(x_2)}{x_3 - x_2};$$

(iii) for every $a \in \operatorname{int} I$, the function $g : I \setminus \{a\} \to \mathbb{R}$ given by

$$g(x) = \frac{f(x) - f(a)}{x - a}$$

is increasing.

Proof We prove the $(i) \Rightarrow (ii)$ implication. Take $\lambda = \dfrac{x_2 - x_1}{x_3 - x_1} \in (0, 1)$. Then the equality $x_2 = \lambda x_3 + (1 - \lambda)x_1$ holds and one must prove now that

$$\frac{f(x_2) - f(x_1)}{\lambda(x_3 - x_1)} \le \frac{f(x_3) - f(x_1)}{x_3 - x_1} \le \frac{f(x_3) - f(x_2)}{(1 - \lambda)(x_3 - x_1)}.$$

After some calculations, one can show that:

$$f(x_2) \le \lambda f(x_3) + (1 - \lambda)f(x_1).$$

The proof of the implication $(ii) \Rightarrow (i)$ follows the inverse path of the proof of $(i) \Rightarrow (ii)$, hence (i) and (ii) are equivalent.

In order to prove $(ii) \Rightarrow (iii)$, we fix $x_1, x_2 \in I \setminus \{a\}$ with $x_1 < x_2$ and we find three situations. If $x_1 < x_2 < a$, then we apply (ii) for the triplet (x_1, x_2, a). If $x_1 < a < x_2$, then we apply (ii) for the triplet (x_1, a, x_2). Finally, if $a < x_1 < x_2$, then we apply (ii) for the triplet (a, x_1, x_2).

We prove now $(iii) \Rightarrow (i)$. Take $x, y \in I$ with $x < y$ and $\lambda \in (0, 1)$. Then $x < \lambda x + (1 - \lambda)y < y$, and by applying (iii) with $a = \lambda x + (1 - \lambda)y$, one deduces

$$\frac{f(x) - f(\lambda x + (1 - \lambda)y)}{x - \lambda x - (1 - \lambda)y} \le \frac{f(y) - f(\lambda x + (1 - \lambda)y)}{y - \lambda x - (1 - \lambda)y}.$$

After some calculations, the relation follows from the definition of convexity. The fact that this relation holds for any $x, y \in I$ with $x < y$ and for any $\lambda \in (0, 1)$ is sufficient to prove the desired assertion. The proof is complete. □

Proposition 2.2.6. *Let $I \subset \mathbb{R}$ be an interval and $f : I \to \mathbb{R}$ be a convex function. Then f admits lateral derivatives in every interior point of I and for every $x, y \in \text{int } I$ with $x < y$, one has*

$$f'_-(x) \le f'_+(x) \le f'_-(y) \le f'_+(y).$$

Proof Fix $a \in \text{int } I$. Since the function $g : I \setminus \{a\} \to \mathbb{R}$ given by

$$g(x) = \frac{f(x) - f(a)}{x - a}$$

is increasing (see the previous result) one deduces that g admits finite lateral limits, which implies the existence of the lateral derivatives of f at a. Moreover, $f'_-(a) \le f'_+(a)$. For $x, y \in \text{int } I$, $x < y$ and for any $u, v \in (x, y)$, $u \le v$, by using again the argument given by the last conclusion of Proposition 2.2.5, one deduces that

$$\frac{f(u) - f(x)}{u - x} \le \frac{f(v) - f(x)}{v - x} = \frac{f(x) - f(v)}{x - v} \le \frac{f(y) - f(v)}{y - v} = \frac{f(v) - f(y)}{v - y}.$$

Passing to the limit for $u \to x$ and $v \to y$, one gets $f'_+(x) \le f'_-(y)$. □

Here we characterize differentiable convex functions of one variable.

Theorem 2.2.7. *Let I be an open interval and $f : I \to \mathbb{R}$ be a function.*

(i) If f is differentiable on I, then f is convex if and only if f' is increasing on I.

(ii) If f is twice differentiable on I, then f is convex if and only if $f''(x) \ge 0$ for every $x \in I$.

Proof In the case of real functions of one variable, the equivalence between the monotonicity of f' and the sign of f'' is sufficient to prove that f is convex if and only if f' is increasing on I. If f is convex, the monotonicity of the derivative follows from Proposition 2.2.6. Conversely, suppose that f' is increasing and we prove that f is convex. Take $a, b \in I$. Define $g : [a, b] \to \mathbb{R}$ given by

$$g(x) = f(x) - f(a) - (x - a)\frac{f(b) - f(a)}{b - a}.$$

Obviously, $g(a) = g(b) = 0$, and

$$g'(x) = f'(x) - \frac{f(b) - f(a)}{b - a}.$$

The function f satisfies the conditions of Lagrange Theorem on $[a, b]$, hence there exists $c \in (a, b)$ such that

$$\frac{f(b) - f(a)}{b - a} = f'(c).$$

Consequently, $g'(x) = f'(x) - f'(c)$. From the monotonicity of f', we deduce that g is decreasing on (a, c) and increasing on (c, b), and since $g(a) = g(b) = 0$, we know that g is negative on the whole interval $[a, b]$. Take $x \in (a, b)$. Then there exists $\lambda \in (0, 1)$ such that

$$x = \lambda a + (1 - \lambda)b.$$

By replacing x in the expression of g and taking into account that $g(x) \le 0$, we deduce that

$$f(\lambda a + (1 - \lambda)b) - f(a) - (1 - \lambda)(b - a)\frac{f(b) - f(a)}{b - a} \le 0,$$

relation which reduces to the definition of the convexity. $\qquad\square$

Example 2.2.8. *Based on the above result, one deduces the convexity of the following functions:* $f : \mathbb{R} \to \mathbb{R}$, $f(x) = ax + b$, *with* $a, b \in \mathbb{R}$; $f : (0, \infty) \to \mathbb{R}$, $f(x) = -\ln x$; $f : (0, \infty) \to \mathbb{R}$, $f(x) = x \ln x$; $f : (0, \infty) \to \mathbb{R}$, $f(x) = x^a, a \ge 1$; $f : \mathbb{R} \to \mathbb{R}$, $f(x) = e^x$; $f : (-1, 1) \to \mathbb{R}$, $f(x) = -\sqrt{1 - x^2}$; $f : (0, \pi) \to \mathbb{R}$, $f(x) = \sin^{-1} x$.

Another example is given by the next result.

Proposition 2.2.9. *Let* $D \subset \mathbb{R}^p$ *be a nonempty convex set. Then the function* $d_D : \mathbb{R}^p \to \mathbb{R}$ *given by* $d_D(x) = d(x, D)$ *is convex.*

Proof Take $x, y \in \mathbb{R}^p$ and $\alpha \in [0, 1]$. For any $\varepsilon > 0$, there exist $d_{x,\varepsilon}, d_{y,\varepsilon} \in D$ such that

$$\|d_{x,\varepsilon} - x\| < d_D(x) + \varepsilon$$
$$\|d_{y,\varepsilon} - y\| < d_D(y) + \varepsilon.$$

Using the convexity of D, one knows:

$$d_D(\alpha x + (1 - \alpha)y) \leq \left\| \alpha x + (1 - \alpha)y - (\alpha d_{x,\varepsilon} + (1 - \alpha)d_{y,\varepsilon}) \right\|$$
$$\leq \alpha \left\| d_{x,\varepsilon} - x \right\| + (1 - \alpha) \left\| d_{y,\varepsilon} - y \right\|$$
$$< \alpha d_D(x) + (1 - \alpha)d_D(y) + \varepsilon.$$

As ε is arbitrarily chosen, we may pass to the limit for $\varepsilon \to 0$ and the conclusion follows. $\qquad\square$

We now characterize differentiable convex functions in the general case.

Theorem 2.2.10. *Let $D \subset \mathbb{R}^p$ be an open convex set and $f : D \to \mathbb{R}$ be a function.*
(i) If f is differentiable on D, then f is convex if and only if for any $x, y \in D$,

$$f(y) \geq f(x) + \nabla f(x)(y - x).$$

(ii) If f is twice differentiable on D, then f is convex if and only if for every $x \in D$ and $y \in \mathbb{R}^p$, one has

$$\nabla^2 f(x)(y, y) \geq 0.$$

Proof (i) Consider first the case when $p = 1$. Fix $x \in D$ with $y \neq x$ and take $\lambda \in (0, 1]$. Since f is convex, we get

$$f(x + \lambda(y - x)) = f((1 - \lambda)x + \lambda y)$$
$$\leq (1 - \lambda)f(x) + \lambda f(y) = f(x) + \lambda(f(y) - f(x)).$$

Consequently,

$$\frac{f(x + \lambda(y - x)) - f(x)}{\lambda(y - x)} \cdot (y - x) \leq f(y) - f(x).$$

Passing to the limit for $\lambda \to 0$, one gets $f'(x)(y - x) \leq f(y) - f(x)$.

We pass now to the general case. For $x \in D$, consider the function $\varphi_{y,x}$ from Proposition 2.2.3, which we already know that is convex. Moreover, $\varphi_{y,x}$ is differentiable on the open interval $I_{y,x}$, and

$$\varphi'_{y,x}(t) = \nabla f(ty + (1 - t)x)(y - x).$$

According to the preceding step,

$$\varphi_{y,x}(1) \geq \varphi_{y,x}(0) + \varphi'_{y,x}(0)$$

which means that

$$f(y) \geq f(x) + \nabla f(x)(y - x).$$

Conversely, fix $x, y \in D$ and $\lambda \in [0, 1]$. Therefore, by assumption,

$$f(x) \geq f(\lambda x + (1 - \lambda)y) + (1 - \lambda)\nabla f(\lambda x + (1 - \lambda)y)(x - y)$$

and

$$f(y) \geq f(\lambda x + (1 - \lambda)y) + \lambda \nabla f(\lambda x + (1 - \lambda)y)(y - x).$$

Multiplying the first inequality by λ, the second one by $(1 - \lambda)$, and summing up the new inequalities, one obtains

$$\lambda f(x) + (1 - \lambda)f(y) \geq f(\lambda x + (1 - \lambda)y),$$

which proves that f is convex.

(ii) The case $p = 1$ is proved in Theorem 2.2.7. Now, in order to pass to the general case, take $x \in D$, $y \in \mathbb{R}^p$. Suppose f is convex. Since D is open, there exists an $\alpha > 0$ such that $u := x + \alpha y \in D$. According to the assumption, $\varphi_{u,x}$ is convex, and taking into account the case that we have already studied, $\varphi_{u,x}''(t) \geq 0$ for any $t \in I_{u,x}$. For $t = 0$, one deduces that

$$0 \leq \varphi_{u,x}''(0) = \nabla^2 f(x)(u - x, u - x),$$

and the conclusion follows. Conversely, for $x, y \in D$ and $t \in I_{x,y}$, $\varphi_{x,y}''(t) \geq 0$. From the case $p = 1$, we get that $\varphi_{x,y}$ is convex, hence f is convex. The proof is now complete.□

From these results, one may observe that some properties of the convex functions have a global character, an aspect which will persist in subsequent sections.

At the end of this subsection, we will discuss a property which is stronger than convexity.

Definition 2.2.11. *Let $D \subset \mathbb{R}^p$ be a convex set. One says that a function $f : D \to \mathbb{R}$ is strictly convex if*

$$f(\lambda x + (1 - \lambda)y) < \lambda f(x) + (1 - \lambda)f(y), \quad \forall x, y \in D, \ x \neq y, \ \forall \lambda \in (0, 1).$$

Definition 2.2.12. *Let $D \subset \mathbb{R}^p$ be a convex set. One says that a function $f : D \to \mathbb{R}$ is strictly concave if $-f$ is strictly convex.*

Every strictly convex function is convex, but the converse is false. To see this, consider a convex function which is constant on an interval. Again, the properties of the strictly concave functions easily follow from the corresponding ones of the strictly convex functions.

By the use of very similar arguments as in the proofs of preceding results, one can deduce the next characterizations.

Theorem 2.2.13. *Let I be an open interval and* $f : I \to \mathbb{R}$ *be a differentiable function. The next assertions are equivalent:*

    (i) f *is strictly convex;*

    (ii) $f(x) > f(a) + f'(a)(x - a)$, *for any* $x, a \in I$, $x \neq a$;

    (iii) f' *is strictly increasing.*

    If, moreover, f is twice differentiable (on I), then one more equivalence holds :

    (iv) $f''(x) \geq 0$ *for any* $t \in I$ *and* $\{x \in I \mid f''(x) = 0\}$ *does not contain any proper interval.*

Example 2.2.14. *Using this result, one gets the strict convexity of the following functions:* $f : (0, \infty) \to \mathbb{R}$, $f(x) = -\ln x$; $f : (0, \infty) \to \mathbb{R}$, $f(x) = x \ln x$; $f : (0, \infty) \to \mathbb{R}$, $f(x) = x^a$, $a > 1$; $f : \mathbb{R} \to \mathbb{R}$, $f(x) = e^x$; $f : (0, \infty) \to \mathbb{R}$, $f(x) = (1 + x^p)^{\frac{1}{p}}$, $p > 1$.

Theorem 2.2.15. *Let* $D \subset \mathbb{R}^p$ *an open convex set and* $f : D \to \mathbb{R}$ *be a differentiable function. The next assertions are equivalent:*

    (i) f *is strictly convex;*

    (ii) $f(x) > f(a) + \nabla f(a)(x - a)$, *for any* $x, a \in D$, $x \neq a$.

    If, moreover, f is twice differentiable (on D), then the preceding two items are implied by the relation:

    (iii) $\nabla^2 f(x)(y, y) > 0$ *for any* $x \in D$ *and* $y \in \mathbb{R}^p \setminus \{0\}$.

2.2.2 Convex Functions of One Variable

In this subsection we will focus on some properties and applications of convex functions defined on real intervals, even if some of the results hold in more general situations.

The class of convex functions is stable under several algebraic operations, an aspect which makes it very useful. Here are some of these operations:

Proposition 2.2.16. *Let* $I, J \subset \mathbb{R}$ *be intervals.*

    (i) Let $n \in \mathbb{N}^*$ *and* $f_1, f_2, \dots, f_n : I \to \mathbb{R}$ *be convex functions, and* $\lambda_1, \lambda_2, \dots, \lambda_n \geq 0$. *Then* $\sum_{i=1}^{n} \lambda_i f_i$ *is convex. If at least one of the functions is strictly convex, and the corresponding scalar is not zero, then* $\sum_{i=1}^{n} \lambda_i f_i$ *is strictly convex.*

    (ii) Let $f : I \to J$ *be (strictly) convex,* $g : J \to \mathbb{R}$ *be convex and (strictly) increasing. Then* $g \circ f$ *is (strictly) convex.*

    (iii) Let $f : I \to J$ *be strictly decreasing, (strictly) convex, and surjective. Then* f^{-1} *is (strictly) convex.*

The next result emphasizes some monotonicity properties of convex functions of one variable.

Theorem 2.2.17. *Let I be a nondegenerate interval (i.e., not a singleton set) and $f : I \to \mathbb{R}$ be convex. Then either f is monotone on $\operatorname{int} I$, or there exists $\overline{x} \in \operatorname{int} I$ such that f is decreasing on $I \cap (-\infty, \overline{x}]$, and increasing on $I \cap [\overline{x}, \infty)$.*

Proof Because of relation (2.2.2), it is sufficient to restrict our attention to the case when I is open, i.e., $I = \operatorname{int} I$. Suppose that f is not monotone on I. Then there exist $a, b, c \in I$, $a < b < c$ such that $f(a) > f(b) < f(c)$ or $f(a) < f(b) > f(c)$. The second situation cannot hold, since in that case, using (2.2.2), one would have

$$f(b) \leq f(a) + \frac{f(c) - f(a)}{c - a}(b - a) = \frac{f(a)(c - b) + f(c)(b - a)}{c - a} < f(b).$$

Hence, $f(a) > f(b) < f(c)$. As f is continuous on $[a, c]$, its minimum on this interval must be attained at a point $\overline{x}$ (from the Weierstrass Theorem). Take $x \in I \cap (-\infty, a)$. According to Proposition 2.2.5,

$$\frac{f(x) - f(\overline{x})}{x - \overline{x}} \leq \frac{f(a) - f(\overline{x})}{a - \overline{x}},$$

i.e.,

$$(\overline{x} - a)f(x) \geq (x - a)f(\overline{x}) + (\overline{x} - x)f(a) \geq (\overline{x} - a)f(\overline{x}),$$

hence $f(\overline{x}) \leq f(x)$. Similarly, one can prove that $f(\overline{x}) \leq f(x)$ for $x \in I \cap (c, \infty)$. It follows that $f(\overline{x}) = \inf f(I)$. We next prove that f is decreasing on $I \cap (-\infty, \overline{x})$. Take $u, v \in I \cap (-\infty, \overline{x})$, $u < v$. On one hand,

$$\frac{f(v) - f(u)}{v - u} \leq \frac{f(\overline{x}) - f(u)}{\overline{x} - u},$$

and on the other hand,

$$\frac{f(\overline{x}) - f(u)}{\overline{x} - u} = \frac{f(u) - f(\overline{x})}{u - \overline{x}} \leq \frac{f(v) - f(\overline{x})}{v - \overline{x}} \leq 0,$$

hence $f(v) - f(u) \leq 0$, which is exactly what we wanted to prove. Similarly, one can show that f is increasing on $I \cap (\overline{x}, +\infty)$. The continuity of f on I finalizes the proof. $\square$

As we saw before, a convex function defined on an interval can have discontinuities only at the extremities of the interval. The previous theorem allows us to consider a different function in those eventual discontinuity points, without losing the convexity. In this way one obtains the next consequence.

Corollary 2.2.18. *Let $a, b \in \mathbb{R}$, $a < b$ and $f : [a, b] \to \mathbb{R}$ be a convex function. Then there exist $\lim_{x \to a+} f(x)$ and $\lim_{x \to b-} f(x)$, and the function*

$$\overline{f}(x) = \begin{cases} \lim_{x \to a+} f(x), & x = a \\ f(x), & x \in (a, b) \\ \lim_{x \to b-} f(x), & x = b \end{cases}$$

is convex and continuous on $[a, b]$.

From the proof of Theorem 2.2.17 one can also obtain the next result, which will be restated in a more general framework in the next chapter.

Corollary 2.2.19. *Let $I \subset \mathbb{R}$ be a nondegenerate interval. If $f : I \to \mathbb{R}$ is a convex and non-monotone function, then it has a global minimum on* int *I*.

The next proposition, sometimes called the Jensen inequality, follows by applying the definitions and the mathematical induction principle.

Proposition 2.2.20. *Let $D \subset \mathbb{R}^p$ be a convex set and $f : D \to \mathbb{R}$. If the function f is convex, then*

$$f(\lambda_1 x_1 + \ldots + \lambda_m x_m) \le \lambda_1 f(x_1) + \ldots + \lambda_m f(x_m)$$

for any $m \in \mathbb{N}^*$, $x_1, \ldots, x_m \in D$, $\lambda_1, \ldots, \lambda_m \ge 0$, $\lambda_1 + \ldots + \lambda_m = 1$. The inequality is strict if the function f is strictly convex, at least two of the points (x_k) are different and the corresponding scalars (λ_k) are strictly positive.

Actually, for the convexity of a continuous function it is sufficient that the inequality from the definition is satisfied for $\lambda = 2^{-1}$.

Theorem 2.2.21. *Let $I \subset \mathbb{R}$ be an interval and $f : I \to \mathbb{R}$ be a continuous function. The function f is convex if and only if*

$$f\left(\frac{x+y}{2}\right) \le \frac{f(x) + f(y)}{2}, \quad \forall x, y \in I. \tag{2.2.3}$$

Proof The necessity of the condition (2.2.3) is obvious. Let us prove that it is also sufficient. Suppose, by contradiction, that f is not convex, which means (2.2.2) is not satisfied. Then there exist $x, y \in I$, $x < y$ and $u \in (x, y)$ such that

$$f(u) > f(x) + \frac{f(y) - f(x)}{y - x}(u - x). \tag{2.2.4}$$

Observe by the relation (2.2.4) that the cases $u = x$ and $u = y$ cannot hold. Consider then $g : [x, y] \to \mathbb{R}$,

$$g(t) = f(t) - f(x) - \frac{f(y) - f(x)}{y - x}(t - x),$$

which is continuous and satisfies the relations $g(x) = g(y) = 0$. By (2.2.4) and Weierstrass' Theorem, there exists $z \in (x, y)$ such that $g(z) = \sup_{t \in [x,y]} g(t) > 0$. Denote by

$$w := \inf\{z \in (x, y) \mid g(z) = \sup_{t \in [x,y]} g(t)\}.$$

By the continuity of g, it follows that $g(w) = \sup_{t \in [a,b]} g(t) > 0$, and hence $w \in (x, y)$. Consequently, there exists $h > 0$ such that $w + h, w - h \in (x, y)$. But $w = 2^{-1}(w + h) +$

$2^{-1}(w - h)$, and also $g(w) \geq g(w + h)$ and $g(w) > g(w - h)$. Accordingly,

$$\frac{g(w - h) + g(w + h)}{2} < g(w) = f(w) - f(x) - \frac{f(y) - f(x)}{y - x}(w - x)$$

$$\leq \frac{f(w + h) + f(w - h)}{2} - f(x) - \frac{f(y) - f(x)}{y - x}\left(\frac{w + h + w - h}{2} - x\right)$$

$$= \frac{g(w - h) + g(w + h)}{2},$$

which is a contradiction. It follows that the assumption made is false, hence f is convex. □

Corollary 2.2.22. *Let $I \subset \mathbb{R}$ be an interval and $f : I \to \mathbb{R}$ be a continuous function. The function f is convex if and only if for any $x \in I$ and $h > 0$ with $x + h, x - h \in I$, one has*

$$f(x + h) + f(x - h) - 2f(x) \geq 0.$$

In the case of strictly convex functions, the results are similar.

Proposition 2.2.23. *Let $I \subset \mathbb{R}$ be an interval and $f : I \to \mathbb{R}$ be a continuous function. The function f is strictly convex if and only if*

$$f\left(\frac{x + y}{2}\right) < \frac{f(x) + f(y)}{2}, \ \forall x, y \in I, \ x \neq y.$$

Corollary 2.2.24. *Let $I \subset \mathbb{R}$ be an interval and $f : I \to \mathbb{R}$ be a continuous function. The function f is strictly convex if and only if for any $x \in I$ and $h > 0$ with $x + h, x - h \in I$, one has*

$$f(x + h) + f(x - h) - 2f(x) > 0.$$

For the case of triplets, a result was proved in 1965 by the Romanian mathematician Tiberiu Popoviciu.

Theorem 2.2.25. *Let $I \subset \mathbb{R}$ be an interval and $f : I \to \mathbb{R}$ be a continuous function. The function f is convex if and only if for any $x, y, z \in I$, one has*

$$\frac{2}{3}\left[f\left(\frac{x + y}{2}\right) + f\left(\frac{y + z}{2}\right) + f\left(\frac{z + x}{2}\right)\right] \leq f\left(\frac{x + y + z}{3}\right) + \frac{f(x) + f(y) + f(z)}{3}. \quad (2.2.5)$$

Proof We prove first the necessity of condition (2.2.5). As in the case of Theorem 2.2.21, continuity is not involved in this step. Without losing the generality, suppose that $x \leq y \leq z$. If $y \leq 3^{-1}(x + y + z)$, then

$$\frac{x + y + z}{3} \leq \frac{x + z}{2} \leq z \text{ and } \frac{x + y + z}{3} \leq \frac{y + z}{2} \leq z,$$

hence there exist $s, t \in [0, 1]$ such that

$$\frac{x + z}{2} = s\frac{x + y + z}{3} + (1 - s)z$$

$$\frac{y+z}{2} = t\frac{x+y+z}{3} + (1-t)z.$$

By summation, one gets

$$(x+y-2z)(s+t-2^{-1}3) = 0.$$

If $x+y=2z$, then $x=y=z$ and (2.2.5) is obvious. If $s+t=2^{-1}3$, then by summing the inequalities

$$f\left(\frac{x+z}{2}\right) \leq sf\left(\frac{x+y+z}{3}\right) + (1-s)f(z)$$

$$f\left(\frac{y+z}{2}\right) \leq tf\left(\frac{x+y+z}{3}\right) + (1-t)f(z)$$

$$f\left(\frac{x+y}{2}\right) \leq \frac{f(x)+f(y)}{2}$$

and by multiplying with $3^{-1}2$, one gets (2.2.5). The case $y > 3^{-1}(x+y+z)$ is similar.

In order to prove the sufficiency, observe that for $y=z$ in (2.2.5), one gets

$$\frac{1}{4}f(x) + \frac{3}{4}f\left(\frac{x+2y}{3}\right) \geq f\left(\frac{x+y}{2}\right), \; \forall x, y \in I.$$

From now on, one can follow the arguments from the proof of the sufficiency in Theorem 2.2.21. □

If f is strictly convex, then the Popoviciu inequality (2.2.5) is strict, except the case when $x=y=z$.

2.2.3 Inequalities

We shall now formulate several inequalities which follow from the general results previously given. In many cases, these inequalities can be seen as examples of discrete optimization and have, in general, a wide applicability in different mathematical areas.

We begin with a refinement of the inequality from Theorem 2.2.21, widely known under the name of Hermite-Hadamard inequality.

Theorem 2.2.26 (Hermite-Hadamard). *Let $a, b \in \mathbb{R}$, $a < b$ and $f : [a, b] \to \mathbb{R}$ be a convex function. Then the next inequality holds:*

$$f\left(\frac{a+b}{2}\right) \leq \frac{1}{b-a}\int_a^b f(x)dx \leq \frac{f(a)+f(b)}{2}.$$

The equality is obtained if and only if f is affine.

Proof Observe first that, because of the continuity of f on (a, b), f is Riemann integrable on $[a, b]$. Moreover, from convexity, one has, for any $\lambda \in [0, 1]$,

$$f(\lambda a + (1 - \lambda)b) \le \lambda f(a) + (1 - \lambda)f(b).$$

By integration with respect to λ, one gets

$$\int_0^1 f(\lambda a + (1 - \lambda)b)d\lambda \le \frac{f(a) + f(b)}{2}.$$

On the other hand, for any λ,

$$f\left(\frac{a + b}{2}\right) = f\left(\frac{\lambda a + (1 - \lambda)b}{2} + \frac{(1 - \lambda)a + \lambda b}{2}\right)$$

$$\le \frac{1}{2}f(\lambda a + (1 - \lambda)b) + \frac{1}{2}f((1 - \lambda)a + \lambda b).$$

By integrating again with respect to λ and changing the variable, it follows that

$$f\left(\frac{a + b}{2}\right) \le \frac{1}{2}\int_0^1 f(\lambda a + (1 - \lambda)b)d\lambda + \frac{1}{2}\int_0^1 f((1 - \lambda)a + \lambda b)d\lambda$$

$$= \int_0^1 f(\lambda a + (1 - \lambda)b)d\lambda.$$

Therefore,

$$f\left(\frac{a + b}{2}\right) \le \int_0^1 f(\lambda a + (1 - \lambda)b)d\lambda \le \frac{f(a) + f(b)}{2}.$$

Finally, if we change the variable to $\lambda a + (1 - \lambda)b = x$ in the integral, then we get the desired inequalities.

The second inequality can be deduced by the integration of (2.2.2) written for the interval $[a, b]$.

For the second conclusion, define $g : [a, b] \to \mathbb{R}$, where

$$g(x) := f(a) + \frac{f(b) - f(a)}{b - a}(x - a).$$

In general, $f(x) \le g(x)$ for any $x \in [a, b]$. If f is affine, then

$$f(x) = g(x), \ \forall x \in [a, b],$$

and a direct calculation shows the equality in this case. Conversely, suppose that f is not affine. Then there exists $\bar{x} \in (a, b)$ with

$$f(\bar{x}) < g(\bar{x}).$$

Take $\alpha > 0$ such that

$$f(\overline{x}) < g(\overline{x}) - \alpha.$$

By using the continuity of the both parts in $\overline{x}$, there exists $\varepsilon > 0$ such that $(\overline{x}-\varepsilon, \overline{x}+\varepsilon) \subset (a, b)$ and

$$f(x) < g(x) - \alpha, \ \forall x \in (\overline{x} - \varepsilon, \overline{x} + \varepsilon).$$

Hence

$$\int_a^b f(x)dx = \int_a^{\overline{x}-\varepsilon} f(x)dx + \int_{\overline{x}-\varepsilon}^{\overline{x}+\varepsilon} f(x)dx + \int_{\overline{x}+\varepsilon}^b f(x)dx$$

$$\leq \int_a^{\overline{x}-\varepsilon} g(x)dx + \int_{\overline{x}-\varepsilon}^{\overline{x}+\varepsilon} (g(x) - \alpha)dx + \int_{\overline{x}+\varepsilon}^b g(x)dx$$

$$= \int_a^b g(x)dx - 2\alpha\varepsilon = (b - a)\frac{f(a) + f(b)}{2} - 2\alpha\varepsilon.$$

Therefore,

$$\int_a^b f(x)dx < (b - a)\frac{f(a) + f(b)}{2},$$

which is a contradiction. □

By using the Jensen inequality (Theorem 2.2.20) for several functions, we can deduce some classical inequalities. Such an example is provided by the convex function $f : \mathbb{R} \to \mathbb{R}$, $f(x) = e^x$. Take $n \in \mathbb{N}^*$ and $x_1, x_2, ..., x_n \in \mathbb{R}$, $\lambda_1, \lambda_2, ..., \lambda_n > 0$ with $\sum_{k=1}^n \lambda_k = 1$. By applying Theorem 2.2.20, we deduce that

$$e^{\sum_{k=1}^n \lambda_k x_k} \leq \sum_{k=1}^n \lambda_k e^{x_k}, \tag{2.2.6}$$

with equality only in the case $x_1 = x_2 = ... = x_n$. Take $a_1, a_2, ..., a_n > 0$ and $x_k = \ln a_k$ for any $k \in \overline{1, n}$. Then, from (2.2.6), we deduce that

$$a_1^{\lambda_1} a_2^{\lambda_2} ... a_n^{\lambda_n} \leq \sum_{k=1}^n \lambda_k a_k. \tag{2.2.7}$$

Again, equality holds if and only if $a_1 = a_2 = ... = a_n$. This inequality is sometimes called the general means inequality, because if one takes $\lambda_1 = \lambda_2 = ... = \lambda_n = n^{-1}$ in (2.2.7), one recovers the well-known inequality between the geometric and the arithmetic means:

$$(a_1 a_2 ... a_n)^{\frac{1}{n}} \leq \frac{a_1 + a_2 + ... + a_n}{n}.$$

By the transform $x_k \to x_k^{-1}$, instead of (2.2.7), one gets

$$a_1^{\lambda_1} a_2^{\lambda_2} \ldots a_n^{\lambda_n} \geq \frac{1}{\sum_{k=1}^{n} \frac{\lambda_k}{a_k}},$$

and from here (again for $\lambda_1 = \lambda_2 = \ldots = \lambda_n = n^{-1}$) the inequality between the harmonic and the geometric means:

$$\frac{n}{\frac{1}{a_1} + \frac{1}{a_2} + \ldots + \frac{1}{a_n}} \leq (a_1 a_2 \ldots a_n)^{\frac{1}{n}}.$$

Also from (2.2.7), for $n = 2$, $a_1 = u > 0$, $a_2 = v > 0$, $\lambda_1 = \frac{1}{p}$, $\lambda_2 = \frac{1}{q}$ with $p, q > 1$, $\frac{1}{p} + \frac{1}{q} = 1$, one deduces that

$$u^{\frac{1}{p}} v^{\frac{1}{q}} \leq \frac{u}{p} + \frac{v}{q}. \tag{2.2.8}$$

By taking $u = \varepsilon a^p$, $v = \frac{1}{\varepsilon} b^q$ with $a, b, \varepsilon > 0$, one has

$$ab \leq \varepsilon \frac{a^p}{p} + \frac{1}{\varepsilon} \frac{b^q}{q},$$

and for $\varepsilon = 1$,

$$ab \leq \frac{a^p}{p} + \frac{b^q}{q}, \quad \forall p, q > 1, \quad \frac{1}{p} + \frac{1}{q} = 1, \quad a, b > 0. \tag{2.2.9}$$

The equality holds if and only if $a^p = b^q$. The relation (2.2.9) is called the Young inequality.

The Hölder inequality can be proved in several ways. We now deduce it as a consequence of (2.2.8). Take $p, q > 1$, $\frac{1}{p} + \frac{1}{q} = 1$, $x_1, x_2, \ldots, x_n > 0$ and $y_1, y_2, \ldots, y_n > 0$. One takes, for a natural k arbitrarily fixed, $u := \frac{x_k^p}{\sum_{k=1}^{n} x_k^p}$ and $v := \frac{y_k^q}{\sum_{k=1}^{n} y_k^q}$. From (2.2.8), one gets

$$\frac{x_k}{\left(\sum_{k=1}^{n} x_k^p\right)^{\frac{1}{p}}} \frac{y_k}{\left(\sum_{k=1}^{n} y_k^q\right)^{\frac{1}{q}}} \leq \frac{1}{p} \frac{x_k^p}{\sum_{k=1}^{n} x_k^p} + \frac{1}{q} \frac{y_k^q}{\sum_{k=1}^{n} y_k^q}.$$

We write this relation for $k \in \overline{1, n}$ and make the sum. We find the Hölder inequality:

$$\sum_{k=1}^{n} x_k y_k \leq \left(\sum_{k=1}^{n} x_k^p\right)^{\frac{1}{p}} \left(\sum_{k=1}^{n} y_k^q\right)^{\frac{1}{q}}.$$

As one may observe, the equality holds if and only if the elements $(x_k^p)_{k \in \overline{1,n}}$ and $(y_k^q)_{k \in \overline{1,n}}$ are proportional.

From here one can deduce the Minkowski inequality. Take $p > 1$ and $x_1, x_2, \ldots, x_n > 0$, $y_1, y_2, \ldots, y_n > 0$. Then, by applying the previous inequality (and by taking $q = \frac{p}{p-1}$), we obtain

$$\sum_{k=1}^{n} (x_k + y_k)^p = \sum_{k=1}^{n} x_k (x_k + y_k)^{p-1} + \sum_{k=1}^{n} y_k (x_k + y_k)^{p-1}$$

$$\leq \left(\sum_{k=1}^{n} x_k^p\right)^{\frac{1}{p}} \left(\sum_{k=1}^{n}(x_k+y_k)^{(p-1)q}\right)^{\frac{1}{q}} + \left(\sum_{k=1}^{n} y_k^p\right)^{\frac{1}{p}} \left(\sum_{k=1}^{n}(x_k+y_k)^{(p-1)q}\right)^{\frac{1}{q}}$$

$$= \left(\left(\sum_{k=1}^{n} x_k^p\right)^{\frac{1}{p}} + \left(\sum_{k=1}^{n} y_k^p\right)^{\frac{1}{p}}\right) \left(\sum_{k=1}^{n}(x_k+y_k)^p\right)^{\frac{1}{q}}.$$

From this, one gets

$$\left(\sum_{k=1}^{n}(x_k+y_k)^p\right)^{\frac{1}{p}} \leq \left(\sum_{k=1}^{n} x_k^p\right)^{\frac{1}{p}} + \left(\sum_{k=1}^{n} y_k^p\right)^{\frac{1}{p}}.$$

Several inequalities can also be obtained based on the following observation, which follows from the Jensen inequality: if f is a convex function defined on $(0, \infty)$, and $x_1, x_2, \ldots, x_n > 0$, $y_1, y_2, \ldots, y_n > 0$, then

$$f\left(\frac{\sum_{k=1}^{n} x_k y_k}{\sum_{k=1}^{n} x_k}\right) \leq \frac{\sum_{k=1}^{n} x_k f(y_k)}{\sum_{k=1}^{n} x_k}.$$

Another example is provided by the inequality

$$(x_1 x_2 \ldots x_n)^{\frac{x_1+x_2+\ldots+x_n}{n}} \leq x_1^{x_1} x_2^{x_2} \ldots x_n^{x_n}, \quad \forall x_1, x_2, \ldots, x_n > 0,$$

which can be obtained by combining the means inequality and the Jensen inequality for the convex function $f : (0, \infty) \to \mathbb{R}$, $f(x) = x \ln x$.

Another result is obtained on the same basis as the previous ones, but is formulated in a slightly different way.

Theorem 2.2.27. *Take $p > 1$ and $a, b > 0$. Then*
(i) $\inf_{t>0}\left[\frac{1}{p} t^{\frac{1}{p}-1} a + \left(1 - \frac{1}{p}\right) t^{\frac{1}{p}} b\right] = a^{\frac{1}{p}} b^{1-\frac{1}{p}}.$
(ii) $\inf_{0<t<1}\left[t^{1-p} a^p + (1-t)^{1-p} b^p\right] = (a+b)^p.$

Proof (i) The function e^x is (strictly) convex on $\mathbb{R}$. Therefore, for every $t > 0$,

$$a^{\frac{1}{p}} b^{1-\frac{1}{p}} = \left[t^{\frac{1}{p}-1} a\right]^{\frac{1}{p}} \left[t^{\frac{1}{p}} b\right]^{1-\frac{1}{p}}$$

$$= e^{\frac{1}{p} \ln\left(t^{\frac{1}{p}-1} a\right) + \left(1-\frac{1}{p}\right) \ln\left(t^{\frac{1}{p}} b\right)}$$

$$\leq \frac{1}{p} e^{\ln\left(t^{\frac{1}{p}-1} a\right)} + \left(1 - \frac{1}{p}\right) e^{\ln\left(t^{\frac{1}{p}} b\right)}$$

$$= \frac{1}{p} t^{\frac{1}{p}-1} a + \left(1 - \frac{1}{p}\right) t^{\frac{1}{p}} b.$$

The equality is obtained when $t = \frac{a}{b}$.

(ii) For $p > 1$, the function $f : (0, \infty) \to \mathbb{R}$, $f(u) = u^p$ is strictly convex. Consequently,

$$(a + b)^p = \left[t\frac{a}{t} + (1 - t)\frac{b}{1 - t} \right]^p$$

$$\leq t\left(\frac{a}{t}\right)^p + (1 - t)\left(\frac{b}{1 - t}\right)^p = t^{1-p} a^p + (1 - t)^{1-p} b^p$$

for any $0 < t < 1$. The equality holds for $t = \frac{a}{a+b}$. $\qquad\square$

Besides these well-known inequalities, we present some refinements and generalizations. The next result is an improvement of Young inequality.

Proposition 2.2.28. *Take $p \in (1, 2]$ and $q \in \mathbb{R}$ such that $p^{-1} + q^{-1} = 1$. Then for every $x, y > 0$, one has*

$$\frac{1}{q}\left(x^{\frac{p}{2}} - y^{\frac{q}{2}}\right)^2 \leq \frac{x^p}{p} + \frac{x^q}{q} - xy \leq \frac{1}{p}\left(x^{\frac{p}{2}} - y^{\frac{q}{2}}\right)^2.$$

Proof For $p = 2$, one has equalities. Suppose $p \in (1, 2)$. Then one must have $q > 2$. Let us prove the first inequality, which reduces to

$$\frac{2 - p}{p}x^p + \frac{2}{q}x^{\frac{p}{2}}y^{\frac{q}{2}} - xy \geq 0.$$

Consider the function $f : [0, \infty) \to \mathbb{R}$, $f(y) = \frac{2-p}{p}x^p + \frac{2}{q}x^{\frac{p}{2}}y^{\frac{q}{2}} - xy$. Its only critical point is $y = x^{p-1}$. Since $f''(y) = x^{\frac{p}{2}}y^{\frac{q}{2}-1}\left(\frac{q}{2} - 1\right) > 0$ for any $y > 0$, one deduces that $y = x^{p-1}$ is a global minimum for f. From $f(x^{p-1}) = 0$, one gets the conclusion. The other inequality can be similarly proved. $\qquad\square$

We present next one of the most important and profound numerical inequalities, due to Hardy.

Theorem 2.2.29 (Hardy Inequality). *Take $p > 1$ and a sequence $(a_n)_{n \in \mathbb{N}^*} \subset [0, \infty)$. Then the next inequality holds:*

$$\sum_{n=1}^{\infty}\left(\frac{a_1 + a_2 + \dots + a_n}{n}\right)^p \leq \left(\frac{p}{p - 1}\right)^p \sum_{n=1}^{\infty} a_n^p. \qquad (2.2.10)$$

If the series $\sum_{n=1}^{\infty} a_n^p$ is convergent, the equality holds if and only if (a_n) is constantly equal to 0. Moreover, the constant of the right-hand side is sharp (cannot be made smaller).

Proof If all the terms a_n are equal to zero, the conclusion is obvious. Suppose that at least one of these terms is strictly positive. Moreover, suppose that the series $\sum_{n=1}^{\infty} a_n^p$

is convergent, because otherwise the result is again trivial. Denote, for every $n \in \mathbb{N}^*$, the partial sum of the sequence (a_n) by S_n, i.e., $S_n = a_1 + a_2 + \ldots + a_n$. Also, denote

$$A_n := \frac{S_n}{n}, \ \forall n \in \mathbb{N}^*.$$

For completeness, take $S_0 = A_0 = 0$. We will use the following elementary inequality:

$$(n + 1)xy^n \le x^{n+1} + ny^{n+1}, \ \forall x, y \ge 0, \ n \in \mathbb{N}^*, \tag{2.2.11}$$

where equality holds if and only if $x = y$.

Then, for a fixed $n \in \mathbb{N}^*$, one has

$$
\begin{aligned}
A_n^p - \frac{p}{p-1}A_n^{p-1}a_n &= A_n^p - \frac{p}{p-1}(nA_n - (n-1)A_{n-1})A_n^{p-1} \\
&= A_n^p \left(1 - \frac{np}{p-1}\right) + \frac{(n-1)p}{p-1}A_n^{p-1}A_{n-1} \\
&\le A_n^p \left(1 - \frac{np}{p-1}\right) + \frac{(n-1)}{p-1}((p-1)A_n^p + A_{n-1}^p) \\
&= \frac{1}{p-1}\left[(n-1)A_{n-1}^p - nA_n^p\right].
\end{aligned}
$$

Therefore,

$$A_n^p - \frac{p}{p-1}A_n^{p-1}a_n \le \frac{1}{p-1}\left[(n-1)A_{n-1}^p - nA_n^p\right], \ \forall n \in \mathbb{N}^*. \tag{2.2.12}$$

Fix $N \in \mathbb{N}^*$. For $n = 1, 2, \ldots, N$, we write the relations (2.2.12) and then compute their sum. We deduce then

$$\sum_{n=1}^{N} A_n^p - \frac{p}{p-1}\sum_{n=1}^{N} A_n^{p-1}a_n \le -\frac{NA_N^p}{p-1} \le 0,$$

from where

$$\sum_{n=1}^{N} A_n^p \le \frac{p}{p-1}\sum_{n=1}^{N} A_n^{p-1}a_n.$$

For the right-hand side of this relation, we use the Hölder inequality to deduce

$$\sum_{n=1}^{N} A_n^p \le \frac{p}{p-1}\left(\sum_{n=1}^{N} a_n^p\right)^{\frac{1}{p}}\left(\sum_{n=1}^{N} A_n^p\right)^{\frac{p-1}{p}}.$$

From here we get, after taking the power p,

$$\sum_{n=1}^{N} A_n^p \le \left(\frac{p}{p-1}\right)^p\sum_{n=1}^{N} a_n^p.$$

By making $N \to \infty$, one gets (2.2.10).

In order to have equality in the final relation, one must also have equality when inequality (2.2.11) is applied, which reduces to $A_n = A_{n-1}$, for any $n \in \mathbb{N}^*$. This proves that the sequence (a_n) is constantly equal to 0.

The same conclusion follows when one analyzes the case of equality in the Hölder inequality when it is applied at a particular point. Therefore, from this analysis, we must have that the sequences $(a_n^p)_{n \in \mathbb{N}^*}$ and $(A_n^p)_{n \in \mathbb{N}^*}$ are proportional. If a finite number of terms a_n are nonzero, this proportionality cannot hold, hence the inequality is strict. Otherwise, if all the terms a_n are strictly positive, then there must exist a $c \in \mathbb{R}$ such that $a_n = cA_n$ for any $n \in \mathbb{N}^*$. We deduce from this fact that $c = 1$ and, moreover, the sequence (a_n) is constant, but in this case the series $\sum_{n=1}^{\infty} a_n^p$ diverges. Therefore, if the series $\sum_{n=1}^{\infty} a_n^p$ converges, the equality holds if and only if (a_n) is constantly equal to 0.

In order to show that one cannot decrease the constant $\left(\frac{p}{p-1}\right)^p$ from the right-hand side, we use the following argument. Take $N \in \mathbb{N}^*$ and $a_n = n^{-\frac{1}{p}}$, for $1 \leq n \leq N$ and $a_n = 0$ for $n > N$. Then

$$\sum_{n=1}^{\infty} a_n^p = \sum_{n=1}^{N} \frac{1}{n},$$

and for $n \leq N$,

$$S_n = \sum_{i=1}^{n} i^{-\frac{1}{p}} > \int_1^n x^{-\frac{1}{p}} dx = \frac{p}{p-1} \left(n^{\frac{p-1}{p}} - 1 \right).$$

Therefore

$$\left(\frac{S_n}{n} \right)^p > \left(\frac{p}{p-1} \right)^p \left(\frac{n^{\frac{p-1}{p}} - 1}{n} \right)^p = \left(\frac{p}{p-1} \right)^p \left(n^{-\frac{1}{p}} - \frac{1}{n} \right)^p.$$

Then there exists $(\varepsilon_n) \subset [0, \infty)$ such that $\varepsilon_n \to 0$ and

$$\left(\frac{S_n}{n} \right)^p > \left(\frac{p}{p-1} \right)^p \left(\frac{1 - \varepsilon_n}{n} \right),$$

an assertion which follows based on the fact that

$$\lim_{n \to \infty} n \left(n^{-\frac{1}{p}} - \frac{1}{n} \right)^p = \lim_{n \to \infty} \frac{\left(\left(\frac{1}{n} \right)^{\frac{1}{p}} - \frac{1}{n} \right)^p}{\frac{1}{n}} = \lim_{x \to 0+} \frac{\left(x^{\frac{1}{p}} - x \right)^p}{x}$$

$$= \lim_{x \to 0+} \frac{\left(x^{\frac{1}{p}} - x \right)^p}{\left(x^{\frac{1}{p}} \right)^p} = \lim_{x \to 0+} \left(1 - x^{1 - \frac{1}{p}} \right)^p = 1.$$

Consequently,

$$\sum_{n=1}^{\infty} \left(\frac{S_n}{n} \right)^p > \sum_{n=1}^{N} \left(\frac{S_n}{n} \right)^p > \left(\frac{p}{p-1} \right)^p \sum_{n=1}^{N} \left(\frac{1 - \varepsilon_n}{n} \right) = \left(\frac{p}{p-1} \right)^p \sum_{n=1}^{N} (1 - \varepsilon_n) a_n^p.$$

Then there exists $v_N \overset{N \to \infty}{\to} 0$ such that

$$\sum_{n=1}^{\infty} \left(\frac{S_n}{n} \right)^p > (1 - v_N) \left(\frac{p}{p-1} \right)^p \sum_{n=1}^{N} a_n^p. \tag{2.2.13}$$

In order to justify this relation, we remark first that it reduces to the fact that there exists $v_N \overset{N \to \infty}{\to} 0$ such that

$$v_N \sum_{n=1}^{N} \frac{1}{n} > \sum_{n=1}^{N} \frac{\varepsilon_n}{n}.$$

On the other hand, this becomes clear by the use of the Stolz-Cesàro criterion (Proposition 1.1.24), one has

$$\lim_{N \to \infty} \frac{\sum_{n=1}^{N} \frac{\varepsilon_n}{n}}{\sum_{n=1}^{N} \frac{1}{n}} = \lim_{N \to \infty} \frac{\frac{\varepsilon_{N+1}}{N+1}}{\frac{1}{N+1}} = 0.$$

Therefore, the relation (2.2.13) is true and, by taking into account the last part of the proof, we deduce that the value of the constant $\left(\frac{p}{p-1} \right)^p$ cannot be smaller. The proof is now complete. $\qquad \square$

We now present the Carleman inequality.

Theorem 2.2.30 (Carleman Inequality). *Let $(a_n)_{n \in \mathbb{N}^*} \subset [0, \infty)$ be a sequence. Then the next inequality holds:*

$$\sum_{n=1}^{\infty} (a_1 a_2 \ldots a_n)^{\frac{1}{n}} \le e \sum_{n=1}^{\infty} a_n.$$

Proof Again, without losing the generality, we may suppose that all the terms are strictly positive. We use the Hardy inequality, where a_n is replaced by $a_n^{\frac{1}{p}}$. Then we can write

$$\sum_{n=1}^{\infty} \left(\frac{a_1^{\frac{1}{p}} + a_2^{\frac{1}{p}} + \ldots + a_n^{\frac{1}{p}}}{n} \right)^p \le \left(\frac{p}{p-1} \right)^p \sum_{n=1}^{\infty} a_n.$$

But

$$\left(\frac{a_1^{\frac{1}{p}} + a_2^{\frac{1}{p}} + \ldots + a_n^{\frac{1}{p}}}{n} \right)^p = e^{\frac{\ln \left(a_1^{\frac{1}{p}} + a_2^{\frac{1}{p}} + \ldots + a_n^{\frac{1}{p}} \right) - \ln n}{\frac{1}{p}}} \overset{p \to \infty}{\to} (a_1 a_2 \ldots a_n)^{\frac{1}{n}}.$$

For passing to the limit, we have utilized the fact that

$$\lim_{x \to 0} \frac{\ln(a_1^x + a_2^x + \ldots + a_n^x) - \ln n}{x} = \frac{\ln(a_1 a_2 \ldots a_n)}{n},$$

which easily follows based on L'Hôpital Rule. Since, on the other hand, $\left(\frac{p}{p-1} \right)^p \overset{p \to \infty}{\to} e$ (increasingly), we get the desired inequality. $\qquad \square$

We end this section by presenting the Kantorovici inequality.

Theorem 2.2.31 (Kantorovici Inequality). *Let A be a symmetric and positive definite square matrix of dimension p. Then for every $x \in \mathbb{R}^p$, one has*

$$\|x\|^4 \le \left\langle (Ax^t)^t, x \right\rangle \cdot \left\langle (A^{-1}x^t)^t, x \right\rangle \le \frac{1}{4} \left(\sqrt{\frac{\lambda_1}{\lambda_p}} + \sqrt{\frac{\lambda_p}{\lambda_1}} \right)^2 \|x\|^4,$$

where λ_1 and λ_p are respectively the greatest and the smaller eigenvalue of A.

Proof It is sufficient to prove the inequality for any x of unit norm. As we observed in the first chapter, the eigenvalues of A are strictly positive reals and, without loss of generality, one may decreasingly arrange them: $\lambda_1 \ge \lambda_2 \ge \dots \ge \lambda_p$. We denote the diagonal matrix having the eigenvalues on its main diagonal (in the mentioned order) by $D := \operatorname{diag}(\lambda_1, \lambda_2, \dots, \lambda_p)$. We also know that there exists an orthogonal matrix B such that $A = B^t D B$. Then $A^{-1} = (B^t D B)^{-1} = B^t D^{-1} B$, and $D^{-1} = \operatorname{diag}(\lambda_1^{-1}, \lambda_2^{-1}, \dots, \lambda_p^{-1})$. Hence

$$\left\langle (Ax^t)^t, x \right\rangle = \left\langle \left(B^t D B x^t \right)^t, x \right\rangle = \left\langle (DBx^t)^t, (Bx^t)^t \right\rangle$$

and

$$\left\langle (A^{-1}x^t)^t, x \right\rangle = \left\langle \left(B^t D^{-1} B x^t \right)^t, x \right\rangle = \left\langle (D^{-1}Bx^t)^t, (Bx^t)^t \right\rangle.$$

On the other hand, the mapping $x \mapsto (Bx^t)^t$ is a bijection from the unit sphere of $\mathbb{R}^p$ into itself, hence in order to get the desired conclusion it is sufficient to prove that for every $u \in \mathbb{R}^p$ with $\|u\| = 1$, one has

$$1 \le \left\langle (Du^t)^t, u \right\rangle \cdot \left\langle (D^{-1}u^t)^t, u \right\rangle \le \frac{1}{4} \left(\sqrt{\frac{\lambda_1}{\lambda_p}} + \sqrt{\frac{\lambda_p}{\lambda_1}} \right)^2.$$

If $\lambda_1 = \lambda_p$, then one has the equality. Suppose that $\lambda_p < \lambda_1$. One has

$$\left\langle (Du^t)^t, u \right\rangle = \sum_{i=1}^{p} u_i^2 \lambda_i; \quad \left\langle (Du^t)^t, u \right\rangle = \sum_{i=1}^{p} u_i^2 \frac{1}{\lambda_i}.$$

Then the first inequality becomes

$$1 \le \left(\sum_{i=1}^{p} u_i^2 \frac{1}{\lambda_i} \right) \left(\sum_{i=1}^{p} u_i^2 \lambda_i \right),$$

i.e.,

$$\frac{1}{\left(\sum_{i=1}^{p} u_i^2 \lambda_i \right)} \le \sum_{i=1}^{p} u_i^2 \frac{1}{\lambda_i}.$$

Since $\sum_{i=1}^{p} u_i^2 = 1$, the previous inequality follows from Jensen inequality applied to the convex function $(0, \infty) \ni x \mapsto \frac{1}{x}$.

For the second inequality, observe that for any $i \in \overline{1, p}$,

$$\frac{1}{\lambda_i} \leq \frac{1}{\lambda_1} + \frac{1}{\lambda_p} - \frac{\lambda_i}{\lambda_1 \lambda_p},$$

because $\lambda_p \leq \lambda_i \leq \lambda_1$. Hence

$$\sum_{i=1}^{p} u_i^2 \frac{1}{\lambda_i} \leq \frac{1}{\lambda_1} + \frac{1}{\lambda_p} - \frac{\sum_{i=1}^{p} u_i^2 \lambda_i}{\lambda_1 \lambda_p}.$$

One gets

$$\left(\sum_{i=1}^{p} u_i^2 \frac{1}{\lambda_i} \right) \left(\sum_{i=1}^{p} u_i^2 \lambda_i \right) \leq \left(\sum_{i=1}^{p} u_i^2 \lambda_i \right) \left(\frac{1}{\lambda_1} + \frac{1}{\lambda_p} - \frac{\sum_{i=1}^{p} u_i^2 \lambda_i}{\lambda_1 \lambda_p} \right)$$

$$= \frac{\left(\sum_{i=1}^{p} u_i^2 \lambda_i \right) \left(\lambda_1 + \lambda_p - \sum_{i=1}^{p} u_i^2 \lambda_i \right)}{\lambda_1 \lambda_p}.$$

The second-degree polynomial

$$\lambda \mapsto \frac{\lambda(\lambda_1 + \lambda_p - \lambda)}{\lambda_1 \lambda_p}$$

attains its maximum for $\lambda = \frac{\lambda_1 + \lambda_p}{2}$, hence

$$\left(\sum_{i=1}^{p} u_i^2 \frac{1}{\lambda_i} \right) \left(\sum_{i=1}^{p} u_i^2 \lambda_i \right) \leq \left(\frac{\lambda_1 + \lambda_p}{2} \right)^2 \frac{1}{\lambda_1 \lambda_p} = \frac{1}{4} \left(\sqrt{\frac{\lambda_1}{\lambda_p}} + \sqrt{\frac{\lambda_p}{\lambda_1}} \right)^2.$$

The proof is now complete. $\qquad\square$

2.3 Banach Fixed Point Principle

This section is dedicated to the Banach fixed point theorem (also known as Contraction Principle, or Banach Principle), which is one of the fundamental results in nonlinear analysis.

Take $f : \mathbb{R}^p \to \mathbb{R}^p$. A point $x \in \mathbb{R}^p$ for which $f(x) = x$ is called a fixed point of f. By a fixed point result we understand a result concerning the existence of the fixed points for a given function.

Banach fixed point theorem is a result which provides, under certain conditions, both the existence and the uniqueness of the fixed point. This theorem also creates the basis for getting several other remarkable mathematical results, such as, to name a few, the implicit function theorem, or theorems about the existence and the uniqueness of solution for differential equations or systems of differential equations.

2.3.1 Contractions and Fixed Points

We begin with a definition:

Definition 2.3.1. *Let $A \subset \mathbb{R}^p$. A mapping $f : A \to \mathbb{R}^q$ is a contraction on A if there exists a real constant $\lambda \in (0, 1)$ such that $\|f(x) - f(y)\| \leq \lambda \|x - y\|$ for any $x, y \in A$.*

Remark that λ does not depend on x and y, and by applying the function f to a pair of points from A, the distance between them shrinks (contracts). The contraction notion is a particular case of the concept of Lipschitz function (Definition 1.2.25), hence, in particular, every contraction is a (uniformly) continuous function (Proposition 1.2.26).

The next assertion, often used in order to prove that a function is Lipschitz, holds: if $A \subset \mathbb{R}^p$ is an open convex set, $f : A \to \mathbb{R}$ is of class C^1 on A, and there exists $M > 0$ such that $\|\nabla f(x)\| \leq M$ for any $x \in A$, then f is Lipschitz on A. The proof of this fact relies on Taylor formula: for every two points $x, y \in A$ we can apply Theorem 1.3.4 to the function f on $[x, y]$, hence we deduce the existence of a point $c_{x,y} \in (x, y) \subset A$ such that

$$\left| f(x) - f(y) \right| = \left| \nabla f(c_{x,y})(x - y) \right| \leq M \|x - y\|.$$

Accordingly, f is Lipschitz on A. We now formulate a particular case of this observation.

Proposition 2.3.2. *Let $A \subset \mathbb{R}^p$ be an open convex set and $f : A \to \mathbb{R}$ of class C^1 on A, satisfying the condition that there exists $M \in (0, 1)$ such that $\|\nabla f(x)\| \leq M$ for any $x \in A$. Then f is a contraction on A.*

Obviously, for $p = 1$ we may consider in the previous proposition also closed, or half-open intervals.

Example 2.3.3. *The function $f : \mathbb{R}^p \to \mathbb{R}^p$ defined by $f(x) = \frac{1}{2}x$ is a contraction of $\mathbb{R}^p$ into itself, because:*

$$\|f(x) - f(y)\| = \frac{1}{2} \|x - y\|.$$

Example 2.3.4. *Take $A = [0, \infty)$ and the function $f : A \to A$, given by $f(x) = \frac{1}{1+x^2}$. The function f is a contraction on A, and this fact can be proven using Proposition 2.3.2. The derivative of the function f is*

$$f'(x) = \left(\frac{1}{1 + x^2} \right)' = -\frac{2x}{(1 + x^2)^2},$$

hence

$$\left| f'(x) \right| = \frac{2x}{(1 + x^2)^2}.$$

Define the auxiliary function $g : [0, \infty) \to [0, \infty)$ given by $g(x) = \frac{2x}{(1+x^2)^2}$. Its derivative is

$$
\begin{aligned}
g'(x) &= \frac{2(1+x^2)^2 - 2x \cdot 2(1+x^2) \cdot 2x}{(1+x^2)^4} \\
&= \frac{2(1+x^2)(1+x^2 - 4x^2)}{(1+x^2)^4} = 2\frac{1 - 3x^2}{(1+x^2)^3}.
\end{aligned}
$$

Remark that $x = \frac{1}{\sqrt{3}}$ is the positive maximum for the function g and $g\left(\frac{1}{\sqrt{3}}\right) = \frac{9}{8\sqrt{3}} < 1$. Hence $\left|f'(x)\right| \le \frac{9}{8\sqrt{3}} < 1$ for every $x \in [0, \infty)$. It follows from Proposition 2.3.2 that f is a contraction.

We now formulate the announced principle. We mention here that the Banach fixed point theorem was proved by the Polish mathematician Stefan Banach in 1922 in the framework of complete normed vector spaces.

Theorem 2.3.5 (Banach Principle). *Let $A \subset \mathbb{R}^p$ be a closed nonempty set and $f : A \to A$ be a contraction. Then f admits a unique fixed point.*

Proof As usual in the case of existence and uniqueness results, we divide the proof into two steps. We prove first the existence, and then the uniqueness of the fixed point. According to the definition, there exists a real number $\lambda \in (0, 1)$ such that

$$
\left\|f(x) - f(y)\right\| \le \lambda \left\|x - y\right\|
$$

for any $x, y \in A$. Consider $x_0 \in A$ as an arbitrary point and denote $x_1 = f(x_0)$, $x_2 = f(x_1), \dots, x_n = f(x_{n-1})$, an operation which can be made for every natural n. Observe that $x_2 = f(f(x_0)) = f^2(x_0)$ and, in general, $x_n = f^n(x_0)$ (we have denoted f^2 instead of $f \circ f$, and, in general, f^n instead of n times $f \circ f \circ \dots \circ f$). We prove that $(x_n)_{n \in \mathbb{N}}$ is a Cauchy sequence. The next relations hold

$$
\|x_2 - x_1\| = \left\|f(x_1) - f(x_0)\right\| \le \lambda \|x_1 - x_0\|,
$$

$$
\|x_3 - x_2\| = \left\|f(x_2) - f(x_1)\right\| \le \lambda \|x_2 - x_1\| \le \lambda^2 \|x_1 - x_0\|.
$$

Using an inductive procedure, one gets the inequality

$$
\|x_{n+1} - x_n\| \le \lambda^n \|x_1 - x_0\|
$$

for any $n \in \mathbb{N}^*$. For $m, n \in \mathbb{N}^*$ arbitrarily taken, one can successively write:

$$
\begin{aligned}
\|x_{m+n} - x_n\| &\le \|x_{n+1} - x_n\| + \|x_{n+2} - x_{n+1}\| + \dots + \|x_{n+m} - x_{n+m-1}\| \\
&\le \lambda^n \|x_1 - x_0\| + \lambda^{n+1} \|x_1 - x_0\| + \dots + \lambda^{n+m-1} \|x_1 - x_0\| \\
&= \|x_1 - x_0\| (\lambda^n + \lambda^{n+1} + \dots + \lambda^{n+m-1}) = \|x_1 - x_0\| \lambda^n \frac{1 - \lambda^m}{1 - \lambda} \\
&\le \|x_1 - x_0\| \frac{\lambda^n}{1 - \lambda}.
\end{aligned}
$$

Hence,

$$\|x_{n+m} - x_n\| \le \frac{\lambda^n}{1 - \lambda} \|x_1 - x_0\| \tag{2.3.1}$$

for any $n, m \in \mathbb{N}^*$. If $\|x_1 - x_0\| = 0$, it follows that $f(x_0) = x_0$, i.e., x_0 is a fixed point and then the existence of the fixed point is assured. If $\|x_1 - x_0\| \ne 0$, then, using the fact that $\lambda \in (0, 1)$, one deduces that

$$\lim_{n \to \infty} \frac{\lambda^n}{1 - \lambda} = 0,$$

hence

$$\lim_{n \to \infty} \frac{\lambda^n}{1 - \lambda} \|x_1 - x_0\| = 0.$$

By using the ε characterization of this convergence, it follows that for any $\varepsilon > 0$, there exists $n_\varepsilon \in \mathbb{N}^*$, such that for any $n \ge n_\varepsilon$, one has

$$\frac{\lambda^n}{1 - \lambda} \|x_1 - x_0\| < \varepsilon.$$

By combining this relation with (2.3.1), it follows that for any $\varepsilon > 0$, there exists $n_\varepsilon \in \mathbb{N}^*$, such that for every $n \ge n_\varepsilon$, and every $m \in \mathbb{N}$, one has

$$\|x_{n+m} - x_n\| < \varepsilon.$$

This proves that (x_n) is a Cauchy sequence, and because $\mathbb{R}^p$ is a complete space, the sequence $(x_n)_{n \in \mathbb{N}}$ is convergent, so there exists $\bar{x} \in \mathbb{R}^p$ with $\lim_{n \to \infty} x_n = \bar{x}$. Since $(x_n) \subset A$ and A is closed, we deduce that $\bar{x} \in A$. Recall that the sequence is given by the relation

$$x_0 \in A, \ f(x_n) = x_{n+1}, \ \forall n \in \mathbb{N}. \tag{2.3.2}$$

The function f is continuous (since every contraction has this property, see Proposition 1.2.26). Hence, from the properties of the continuous functions, the limit of the sequence $(f(x_n))_n$ exists and equals $f(\bar{x})$. In the relation (2.3.2), we pass to the limit for $n \to \infty$ and we get

$$\lim_{n \to \infty} f(x_n) = \lim_{n \to \infty} x_{n+1},$$

i.e.,

$$f(\bar{x}) = \bar{x},$$

hence $\bar{x}$ is a fixed point. The existence is proved.

In order to prove the uniqueness, suppose there are two different fixed points x and y. Then

$$\|x - y\| = \|f(x) - f(y)\| \le \lambda \|x - y\|.$$

Since $\|x - y\| > 0$, it follows that $1 \le \lambda$, which is absurd. Therefore, there exists a unique fixed point for f. □

The above result is very important and deserves an extended comment. The analysis of the statement, and also of the proof, provides us with some useful conclusions.

Firstly, observe (in the proof of the existence) how the fixed point was obtained: for every initial point $x_0 \in A$, the sequence given by the relation (2.3.2) converges to the unique fixed point $\overline{x}$ of the mapping f. Coming back to the above inequalities, the next relations hold:

$$
\begin{aligned}
\|x_n - x_0\| &\leq \|x_1 - x_0\| + \|x_2 - x_1\| + \cdots + \|x_n - x_{n-1}\| \\
&\leq \|x_1 - x_0\| + \lambda \|x_1 - x_0\| + \cdots + \lambda^{n-1} \|x_1 - x_0\| \\
&= \|x_1 - x_0\| (1 + \lambda + \lambda^2 + \cdots + \lambda^{n-1}) = \|x_1 - x_0\| \frac{1 - \lambda^n}{1 - \lambda},
\end{aligned}
$$

and, by passing to the limit for $n \to \infty$, we deduce that

$$
\|x_0 - \overline{x}\| \leq \frac{1}{1 - \lambda} \|x_0 - f(x_0)\| .
$$

In this way, we see that for every $x \in A$, one has

$$
\|x - \overline{x}\| \leq \frac{1}{1 - \lambda} \|x - f(x)\| . \tag{2.3.3}
$$

Moreover, one gets the following estimations:

$$
\|x_n - \overline{x}\| \leq \|x_1 - x_0\| \frac{\lambda^n}{1 - \lambda} \tag{2.3.4}
$$

for any $n \in \mathbb{N}^*$, which follows from (2.3.1) passing to the limit for $m \to \infty$.

The "a priori" estimation we get in this way is useful in determining the maximum number of steps of the iteration (2.3.2) which one needs, in order to obtain the desired precision in the estimation of the fixed point, by knowing the initial value x_0 and the value $x_1 = f(x_0)$, a fact which we will use in the study of some algorithms. More exactly, for getting an error smaller than $\varepsilon > 0$, one needs that

$$
\|x_1 - x_0\| \frac{\lambda^n}{1 - \lambda} < \varepsilon,
$$

which drives us to the conclusion that we need a number n of iterations greater than the value

$$
\left| \frac{\ln \varepsilon + \ln(1 - \lambda) - \ln(\|x_1 - x_0\|)}{\ln \lambda} \right| . \tag{2.3.5}
$$

For example, if we need ε to be of the form 10^{-m}, the size order of n is of the type

$$
\frac{m}{|\ln \lambda|} + \text{constant},
$$

and for additional decimal places of accuracy, one needs to supplement the number of iterations by $|\ln \lambda|^{-1}$. Therefore, the closer λ approaches to 0, this value is smaller, and we will need to iterate less in order to obtain the desired precision. Also, one observes from the relation (2.3.5) that a smaller value of $\|x_1 - x_0\|$ implies that the decrease of the number of iterations which are necessary for a prescribed precision. Therefore, in case of some algorithms, it is preferable to start from points as close as possible to the initial point. We will later discuss examples on the computational effects described here in the Chapter 6.

One can also get the following estimate:

$$\|x_n - \overline{x}\| \le \frac{\lambda}{1 - \lambda} \|x_n - x_{n-1}\|, \tag{2.3.6}$$

which is obtained by passing to the limit for $m \to \infty$ in the relation

$$\|x_{n+m} - x_n\| \le \|x_{n+1} - x_n\| + \|x_{n+2} - x_{n+1}\| + \cdots + \|x_{n+m} - x_{n+m-1}\|$$
$$\le \lambda \|x_n - x_{n-1}\| + \lambda^2 \|x_n - x_{n-1}\| + \cdots + \lambda^m \|x_n - x_{n-1}\|$$
$$= (\lambda + \lambda^2 + \dots + \lambda^m) \|x_n - x_{n-1}\|.$$

The speed of convergence of the sequence (x_n) is given by the approximation

$$\|x_{n+1} - \overline{x}\| \le \lambda \|x_n - \overline{x}\|,$$

which is deduced from the relations

$$\|x_{n+1} - \overline{x}\| = \|f(x_n) - f(\overline{x})\| \le \lambda \|x_n - \overline{x}\|.$$

More details regarding the speed of convergence will be given in Chapter 6.

Another inequality which follows immediately from the definition of the contraction is

$$\|x_n - \overline{x}\| \le \lambda^n \|x_0 - \overline{x}\|, \ \forall n \in \mathbb{N}.$$

Besides these observations regarding the convergence type of the iterations towards the fixed point, we make some other remarks concerning the assumptions of the Banach Principle.

Remark 2.3.6. *The Banach fixed point theorem shows not only the existence, but the uniqueness of the fixed point and, at the same time, shows us a method to approximate the fixed point $\overline{x}$, and allows us to emphasize an estimation of the error produced by considering this approximation. This approximation method of the solution by the terms of the sequence $x_n = f^n(x_0)$ is called the successive approximations method, or the Picard method, after the name of the French mathematician Charles Émile Picard, which initiated it in 1890.*

Remark 2.3.7. *The assumption $\lambda < 1$ is essential both for the existence, as for the uniqueness of the fixed point. One can see, for instance, that for the identity map $f(x) = x$, for any $x \in \mathbb{R}$, every point of $\mathbb{R}$ is a fixed point, while the map $f(x) = x+1$, for any $x \in \mathbb{R}$, does not have any fixed point. In both cases, $\lambda = 1$.*

Remark 2.3.8. *If A is not closed, one loses the completeness argument and the conclusion of Banach Principle does not hold. For example, the mapping $f : (0, 1] \to (0, 1]$ given by $f(x) = \frac{x}{2}$ does not have any fixed point, although is a contraction.*

Aiming to further illustrate the assumptions of the Banach Principle, we introduce a weaker notion than the one of contraction.

Definition 2.3.9. *A function $f : \mathbb{R}^p \to \mathbb{R}^p$ is called a weak contraction if*

$$\forall x, y \in \mathbb{R}^p, \ x \neq y, \ \|f(x) - f(y)\| < \|x - y\|.$$

Example 2.3.10. *There exist weak contractions, defined on closed sets, without fixed points. We consider the next example. Take $f : \mathbb{R} \to \mathbb{R}$, defined by*

$$f(x) = 1 + x - \frac{x}{1 + |x|}.$$

Obviously, $f(x) > x$ for any $x \in \mathbb{R}$, hence f does not have any fixed point. On the other hand, f is a weak contraction, which one can show by considering the relation

$$\left|f(x) - f(y)\right| = \left|(x - y) - \left(\frac{x}{1 + |x|} - \frac{y}{1 + |y|}\right)\right|$$

in several situations.
    If $x, y \geq 0$, then:

$$\left|f(x) - f(y)\right| = \left|x - y - \left(\frac{x}{1 + x} - \frac{y}{1 + y}\right)\right| = \left|x - y - \frac{x - y}{(1 + x)(1 + y)}\right|$$

$$= \left|(x - y)\left(1 - \frac{1}{(1 + x)(1 + y)}\right)\right| < |x - y|$$

because $x \neq y$ and

$$1 - \frac{1}{(1 + x)(1 + y)} \in (0, 1).$$

If $x, y < 0$, one has:

$$\left|f(x) - f(y)\right| = \left|x - y - \left(\frac{x}{1 - x} - \frac{y}{1 - y}\right)\right| = \left|(x - y) - \frac{x - y}{(1 - x)(1 - y)}\right|$$

$$= \left|(x - y)\left(1 - \frac{1}{(1 - x)(1 - y)}\right)\right| < |x - y|$$

because $x \neq y$ and

$$1 - \frac{1}{(1 - x)(1 - y)} \in (0, 1).$$

If $x > 0$ and $y < 0$, one has:

$$\left|f(x) - f(y)\right| < \left|x - y - \left(\frac{x}{1 + x} - \frac{y}{1 - y}\right)\right| = \left|(x - y) - \frac{(x - y) - 2xy}{(1 + x)(1 - y)}\right|$$

$$= x - y - \frac{(x - y) - 2xy}{(1 + x)(1 - y)} < x - y = |x - y|,$$

where the last inequality can be obtained by direct calculation.

Observe that the Picard iterations associated to f diverge to $+\infty$ for any initial data x_0: if $x_0 \geq 0$, then (x_n) is strictly increasing, has positive values and cannot have a finite limit, while if $x_0 < 0$, then the terms of the sequence become positive from a certain rank, and we are again in the previous framework.

The weak contractions may yet have a fixed point, if at least a sequence of Picard iterations has a convergent subsequence. More precisely, the next assertion holds.

Proposition 2.3.11. *Let $A \subset \mathbb{R}^p$ be a closed set and $f : A \to A$ be a function such that*

$$\|f(x) - f(y)\| < \|x - y\|, \ \forall x, y \in A, \ x \neq y,$$

i.e., f is a weak contraction on A. If there exists $a \in A$ such that the sequence of Picard iterations, having the initial data a, and given by $x_1 = f(a)$, $x_{n+1} = f(x_n)$, $n \geq 1$, has a subsequence which is convergent to a point $\overline{x} \in A$, then $\overline{x}$ is the unique fixed point of f on A.

Proof Suppose, by contradiction, that $f(\overline{x}) \neq \overline{x}$. Consider

$$D := \{(x, y) \in A \times A \mid x \neq y\}$$

and the mapping $g : D \to \mathbb{R}$ given by

$$g(x, y) := \frac{\|f(x) - f(y)\|}{\|x - y\|}.$$

From the assumptions made, we have that $g(x, y) < 1$. But $(\overline{x}, f(\overline{x})) \in D$, hence $g(\overline{x}, f(\overline{x})) < 1$. Let $(x_{n_k})_{k \in \mathbb{N}}$ be a subsequence of $(x_n)_{n \in \mathbb{N}}$ which converges to $\overline{x}$. Then

$$\|f(x_{n_k}) - f(\overline{x})\| < \|x_{n_k} - \overline{x}\|, \ \forall k \in \mathbb{N},$$

from which one deduces that $f(x_{n_k}) \to f(\overline{x})$. It follows that

$$\|x_{n_k} - f(x_{n_k})\| \to \|\overline{x} - f(\overline{x})\| \quad \text{and} \quad g(x_{n_k}, f(x_{n_k})) \to g(\overline{x}, f(\overline{x})) < 1.$$

Take r such that $g(\overline{x}, f(\overline{x})) < r < 1$. Taking into account the continuity of f (in particular, f is Lipschitz), there exists a $k_0 \in \mathbb{N}$ such that for every $k \geq k_0$,

$$\frac{1}{3} \|\overline{x} - f(\overline{x})\| < \|x_{n_k} - f(x_{n_k})\|$$

and $g(x_{n_k}, f(x_{n_k})) < r$ or, equivalently,

$$\|f(x_{n_k}) - f(f(x_{n_k}))\| < r \|x_{n_k} - f(x_{n_k})\|.$$

Then for every $i > k \geq k_0$, one has:

$$\frac{1}{3} \|\overline{x} - f(\overline{x})\| < \|x_{n_i} - f(x_{n_i})\| = \|f(x_{n_i-1}) - f(x_{n_i})\|$$

$$< \left\| x_{n_i-1} - x_{n_i} \right\| < \dots < \left\| f(x_{n_{i-1}}) - f(f(x_{n_{i-1}})) \right\|$$
$$< r \left\| x_{n_{i-1}} - f(x_{n_{i-1}}) \right\| < \dots < r^{i-k} \left\| x_{n_k} - f(x_{n_k}) \right\|.$$

We make $i \to \infty$ and we get $\left\| \overline{x} - f(\overline{x}) \right\| = 0$, which contradicts $(\overline{x}, f(\overline{x})) \in D$. The uniqueness of the fixed point is straightforward. $\qquad\square$

We present next some generalizations and consequences of the Banach Principle.

A first generalization of the Banach Principle mainly says that it is sufficient for one of the iterations of the function to be a contraction to obtain the conclusions of the Banach Principle. We will see that such an assumption is weaker than the condition that f is a contraction. Remark first that if f is a contraction with constant $\lambda < 1$, then f^n is a contraction with constant $\lambda^n < 1$, hence, in the assumptions of Banach Principle, has a unique fixed point. As every fixed point of f is a fixed point for every iteration, one deduces that f and all its iterations have the same fixed point. The role of the next theorem is also to formulate a partial converse of this observation.

Theorem 2.3.12. *Take $f : \mathbb{R}^p \to \mathbb{R}^p$. If there exists $q \in \mathbb{N}^*$ such that f^q is a contraction, then f has a unique fixed point. Moreover, for every initial data $x_0 \in \mathbb{R}^p$, the sequence of the Picard iterations converges to the fixed point of f.*

Proof Let $q \in \mathbb{N}^*$ be such that f^q is a contraction. From the Banach Principle, f^q has a unique fixed point, which we denote by $\overline{x}$. Observe, based on the previous comment, that $\overline{x}$ is the sole candidate to be a fixed point for f (any fixed point for f is a fixed point for any iteration). The next relations hold:

$$f(\overline{x}) = f(f^q(\overline{x})) = f^q(f(\overline{x})),$$

i.e., $f(\overline{x})$ is a fixed point for f^q. As f^q has a unique fixed point, one deduces that $f(\overline{x}) = \overline{x}$. Let us now prove the assertion concerning the convergence of the Picard iterations. We start from a fixed element $x_0 \in \mathbb{R}^p$ and we construct the associated Picard sequence $(f^n(x_0))_n$. We must prove that this sequence converges to $\overline{x}$. Take $r \in \overline{0, q-1}$. Then, the set of the terms of sequence $(f^n(x_0))_n$ is the union of the sets of terms of subsequences of the type $(f^{qk+r}(x_0))_k$. On the other hand, $(f^{qk+r}(x_0))_k$ can be seen as the sequence of Picard iterations associated to f^q, with the initial point $f^r(x_0)$, because:

$$f^{qk+r}(x_0) = (f^q)^k(f^r(x_0)).$$

Since f^q is a contraction, we know from the Banach Principle that all the Picard iterations of f^q converge to the fixed point $\overline{x}$. Hence all the q subsequences which partition the initial sequence have the same limit (i.e., $\overline{x}$), which shows that $\lim_{n\to\infty} f^n(x_0) = \overline{x}$ and the theorem is completely proved. $\qquad\square$

As in the case of Banach Principle, the result holds if $f : A \to A$, where A is a closed (hence, complete) subset of $\mathbb{R}^p$. Observe that in this theorem is not necessary for f to have any special property (even continuity of f is not assumed). For instance, the function $f : [0, 1] \to [0, 1]$ given by

$$f(x) = \begin{cases} 0 \text{ if } x \in [0, \frac{1}{2}] \\ \frac{1}{2} \text{ if } x \in (\frac{1}{2}, 1] \end{cases}$$

is discontinuous, but $f^2(x) = 0$ for any $x \in [0, 1]$, hence is a contraction. Of course, the unique fixed point of f is $\bar{x} = 0$.

Let us give now another example which proves that the assumption that an iterate of f is a contraction is weaker than the condition that f is a contraction. Consider the function $f : \mathbb{R} \to \mathbb{R}$ given by $f(x) = e^{-x}$. This function is not a contraction on $\mathbb{R}$: for instance

$$\left| f(-2) - f(0) \right| = e^2 - 1 > |-2 - 0| = 2.$$

But $f^2(x) = e^{-e^{-x}}$ is a contraction. To prove this, we evaluate the absolute value of the derivative (in order to apply Proposition 2.3.2):

$$\left| (f^2)'(x) \right| = \left| e^{-e^{-x}} e^{-x} \right| = e^{-x - e^{-x}} \le e^{-1} < 1, \ \forall x \in \mathbb{R},$$

based on the observation that

$$1 \le x + e^{-x}, \ \forall x \in \mathbb{R}.$$

Therefore, according to Theorem 2.3.12, there exists a unique fixed point $\bar{x}$ of the function f (hence $\bar{x} = e^{-\bar{x}}$) which can be approximated using the Picard iterations. A numerical calculation (made on a computer, see Chapter 6) suggests the approximate value $\bar{x} \simeq 0.567$. We will return in Chapter 6 to the possibility of approximating the fixed points in some concrete situations.

At the end of this section, we consider another two interesting consequences of the Banach fixed point principle. The first one refers at the continuous dependence of the fixed point with respect to a parameter.

Theorem 2.3.13. *Let $g : \mathbb{R}^p \times \mathbb{R}^q \to \mathbb{R}^p$ be a continuous function. Suppose that there exists $\alpha \in (0, 1)$ such that*

$$\left\| g(x, t) - g(y, t) \right\| \le \alpha \left\| x - y \right\|$$

for every $t \in \mathbb{R}^q$ and $x, y \in \mathbb{R}^p$. For fixed t in $\mathbb{R}^q$, denote by $\mu(t)$ the unique fixed point of the contraction $g(\cdot, t)$. Then the mapping $\mu : \mathbb{R}^q \to \mathbb{R}^p$ is continuous.

Proof Take $t_0 \in \mathbb{R}^q$ and $\varepsilon > 0$. The continuity of g in $(\mu(t_0), t_0)$ implies the existence of a number $\delta > 0$ such that for any fixed t, with $\|t - t_0\| < \delta$, one has

$$\left\| g(\mu(t_0), t) - g(\mu(t_0), t_0) \right\| < \varepsilon(1 - \alpha),$$

which is equivalent to

$$\|g(\mu(t_0), t) - \mu(t_0)\| < \varepsilon(1 - \alpha).$$

From the relation (2.3.3) and the above relations, we deduce that for any t with $\|t - t_0\| < \delta$, one has

$$\|\mu(t) - \mu(t_0)\| \leq \frac{\|\mu(t_0) - g(\mu(t_0), t)\|}{1 - \alpha} < \varepsilon.$$

This proves the theorem. $\qquad\qquad\square$

Theorem 2.3.14. *Let $f : \mathbb{R}^p \to \mathbb{R}^p$ be a contraction. Then the mapping $v : \mathbb{R}^p \to \mathbb{R}^p$ given by $v(x) = x + f(x)$ is a bicontinuous bijection (i.e., is a homeomorphism of $\mathbb{R}^p$).*

Proof It is clear that v is continuous. Moreover, f is injective because the relation $v(x) = v(y)$ and the contraction property of f imply $x = y$. Consider $g : \mathbb{R}^p \times \mathbb{R}^p \to \mathbb{R}^p$ given by

$$g(x, y) = y - f(x).$$

It is clear that g satisfies the property from the assumption of the previous theorem, hence the mapping $x \mapsto g(x, y)$ has a unique fixed point $\mu(y)$ for any $y \in \mathbb{R}^p$. Therefore,

$$\mu(y) = y - f(\mu(y)),$$

i.e.,

$$y = v(\mu(y)), \tag{2.3.7}$$

hence v is surjective. It remains to prove that v^{-1} is continuous. Relation (2.3.7) shows that the functions v and μ are inverse one to each other, and the continuity of μ (hence of v^{-1}) is assured by the previous theorem. Hence, v is a homeomorphism. $\qquad\square$

2.3.2 The Case of One Variable Functions

We now present some fixed point results for real functions of one variable. Geometrically, the fixed points of a function f in this context are those points x for which $(x, f(x))$ lays on the first bisector, i.e., the abscissae of points where the graph of f intersects the first bisector.

Obviously, the first result which we must mention is the particular case of the Banach Principle, which works for contractions which map a closed subset of $\mathbb{R}$ into itself. The first result which is different from the Banach Principle is the Knaster fixed point theorem.

Theorem 2.3.15 (Knaster). *Let $a, b \in \mathbb{R}$, $a < b$ and $f : [a, b] \to [a, b]$ be an increasing function. Then f has at least one fixed point.*

Proof Define the set

$$A := \{x \in [a, b] \mid f(x) \geq x\}.$$

It is clear, on one hand, that A is nonempty (because $a \in A$) and, on the other hand, that A is bounded (being a subset of $[a, b]$). Hence, according to the completeness axiom, A admits a supremum in $\mathbb{R}$. Denote this number by $\overline{x}$. Therefore, $\overline{x} = \sup A$ and it is clear that $\overline{x} \in [a, b]$. As $\overline{x} \geq x$ for any $x \in A$, the monotony of f allows us to write the inequality $f(\overline{x}) \geq f(x) \geq x$ for any $x \in A$. Hence, $f(\overline{x})$ is a majorant for A, so $f(\overline{x}) \geq \overline{x}$. It also follows from monotony that $f(f(\overline{x})) \geq f(\overline{x})$ and hence $f(\overline{x}) \in A$, i.e., $f(\overline{x}) \leq \overline{x}$. Consequently, one has the equality $f(\overline{x}) = \overline{x}$ and $\overline{x}$ is a fixed point. □

Remark 2.3.16. *If in the previous result one takes f to be decreasing, then the conclusion does not hold. One may consider the following counterexample: $f : [0, 1] \to [0, 1]$ given by*

$$f(x) = \begin{cases} 1 - x \ \text{if } x \in [0, \frac{1}{2}) \\ \frac{1}{2} - \frac{x}{2} \ \text{if } x \in [\frac{1}{2}, 1]. \end{cases}$$

It is clear that f is decreasing on $[0, 1]$, but still does not admit any fixed point.

The next result is simple, but will be useful in many situations.

Theorem 2.3.17. *Let $a, b \in \mathbb{R}$, $a < b$ and $f : [a, b] \to [a, b]$ be a continuous function. Then there exists $\overline{x} \in [a, b]$ such that $f(\overline{x}) = \overline{x}$, i.e., f has at least one fixed point.*

Proof Define the function $g : [a, b] \to \mathbb{R}$, given by $g(x) = f(x) - x$. It is obvious that g is continuous it is the difference of continuous functions and, in particular, it has the Darboux property. Obviously, because $f(a), f(b) \in [a, b]$, one has the inequalities

$$g(a) = f(a) - a \geq 0$$
$$g(b) = f(b) - b \leq 0,$$

hence $g(a) \cdot g(b) \leq 0$, and by the use of the Darboux property, one deduces that there exists a point $\overline{x} \in [a, b]$ such that $g(\overline{x}) = 0$. Therefore, $f(\overline{x}) = \overline{x}$ and the proof is now complete. □

Remark 2.3.18. *In this situation, the uniqueness of the fixed point is not ensured. The immediate example is the identity function of the $[a, b]$ interval. It is also essential that the interval is closed. For example, the function $f : [0, 1) \to [0, 1)$ given by $f(x) = \frac{x+1}{2}$ does not have any fixed point. It is equally essential that the interval is bounded: the function $f : [1, +\infty) \to [1, +\infty)$ given by $f(x) = x + x^{-1}$ does not have any fixed point. Finally, the result does not hold anymore in the case where the function is not defined on an interval. For instance, $f : [-2, -1] \cup [1, 2] \to [-2, -1] \cup [1, 2]$, $f(x) = -x$ does not have any fixed point.*

In some situations, we can say something about the structure of the set of fixed points.

Theorem 2.3.19. *Let $f : [0, 1] \to [0, 1]$ be a 1-Lipschitz function on $[0, 1]$. Then the set of the fixed points of f is a (possibly degenerate) interval.*

Proof Let
$$F := \{x \in [0, 1] \mid f(x) = x\}$$
be the set of fixed points of f. It is clear, from the continuity of f, that F is a compact set, so it admits a minimum and a maximum, denoted respectively by a and b. Obviously, $F \subset [a, b]$. If we fix now an arbitrary $\overline{x} \in [a, b]$, it is sufficient to prove that $\overline{x}$ is a fixed point of f, i.e., $\overline{x} \in F$. Since a is a fixed point of f and $a \leq \overline{x}$, one gets
$$f(\overline{x}) - a \leq \left|f(\overline{x}) - a\right| = \left|f(\overline{x}) - f(a)\right| \leq \overline{x} - a,$$
so $f(\overline{x}) \leq \overline{x}$. By a similar argument in the case of b, one has:
$$b - f(\overline{x}) \leq \left|b - f(\overline{x})\right| = \left|f(b) - f(\overline{x})\right| \leq b - \overline{x},$$
which shows that $f(\overline{x}) \geq \overline{x}$. Consequently, $f(\overline{x}) = \overline{x}$ and the proof is complete. $\square$

We now emphasize a supplemental condition, which assures the uniqueness of the fixed point.

Theorem 2.3.20. *Let $a, b \in \mathbb{R}$, $a < b$ and $f : [a, b] \to [a, b]$ be a continuous function on $[a, b]$, which is differentiable on (a, b) and has the property that $f'(x) \neq 1$ for any $x \in (a, b)$. Then f has a unique fixed point.*

Proof Theorem 2.3.17 assures the existence of the fixed point. If, by contradiction, there are two fixed points of f, denoted by x_1 and x_2, then the Lagrange Theorem applied to f on the interval $[x_1, x_2]$, implies the existence of a point $c \in (x_1, x_2) \subset (a, b)$ such that
$$f(x_1) - f(x_2) = f'(c)(x_1 - x_2).$$
But this relation implies $f'(c) = 1$, which contradicts the assumption. This proves the result. $\square$

There is an important difference between the proof of Banach fixed point theorem and the proof of Theorem 2.3.17 above, which is that we are not given any information concerning the position of the fixed point, or a method to approximate it. Recall that the successive approximations method from the Banach's proof works like this: we start with an arbitrary point x_0 (this time, from the interval $[a, b]$), and we construct by recurrence a sequence (the sequence of Picard approximations) from $[a, b]$ by the relation $x_{n+1} = f(x_n)$ for every $n \in \mathbb{N}$. All of the hard work in the proof of the Banach

Principle was to show that (x_n) converges. In case we know this, the continuity of f furnishes the conclusion that the limit of x_n, denoted by $\bar{x}$, is a fixed point for f. Indeed, we may write:

$$f(\bar{x}) = f(\lim_{n\to\infty} x_n) = \lim_{n\to\infty} f(x_n) = \lim_{n\to\infty} x_{n+1} = \bar{x}. \tag{2.3.8}$$

In order to prove that sequence of the successive approximations is convergent, one usually shows that this sequence is fundamental.

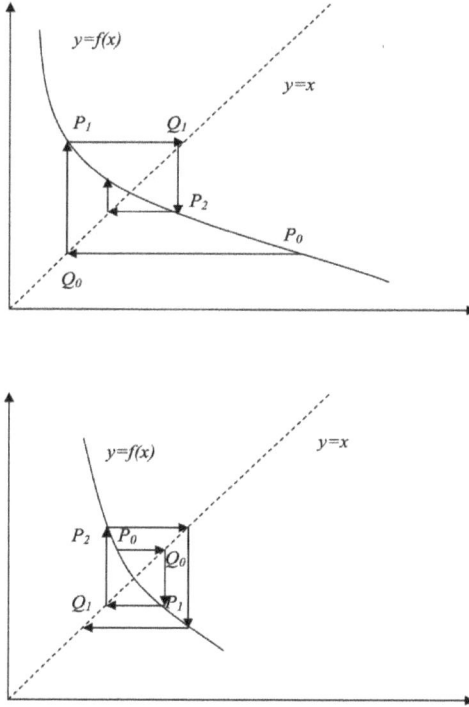

Figure 2.2: Behavior of the Picard iterations.

We can show that in certain conditions, the Picard iterations are convergent to a fixed point, outside the standard assumptions of the Banach Principle. This result is called the Picard convergence theorem, and its geometric meaning stems from the following discussion. Take $x_0 \in [a, b]$ arbitrary and consider the Picard iteration associated with the function f and the initial point x_0, $x_{n+1} = f(x_n)$, for any $n \in \mathbb{N}$. We may describe this procedure of determining the Picard sequence geometrically. Consider first $P_0(x_0, f(x_0))$ on the curve of equation $y = f(x)$ (the graph of f). Take the point $Q_0(f(x_0), f(x_0))$ on the first bisector $y = x$, and we project it vertically on the curve $y = f(x)$ in order to find the point $P_1(x_1, f(x_1))$. We project the point P_1 horizontally in Q_1, which lies on the first bisector, and continue this procedure by the vertical projection of Q_1 on $y = f(x)$ in the point $P_2(x_2, f(x_2))$. One can repeat this procedure in such a way that a "spider net" is obtained, where the fixed point of f is "captured" (i.e., the intersection of the curves $y = f(x)$ and $y = x$). Actually, one may observe that this "capturing" happens if the slopes of the tangents to the curve $y = f(x)$ are smaller in absolute value than the slope of the line $y = x$ (which equals 1). If this condition concerning the slopes is not satisfied, then the points $Q_0, Q_1, \ldots$ move away from the fixed point (see Figure 2.2 for the two different situations).

Therefore, the Picard Theorem analytically transposes this geometric observation.

Theorem 2.3.21 (Picard Theorem). *Let $a, b \in \mathbb{R}$, $a < b$ and $f : [a, b] \to [a, b]$ be a continuous function on $[a, b]$ and differentiable on (a, b), having the property that $|f'(x)| < 1$ for any $x \in (a, b)$. Then f has a unique fixed point, and the sequences of Picard iterations are convergent to the unique fixed point of f.*

Proof Theorem 2.3.20 assures the existence and the uniqueness of the fixed point. Denote it by $\overline{x}$. Take $x_0 \in [a, b]$ and let (x_n) be the Picard sequence which is generated starting at x_0. If for $n \in \mathbb{N}$ we have $x_n = x_{n+1}$, then x_n is the fixed point, and the sequence becomes stationary. In this case the convergence is obvious. Suppose then that (x_n) is not stationary. From the Lagrange Theorem (applied to the function f on an interval having endpoints $\overline{x}$ and x_n) one deduces the existence of an element $c_n \in (a, b)$ for which we can write the relation:

$$x_{n+1} - \overline{x} = f(x_n) - f(\overline{x}) = f'(c_n)(x_n - \overline{x}).$$

This inequality and the assumption on the derivative of f imply collectively that

$$|x_{n+1} - \overline{x}| < |x_n - \overline{x}|. \tag{2.3.9}$$

We prove now that (x_n) converges to $\overline{x}$. Since (x_n) is a bounded sequence (in $[a, b]$), it is sufficient to prove that every convergent subsequence has its limit equal to $\overline{x}$. Consider then a convergent subsequence $(x_{n_k})_k$ of (x_n) and denote by $x \in [a, b]$ its limit. Since (n_k) is strictly increasing, by applying inductively the inequality (2.3.9), one gets:

$$|x_{n_{k+1}} - \overline{x}| \le |x_{n_k+1} - \overline{x}| < |x_{n_k} - \overline{x}|. \tag{2.3.10}$$

But $|x_{n_k} - \overline{x}| \overset{k\to\infty}{\to} |x - \overline{x}|$, and $|x_{n_{k+1}} - \overline{x}| \overset{k\to\infty}{\to} |x - \overline{x}|$, and also

$$|x_{n_k+1} - \overline{x}| = |f(x_{n_k}) - f(\overline{x})| \overset{k\to\infty}{\to} |f(x) - f(\overline{x})|.$$

Passing to the limit in (2.3.10), one gets then

$$|f(x) - f(\overline{x})| = |x - \overline{x}|.$$

If $x \ne \overline{x}$, by applying again the Lagrange Theorem, we have

$$|f(x) - f(\overline{x})| < |x - \overline{x}|,$$

which provides a contradiction. Therefore, $\bar{x} = x$, and the theorem is proved. $\square$

We want to emphasize now that if the condition $|f'(x)| < 1$ does not hold for any $x \in (a, b)$, then, even if f has a unique fixed point, the Picard sequence may not converge to this fixed point. For instance, consider the function $f : [-1, 1] \to [-1, 1]$ given by $f(x) = -x$. Obviously, f is differentiable on its domain, but the absolute value of its derivative equals 1. The sole fixed point of f is $\bar{x} = 0$. By construction the Picard iteration starting from $x_0 \neq 0$, the Picard sequence has the form $x_0, -x_0, x_0, -x_0, \dots,$ hence is not convergent. Remark also that a function which satisfies the assumptions of the previous theorem is not necessarily a contraction, hence the Picard Theorem cannot be obtained from the Banach Principle. To illustrate this aspect, we have the following example. Take $f : [0, 1] \to [0, 1]$, $f(x) = \frac{1}{1+x}$. Obviously, $|f'(x)| \in (0, 1)$ for any $x \in (0, 1)$, but f is not a contraction because

$$\lim_{(x,y)\to(0,0),x\neq y} \frac{|f(x) - f(y)|}{|x - y|} = \lim_{(x,y)\to(0,0),x\neq y} \frac{1}{(x + 1)(y + 1)} = 1.$$

However, we could apply the Banach Principle for the function f if we restrict it to an interval on which it is a contraction. Observe that for $x \in \left[\frac{1}{2}, 1\right]$, $f(x) \in \left[\frac{1}{2}, \frac{2}{3}\right]$, so we can define $f : \left[\frac{1}{2}, 1\right] \to \left[\frac{1}{2}, 1\right]$ and remark that this restriction is a contraction, because $\sup_{x \in [\frac{1}{2}, 1]} |f'(x)| = \frac{4}{9} < 1$. Therefore, there exists a unique fixed point of f in $\left[\frac{1}{2}, 1\right]$, and since the initial function has a unique fixed point, the fixed point of the restriction coincides with it.

Also in the framework of this example, one can apply Theorem 2.3.12, because $f^2(x) = \frac{x+1}{x+2}$ is a contraction ($\sup_{x \in [0,1]} |(f^2)'(x)| = \frac{1}{4} < 1$).

In some cases, it may happen that we cannot restrict the function to a contraction, nor can we find an iteration which is a contraction, so one cannot apply any of the two methods from the preceding example. For example, take the function $f : [0, 1] \to [0, 1]$, $f(x) = \frac{x}{1+x}$. Again, $|f'(x)| \in (0, 1)$ for any $x \in (0, 1)$, and for every $n \in \mathbb{N}^*$ and every $x \in [0, 1]$, $f^n(x) = \frac{x}{1+nx}$. As above:

$$\lim_{(x,y)\to(0,0),x\neq y} \frac{|f^n(x) - f^n(y)|}{|x - y|} = \lim_{(x,y)\to(0,0),x\neq y} \frac{1}{(nx + 1)(ny + 1)} = 1$$

and hence f^n is not a contraction. Moreover, it is impossible to find any restriction of the function f which is a contraction from a set into itself, because the fixed point of f is $\bar{x} = 0$, so, such a restriction should be defined on an interval which contains 0, which is impossible in view of relation

$$\lim_{(x,y)\to(0,0),x\neq y} \frac{|f(x) - f(y)|}{|x - y|} = 1.$$

However, the function from the Picard Theorem is necessarily a weak contraction, and we will see in the subsequent sections how one can reobtain the Picard Theorem by using a more general result concerning weak contractions (Theorem 2.3.24 and Corollary 2.3.25).

In the assumptions of the Picard Theorem, the sequence of iterations is convergent to the fixed point of the function f for any possible choice of the initial term x_0 in the interval $[a, b]$. In order to obtain a quicker convergence, it is natural to try to start with this first term as close as we can to $\bar{x}$. Observe that, because of equality $\bar{x} = f(\bar{x})$, the point $\bar{x}$ lies in the image of f, and since $\bar{x}$ is a fixed point, it would be in the images of the functions f^n for every natural nonzero n. Therefore, $\bar{x}$ will lie in the intersection of all the images of the above functions. Hence, if we could calculate $\bigcap_{n \in \mathbb{N}^*} \operatorname{Im} f^n$, we could be close enough to the fixed point. In case this intersection consists of only one point, that point is the desired fixed point. For example, consider the function $f : [0, 1] \to [0, 1]$ given by $f(x) = \frac{x+1}{4}$. Inductively, one can show that for every $n \in \mathbb{N}^*$,

$$f^n(x) = \frac{3x + 4^n - 1}{3 \cdot 4^n},$$

and

$$\operatorname{Im} f^n = \left[\frac{1}{3}(1 - \frac{1}{4^n}), \frac{1}{3}(1 + \frac{2}{4^n}) \right].$$

If we take the intersection of all these intervals, we get

$$\bigcap_{n \in \mathbb{N}^*} \operatorname{Im} f^n = \left\{ \frac{1}{3} \right\},$$

hence $\bar{x} = \frac{1}{3}$ is the fixed point of the function f we are looking for.

The next result highlights the most important assumption of Picard Theorem, by emphasizing a partial converse.

Proposition 2.3.22. *Let $f : \mathbb{R} \to \mathbb{R}$ be a continuous function and $x_0 \in \mathbb{R}$. If the sequence of Picard iterations starting from x_0 converges to a number $l \in \mathbb{R}$, without being stationary, and f is differentiable at l, then $|f'(l)| \leq 1$.*

Proof Suppose by contradiction that $|f'(l)| > 1$. It is clear, from the continuity of f and from a previous comment, that l must be a fixed point of f. Since

$$\lim_{x \to l} \frac{f(x) - f(l)}{x - l} = f'(l)$$

we get

$$\lim_{x \to l} \left| \frac{f(x) - f(l)}{x - l} \right| = |f'(l)|.$$

Take

$$\varepsilon := \frac{|f'(l)| - 1}{2} > 0.$$

Then for this ε, there exists $\delta > 0$ such that for any $x \in (l - \delta, l + \delta) \setminus \{l\}$,

$$\left|f'(l)\right| - \varepsilon < \left|\frac{f(x) - f(l)}{x - l}\right| < \left|f'(l)\right| + \varepsilon$$

or

$$\frac{\left|f'(l)\right| + 1}{2} < \left|\frac{f(x) - l}{x - l}\right|.$$

In particular, since $\left|f'(l)\right| > 1$,

$$|x - l| < |x - l| \frac{\left|f'(l)\right| + 1}{2} < \left|f(x) - f(l)\right|$$

for every $x \in (l - \delta, l + \delta) \setminus \{l\}$. Since the sequence (x_n) of the Picard iterations starting from x_0 converges to l, without being stationary, there exists $n_\delta \in \mathbb{N}$ such that for any $n \geq n_\delta$,

$$x_n \in (l - \delta, l + \delta) \setminus \{l\}.$$

One gets from the above relations that

$$|x_n - l| < \left|f(x_n) - f(l)\right| = |x_{n+1} - l|$$

for every $n \geq n_\delta$. In particular, we get

$$|x_{n_\delta} - l| < |x_{n_\delta+1} - l| < |x_n - l|$$

for every $n > n_\delta + 1$. Passing to the limit in the last relation for $n \to \infty$, we get the contradiction:

$$|x_{n_\delta} - l| < |x_{n_\delta+1} - l| \leq 0.$$

Consequently, the assumption made is false, hence $\left|f'(l)\right| \leq 1$. $\qquad \square$

Remark 2.3.23. *In general, one cannot obtain the inequality on the derivative if the sequence is stationary. If one considers the function $f : \mathbb{R} \to \mathbb{R}$, $f(x) = x^3 + 2x$, the sequence of Picard iterations starting from 0 is convergent (even stationary) to 0, but $f'(0) = 2$.*

We now present a very interesting result due to Beardon, which says that in case of a weak contraction on $\mathbb{R}$ (which, as we already saw in Example 2.3.10, may not have fixed points), the Picard iteration process gives the same limit point (possibly equal to $+\infty$ or $-\infty$), for any choice of the initial data.

Theorem 2.3.24 (Beardon's Theorem). *Let $f : \mathbb{R} \to \mathbb{R}$ be a function which is a weak contraction on $\mathbb{R}$, i.e., for any two distinct real numbers x and y, one has*

$$\left|f(x) - f(y)\right| < |x - y|.$$

Then there exists $\bar{x} \in \overline{\mathbb{R}}$ such that for any $x \in \mathbb{R}$, the Picard sequence generated by

$$x_n = f^n(x), \forall n \in \mathbb{N}^*$$

has the limit $\bar{x}$.

Proof Suppose firstly that f has a fixed point, $\bar{x} \in \mathbb{R}$. Obviously, this will be the unique fixed point of f. Without loss of generality, we may consider $\bar{x} = 0$, hence

$$\left|f(x)\right| < |x|$$

for any $x \neq 0$. Fix $x \in \mathbb{R}$. Then the sequence $(\left|f^n(x)\right|)_n$ is decreasing and hence is convergent to a number $\mu(x) \geq 0$. We will prove that $\mu(x) = 0$ for any initial data x. Suppose by contradiction that $\mu(x) > 0$. Then $f(\mu(x)) =: y_1$, and $f(-\mu(x)) =: y_2$, where $|y_1|, |y_2| < \left|\mu(x)\right|$. By the continuity of f, there are two neighborhoods of $\mu(x)$ and $-\mu(x)$, respectively, which are applied through f in the interval $I := (-\mu(x), \mu(x))$, which contains y_1 and y_2. Then, for n sufficiently large, $f^n(x)$ also lies in I, which contradicts the inequality $\left|f^n(x)\right| \geq \left|\mu(x)\right|$. Therefore, for every $x \in \mathbb{R}$, one has the convergence $f^n(x) \to 0$.

Suppose now f does not have any fixed point. Then $f(x) > x$ or $f(x) < x$ for any real x. We will only prove the case $f(x) > x$, the other situation following analogously. It is clear that for any real x, the sequence $(f^n(x))$ is strictly increasing, so it has the limit in $(-\infty, +\infty]$. If the limit would be a real number l, then $f^n(x) \neq l$ for any natural nonzero n, and we get

$$\left|f^{n+1}(x) - f(l)\right| < \left|f^n(x) - l\right|$$

from where, by passing to the limit, we get that $f(l) = l$, i.e., l is a fixed point, which is a contradiction. Hence, $f^n(x) \to +\infty$. Obviously, in case $f(x) < x$, we get $f^n(x) \to -\infty$. The proof is now complete. □

Remark also from the proof that if $\bar{x} \in \mathbb{R}$, then $\bar{x}$ is necessarily the sole fixed point of f. Actually, if f has a fixed point (which is necessarily unique from the contraction condition), then all the iterations converge to this fixed point, and if f does not have a fixed point, then the Picard iterations converge to $+\infty$ or $-\infty$, according to $f(x) > x$ or $f(x) < x$, respectively.

Based on this result we can deduce the next corollary.

Corollary 2.3.25. *Let $a, b \in \mathbb{R}$, $a < b$ and $f : [a, b] \to [a, b]$ be a weak contraction. Then f has a unique fixed point, and for any initial data $x \in [a, b]$, the sequences of the Picard iterations are convergent to this fixed point.*

Proof The existence and the uniqueness of the fixed point are assured by the Theorem 2.3.17 and by the contraction condition. By repeating the arguments from the first part of the proof of the previous theorem, one gets the conclusion concerning the Picard iterations. □

The preceding corollary one can obtain also the Picard theorem of convergence (Theorem 2.3.21), because in the assumptions of that theorem, the function is a weak contraction.

2.4 Graves Theorem

We dedicate this section to an important result, known as the Graves Theorem, which gives sufficient conditions for a function $f : \mathbb{R}^n \to \mathbb{R}^m$ to be open (i.e., the image through f of an open set in $\mathbb{R}^n$ is open in $\mathbb{R}^m$). If f is linear, this is an well-known and deep result, known in Functional Analysis as the Open Mapping Principle (and is applicable, as well as the Graves Theorem, in a much wider setting). We give this principle next and illustrate it through a very short and elementary proof (in our particular framework).

Theorem 2.4.1 (Open Mapping Principle). *Let $T : \mathbb{R}^p \to \mathbb{R}^m$ be a linear surjective map $(m, p \in \mathbb{N}^*)$. Then T is open.*

Proof First of all, we recall that T is continuous. Let $\{e_1, e_2, ..., e_m\}$ be the canonical base of $\mathbb{R}^m$. Taking into account the surjectivity of T, for any $i \in \overline{1, m}$, there exist $x_i \in \mathbb{R}^p$ with $T(x_i) = e_i$. Now take $\psi : \mathbb{R}^m \to \mathbb{R}^p$, $\psi(\alpha) = \psi(\alpha_1, ..., \alpha_m) = \sum_{i=1}^m \alpha_i x_i$. Clearly, ψ is continuous and, moreover, $T \circ \psi : \mathbb{R}^m \to \mathbb{R}^m$ is the identity of $\mathbb{R}^m$. In particular, ψ is injective. Now, consider $U \subset \mathbb{R}^p$ an open set. Since ψ is continuous, $\psi^{-1}(U)$ is open and since ψ is injective, $U = \psi(\psi^{-1}(U))$. But

$$T(U) = T(\psi(\psi^{-1}(U))) = (T \circ \psi)(\psi^{-1}(U)) = \psi^{-1}(U).$$

So, $T(U)$ is open. □

Remark that, in fact, a stronger property than the usual openness can be deduced from the Open Mapping Principle: there exists $L > 0$ such that $B(Tx, Lr) \subset T(B(x, r))$ for every $x \in \mathbb{R}^p$ and $r > 0$.

To see this, observe that it is sufficient to prove the above relation for $x = 0$. In this case, since $T(B(0, 1))$ is an open set, there exists an $L > 0$ such that $B(0, L) \subset T(B(0, 1))$. The fact that $B(0, Lr) \subset T(B(0, r))$ for every $r > 0$ easily follows.

The property mentioned before is called linear openness, and the corresponding constant $L > 0$ is called the linear openness modulus.

We now present two results given in 1950 by the American mathematician Lawrence Murray Graves. For a general function $f : \mathbb{R}^p \to \mathbb{R}^m$, one says that f is linearly open at x_0 with modulus $L > 0$ if there is $\rho > 0$ such that $B(f(x_0), Lr) \subset f(B(x_0, r))$

for every $r \in (0, \rho)$. Of course, in case of linear operators, the linear openness at a single point is equivalent to the linear openness at every point.

Theorem 2.4.2 (Graves). *Let $T : \mathbb{R}^p \to \mathbb{R}^m$ be a linear surjective map, and denote its linear openness modulus by $L > 0$. Let $M \in (0, L)$, $r > 0$, $U \subset \mathbb{R}^p$ be an open set, $x_0 \in U$, and $f : U \to \mathbb{R}^m$ be a continuous function such that*

$$\left\| f(u) - f(v) - T(u - v) \right\| \leq M \left\| u - v \right\|$$

for any $u, v \in B(x_0, r)$. Then f is linearly open at x_0 with modulus $L - M > 0$.

Proof Take $y \in \mathbb{R}^m$ such that $\left\| y - f(x_0) \right\| < r(L - M)$. We construct inductively a sequence (ξ_n) as follows: take $\xi_0 := 0$ and, for $n \geq 1$, ξ_n satisfies

$$T(\xi_n - \xi_{n-1}) = y - f(x_0 + \xi_{n-1}) \tag{2.4.1}$$

and

$$\left\| \xi_n - \xi_{n-1} \right\| \leq L^{-1} \left\| y - f(x_0 + \xi_{n-1}) \right\|. \tag{2.4.2}$$

We easily deduce

$$T(\xi_n - \xi_{n-1}) = T(\xi_{n-1} - \xi_{n-2}) - f(x_0 + \xi_{n-1}) + f(x_0 + \xi_{n-2}),$$

hence

$$\left\| \xi_n - \xi_{n-1} \right\| \leq L^{-1} \left\| T(\xi_n - \xi_{n-1}) \right\| \leq L^{-1} \left\| T(\xi_{n-1} - \xi_{n-2}) - f(x_0 + \xi_{n-1}) + f(x_0 + \xi_{n-2}) \right\|$$
$$\leq L^{-1} M \left\| \xi_{n-1} - \xi_{n-2} \right\|, \tag{2.4.3}$$

if $\xi_n, \xi_{n-1} \in B(0, r)$. This is true for every n. Indeed, from (2.4.1) and (2.4.2), we obtain that

$$\left\| \xi_1 \right\| \leq L^{-1} \left\| y - f(x_0) \right\| < r(1 - L^{-1}M) < r,$$

and since $\xi_0 = 0$, we obtain $\xi_1, \xi_0 \in B(0, r)$, hence (2.4.3) holds for $n = 2$. From this,

$$\left\| \xi_2 \right\| = \left\| \xi_2 - \xi_1 + \xi_1 \right\| \leq \left\| \xi_1 \right\| (1 + L^{-1}M) < r,$$

hence (2.4.3) holds for $n = 3$. Suppose (2.4.3) holds until n, and deduce that

$$\left\| \xi_n \right\| \leq \left\| \xi_1 \right\| \left(1 + L^{-1}M + \ldots + \left(L^{-1}M \right)^{n-1} \right) < r, \tag{2.4.4}$$

which means that (2.4.3) holds for $n + 1$. Moreover, from (2.4.3) we know that the sequence (ξ_n) is Cauchy, and hence is convergent to an element $\xi \in \mathbb{R}^p$. From (2.4.1) and the continuity of f, $y = f(x_0 + \xi)$, and (2.4.4) shows that

$$\left\| \xi \right\| \leq \left\| \xi_1 \right\| (1 - L^{-1}M)^{-1} \leq L^{-1} \left\| y - f(x_0) \right\| (1 - L^{-1}M)^{-1} < L^{-1} r(L - M)(1 - L^{-1}M)^{-1} = r.$$

But this means exactly that

$$B(f(x_0), (L - M)r) \subset f(B(x_0, r)).$$

Since this is true if one replaces r above with arbitrary $r' \in (0, r)$, we have the conclusion. □

Corollary 2.4.3. *Let $T : \mathbb{R}^p \to \mathbb{R}^m$ be a linear surjective map, with linear openness modulus $L > 0$. If S is a linear map such that $\|S - T\| \le L/2$, then S is surjective, hence linearly open, with the linear openness modulus $L/2$.*

Proof Apply the previous theorem for $f := S$ and $x_0 := 0$. □

On the basis of the previous results, one deduces the celebrated Lyusternik-Graves Theorem.

Theorem 2.4.4 (Lyusternik-Graves). *Let $U \subset \mathbb{R}^p$ be an open set, $x_0 \in U$, and $f : U \to \mathbb{R}^m$ a Fréchet differentiable function, with ∇f continuous at x_0, and $\nabla f(x_0)$ surjective. Then f is linearly open at x_0.*

Proof Denote by L the openness modulus of $\nabla f(x_0)$, given by the Open Mapping Principle. Since ∇f is continuous at x_0, there exists $\delta > 0$ such that for every $x \in B(x_0, \delta)$, one has

$$\|\nabla f(x) - \nabla f(x_0)\| \le L/8 < L/2. \tag{2.4.5}$$

This means, on the basis of Corollary 2.4.3, that for every $x \in B(x_0, \delta)$, $\nabla f(x)$ is linearly open with modulus $L/2$. For such x, one can apply Lagrange Theorem to the function $y \mapsto f(y) + \nabla f(x)y$, to get that

$$\|f(u) - f(v) - \nabla f(x)(u - v)\| \le \|u - v\| \sup_{t \in [x_0, x]} \|\nabla f(t) - \nabla f(x)\| \le \frac{L}{4} \|u - v\|,$$

if $u, v \in B(x_0, \delta)$, where (2.4.5) was used in the last inequality. The conclusion now follows from Theorem 2.4.2, with $L/2$ instead of L, and $L/4$ instead of M. □

The condition that $\nabla f(x_0)$ is surjective, as the iterative procedure from the proof of Theorem 2.4.2, was introduced in 1934 by the Russian mathematician Lazar Aronovich Lyusternik. For more extensions, developments and historical facts, see (Dontchev and Rockafellar, 2009) and (Klatte and Kummer, 2002).

2.5 Semicontinuous Functions

The aim of this section is to introduce and study some generalizations of the continuity concept, which will be useful in the subsequent discussion concerning the existence of the solutions of the optimization problems.

Observe that a function $f : \mathbb{R}^p \to \mathbb{R}$ is continuous at a if and only if the next two conditions simultaneously hold:

$$\forall \lambda \in \mathbb{R}, \ \lambda < f(a), \ \exists U \in \mathcal{V}(a), \ \forall x \in U, \ \lambda < f(x) \tag{2.5.1}$$

and

$$\forall \lambda \in \mathbb{R}, \ \lambda > f(a), \ \exists U \in \mathcal{V}(a), \ \forall x \in U, \ \lambda > f(x). \tag{2.5.2}$$

By taking each of these two conditions separately, we can define lower and upper semicontinuity. Therefore, the function $f : \mathbb{R}^p \to \mathbb{R}$ is lower semicontinuous at $a \in \mathbb{R}$ if the condition (2.5.1) is satisfied, and f is upper semicontinuous at a if the condition (2.5.2) is satisfied. Similarly, if the respective conditions hold in every point of $\mathbb{R}^p$, one says that f is lower semicontinuous, or respectively upper semicontinuous. Obviously, according to the definitions, a function is continuous at $a \in \mathbb{R}^p$ if and only if it is simultaneously upper and lower semicontinuous at a. It is easy to provide examples of functions which are semicontinuous without being continuous. For instance, the function $f : \mathbb{R} \to \mathbb{R}$,

$$f(x) = \begin{cases} 1, x \neq 0 \\ 0, x = 0 \end{cases}$$

is lower semicontinuous (on $\mathbb{R}$), but is discontinuous at 0. Similarly, the function $f : \mathbb{R} \to \mathbb{R}$,

$$f(x) = \begin{cases} 0, x \neq 0 \\ 1, x = 0 \end{cases}$$

is upper semicontinuous (on $\mathbb{R}$), but is discontinuous at 0.

A more elaborate example is the Riemann function $f : [0, 1] \to \mathbb{R}$, given by

$$f(x) = \begin{cases} \frac{1}{n}, \ x \in (0, 1], x = \frac{m}{n}, \ m, n \in \mathbb{N}^*, \ (m, n) = 1 \\ 0, \ x \in [0, 1] \setminus \mathbb{Q} \text{ or } x = 0, \end{cases}$$

which is continuous on $[0, 1] \setminus \mathbb{Q} \cup \{0\}$ and discontinuous on $(0, 1] \cap \mathbb{Q}$. This function has its limit equal to 0 at every point. This shows that it is upper semicontinuous.

One can easily show that f is upper semicontinuous if and only if $-f$ is lower semicontinuous. As consequence, we will restrict our study to lower semicontinuous functions, since the results can be easily reformulated in the case of the upper semicontinuous functions.

Having a function $f : \mathbb{R}^p \to \mathbb{R}$, besides the epigraph of the function previously introduced (Proposition 2.2.3), that is

$$\operatorname{epi} f = \{(x, t) \in \mathbb{R}^p \times \mathbb{R} \mid f(x) \leq t\},$$

we introduce now the level sets: if $v \in \mathbb{R}$,

$$N_v f := \{x \in \mathbb{R}^p \mid f(x) \leq v\} = f^{-1}((-\infty, v]).$$

Those functions which are globally lower semicontinuous, have the following characterization theorem.

Theorem 2.5.1. *Let $f : \mathbb{R}^p \to \mathbb{R}$. The next assertions are equivalent:*
(i) f is lower semicontinuous (on $\mathbb{R}^p$);
(ii) $N_v f$ is closed in $\mathbb{R}^p$ for any $v \in \mathbb{R}$;
(iii) epi f is a closed set in $\mathbb{R}^p \times \mathbb{R}$;
(iv) $\{x \in \mathbb{R}^p \mid f(x) > \beta\}$ is open in $\mathbb{R}^p$ for any $\beta \in \mathbb{R}$.

Proof (i) $\Rightarrow$ (ii) Take $v \in \mathbb{R}$. We prove that $N_v f$ has an open complement. Take $x \notin N_v f$, i.e., $f(x) > v$. Since f is lower semicontinuous at x, there exists U, a neighborhood of x, such that $f(y) > v$ for any $y \in U$. Therefore, $U \cap N_v f = \emptyset$, i.e., $U \subset \mathbb{R}^p \setminus N_v f$. It follows that the complement of $N_v f$ is open, i.e., $N_v f$ is closed.

(ii) $\Rightarrow$ (iii) We prove that $(\mathbb{R}^p \times \mathbb{R}) \setminus \mathrm{epi}\, f$ is open. Take $(x, t) \notin \mathrm{epi}\, f$, which means $f(x) > t$. There exists v such that $f(x) > v > t$. Then $x \notin N_v f$ and according to (ii), there exists U, a neighborhood of x, such that $U \cap N_v f = \emptyset$. It follows that $U \times (-\infty, v] \cap \mathrm{epi}\, f = \emptyset$. Since $U \times (-\infty, v]$ is a neighborhood of (x, t), we obtain the conclusion.

(iii) $\Rightarrow$ (i) Take $x \in \mathbb{R}^p$ and $t \in \mathbb{R}$ such that $f(x) > t$. Then $(x, t) \notin \mathrm{epi}\, f$ and hence there exists U, a neighborhood of x, and $\varepsilon > 0$, such that $U \times (t - \varepsilon, t + \varepsilon) \cap \mathrm{epi}\, f = \emptyset$. Therefore, for every $y \in U$, $(y, t) \notin \mathrm{epi}\, f$, i.e., $f(y) > t$. Accordingly, f is lower semicontinuous.

(ii) $\Leftrightarrow$ (iv) The relation

$$\{x \in \mathbb{R}^p \mid f(x) > \beta\} = \mathbb{R}^p \setminus N_\beta f$$

proves the equivalence between (ii) and (iv). $\qquad\qquad$ □

A function is upper semicontinuous if and only if for any $y \in \mathbb{R}$, the sets of the type $\{x \in \mathbb{R}^p \mid f(x) \geq y\}$ are closed, which is equivalent to the fact that the sets of the type $\{x \in \mathbb{R}^p \mid f(x) < y\}$ are open.

Let $f : A \subset \mathbb{R}^p \to \mathbb{R}$ be a function. Its lower and upper limits at $a \in \mathrm{cl}\, A$ are given, respectively, by

$$\liminf_{x \to a} f(x) := \sup_{U \in \mathcal{V}(a)} \inf_{x \in U \cap A} f(x) \text{ and } \limsup_{x \to a} f(x) := \inf_{U \in \mathcal{V}(a)} \sup_{x \in U \cap A} f(x). \qquad (2.5.3)$$

From their definitions, it is obvious that

$$\liminf_{x \to a} f(x) \leq \limsup_{x \to a} f(x) \text{ and } \limsup_{x \to a} f(x) = -\liminf_{x \to a}(-f)(x).$$

Moreover, for a sequence (x_n), one defines its lower and upper limits as

$$\liminf_{n \to \infty} x_n := \sup_{n \in \mathbb{N}} \inf_{k \geq n} x_k \text{ and and } \limsup_{n \to \infty} x_n := \inf_{n \in \mathbb{N}} \sup_{k \geq n} x_k. \qquad (2.5.4)$$

In fact, the definitions (2.5.4) naturally follow from (2.5.3) for the function $f : \mathbb{N} \to \mathbb{R}$, $f(n) := x_n$, and $a := \infty \in \mathrm{cl}\, \mathbb{N}$.

Theorem 2.5.2. *Let $f : \mathbb{R}^p \to \mathbb{R}$ and $x \in \mathbb{R}^p$. Then:*

(i) f is lower semicontinuous at x if and only if $f(x) = \liminf_{y \to x} f(y)$.

(ii) If f is lower semicontinuous at x, then for every $(x_n) \to x$, one has $\liminf_{n \to \infty} f(x_n) \geq f(x)$.

Proof (i) It is easy to observe that for every $U \in \mathcal{V}(x)$, $\inf_{y \in U} f(y) \leq f(x)$, hence $\liminf_{y \to x} f(y) \leq f(x)$, and this is true for an arbitrary function. Suppose f is lower semicontinuous at x and take $\lambda \in \mathbb{R}$, $\lambda < f(x)$. Then, according to the definition of the lower semicontinuity, there exists $U \in \mathcal{V}(x)$ such that $\lambda < f(y)$, for any $y \in U$. It means that $\lambda \leq \inf_{y \in U} f(y) \leq \liminf_{y \to x} f(y)$. Since λ was taken arbitrary and assumed to be smaller than $f(x)$, it follows that $f(x) \leq \liminf_{y \to x} f(y)$, and hence we have equality. Suppose now $f(x) = \liminf_{y \to x} f(y)$ and take $\lambda \in \mathbb{R}$, $\lambda < f(x)$. According to the definition of $\liminf$, we know that there is $U \in \mathcal{V}(x)$ such that $\lambda < \inf_{y \in U} f(y) < f(z)$, for every $z \in U$, hence f is lower semicontinuous at x.

(ii) Take arbitrary $U \in \mathcal{V}(x)$. Observe that since $(x_n) \to x$, there exists $n_U \in \mathbb{N}$ such that, for every $k \geq n_U$, one has $x_k \in U$. Accordingly, $\inf_{k \geq n_U} f(x_k) \geq \inf_{y \in U} f(y)$. It follows that

$$\sup_{n \in \mathbb{N}} \inf_{k \geq n} f(x_k) \geq \inf_{y \in U} f(y), \ \forall U \in \mathcal{V}(x).$$

Passing to the supremum for $U \in \mathcal{V}(x)$, and taking into account (i), the conclusion follows. □

For upper semicontinuous functions, similar results can be deduced.

The next result is a generalization of the Weierstrass Theorem.

Theorem 2.5.3. *Let $f : \mathbb{R}^p \to \mathbb{R}$ be a lower semicontinuous function and $K \subset \mathbb{R}^p$ be a compact set. Then f is lower bounded on K and it attains its minimum on K.*

Proof Take $(x_k) \subset K$ such that $f(x_k) \to \inf\{f(x) \mid x \in K\}$. From the compactness of K, the sequence (x_k) has a convergent subsequence to an element $\overline{x} \in K$. From the fact that the sets of the type $K \cap N_v f$ are closed for any $v \in \mathbb{R}$, $v > \inf\{f(x) \mid x \in K\}$, we deduce that $\overline{x}$ lies in all these sets (because for every v, the terms x_k are, from a certain rank, in $K \cap N_v f$). Then we deduce that $f(\overline{x}) \leq v$ for any $v > \inf\{f(x) \mid x \in K\}$. Consequently, $f(\overline{x}) \leq \inf\{f(x) \mid x \in K\}$. On one hand, this means that $\inf\{f(x) \mid x \in K\} \in \mathbb{R}$, hence f is lower bounded on K, and, on the other hand, that $\overline{x}$ is the point we are looking for, which realizes the minimum of f on K. The proof is complete. □

Obviously, for upper semicontinuous functions, we will have the other "half" of the Weierstrass Theorem.

Theorem 2.5.4. *Let $f : \mathbb{R}^p \to \mathbb{R}$ be an upper semicontinuous function and $K \subset \mathbb{R}^p$ be a compact set. Then f is upper bounded on K and it attains its maximum on K.*

We know that, in general, the (pointwise) supremum of a family of continuous functions is not a continuous function. For instance, if we define for every $n \in \mathbb{N}$, $f_n : [0, 1] \to \mathbb{R}$, $f_n(x) = -x^n$, then the pointwise supremum is the function

$$f(x) = \sup_{n \in \mathbb{N}} f_n(x) = \begin{cases} 0, & x \in [0, 1) \\ -1, & x = 1, \end{cases}$$

which is not continuous on $[0, 1]$. In turn, f is lower semicontinuous on $[0, 1]$, a fact which is not accidental, as the following result shows.

Theorem 2.5.5. *Let I be a nonempty arbitrary family of indices, and $(f_i)_{i \in I}$ a family of lower semicontinuous functions from $\mathbb{R}^p$ into $\mathbb{R}$. If for any $x \in \mathbb{R}^p$,*

$$\sup\{f_i(x) \mid i \in I\} \in \mathbb{R},$$

then the function $f : \mathbb{R}^p \to \mathbb{R}$ given by $f(x) = \sup\{f_i(x) \mid i \in I\}$ is lower semicontinuous on $\mathbb{R}^p$.

Proof For any $\alpha \in \mathbb{R}$, one has:

$$f^{-1}\left((\alpha, \infty)\right) = \{x \in \mathbb{R}^p \mid f(x) > \alpha\}$$
$$= \bigcup_{i \in I}\{x \in \mathbb{R}^p \mid f_i(x) > \alpha\}.$$

Since the functions f_i are lower semicontinuous, the sets $\{x \in \mathbb{R}^p \mid f_i(x) > \alpha\}$ are open according to Theorem 2.5.1. Since every union of open sets is an open set, we deduce that $f^{-1}\left((\alpha, \infty)\right)$ is open, and by applying again Theorem 2.5.1, we get that f is lower semicontinuous on $\mathbb{R}^p$. $\qquad\square$

3 The Study of Smooth Optimization Problems

This chapter plays a central role in this monograph. In its first section, we present smooth optimization problems and deduce existence conditions for minimality. We take this opportunity to prove and discuss the Ekeland Variational Principle and its consequences. We then obtain necessary conditions for optimality as well as sufficient optimality conditions of the first and second-order for smooth objective functions under geometric restrictions (i.e., restrictions of the type $x \in M$, where M is an arbitrary set). The second section is dedicated to the investigation of optimality conditions under functional restrictions (with equalities and inequalities). The main aim is to deduce Karush-Kuhn-Tucker conditions and to introduce and compare several qualification conditions. Special attention is paid to the case of convex and affine data. Subsequently, we derive second-order optimality conditions for the case of functional restrictions. The last section of this chapter includes two examples which show that, for practical problems, the computational challenges posed by the optimality conditions are sometimes not easy to solve.

3.1 General Optimality Conditions

Let $U \subset \mathbb{R}^p$ be a nonempty open set, $f : U \to \mathbb{R}$ be a function and $M \subset U$ a nonempty set. We are interested in studying the minimization problem for the function f when its argument belongs to M. Formally, we write this problem in the following form:

$$(P) \ \min f(x), \text{ subject to } x \in M.$$

The function f is called the objective function, or cost function, and the set M is called the set of feasible points of the problem (P), or the set of constraints, or the set of restrictions.

We should say from the very beginning that we shall study the minimization of f, but, by virtue of the relation $\max f = -\min(-f)$, similar results for its maximization could be obtained. Let us start by defining the notion of a solution associated to the problem (P).

Definition 3.1.1. *One says that $\overline{x} \in M$ is a local solution (or, simply, solution) of the problem (P) or minimum point of the function f on the set M if there exists a neighborhood V of $\overline{x}$ such that $f(\overline{x}) \leq f(x)$ for every $x \in M \cap V$. If $V = \mathbb{R}^p$, one says that $\overline{x}$ is a global solution of (P) or minimal global point of f on M.*

Of course, for the maximization problem, the corresponding solution is clear. Let us mention that in this chapter we will only deal with the case of smooth functions (up to the order two, i.e., $f \in C^2$).

Remark 3.1.2. *We shall distinguish two main situations for the study of problem* (P) : (i) *the case where M = U and* (ii) *the case where M appears as an intersection between a closed set of* $\mathbb{R}^p$ *and U. In the former case, we say that the optimization problem* (P) *has no constraints (or restrictions), while in the latter, we call this a problem with constraints. In the case of a problem without constraints, we use as well the term "local minimum point of f" instead of local solution. Let us observe that in Definition 3.1.1, if* $\overline{x} \in$ int M, *then* $\overline{x}$ *is a local solution of the unconstrained problem (it is enough to take a smaller neighborhood V such that* $V \subset M$). *So, in the case of problems with restrictions, the interesting situation is when* $\overline{x} \in$ bd M. *If* $\overline{x} \in$ int M *we say that the restriction is inactive.*

In the next sections, the two cases mentioned above will be treated together, but afterwards the discussion will split.

The basis of Optimization Theory consists of two fundamental results: the Weierstrass Theorem which ensures the existence of extrema and the Fermat Theorem (on stationary, or critical points) which gives a necessary condition for a point to be an extremum (without constraints) of a function. The theory follows the main trajectory of these fundamental results: on one hand, the study of existence conditions, and, on the other hand, the study of the (necessary and sufficient) optimality conditions.

We now recall these basic results. The classical Weierstrass Theorem, also given in the first chapter, states that a continuous function on a compact interval has a global minimum point on that interval. We have shown already that some conditions can be relaxed. Here is the theorem again.

Theorem 3.1.3 (Weierstrass Theorem). *If* $f : K \subset \mathbb{R}^p \to \mathbb{R}$ *is a lower (upper) semicontinuous function and K is a compact set, then f is lower (upper) bounded on K (i.e., f(K) is a bounded below (above) set) and f attains its minimum (maximum) on K, i.e., there exists* $\overline{x} \in K$ *such that* $\inf_{x \in K} f(x) = f(\overline{x})$ ($\sup_{x \in K} f(x) = f(\overline{x})$, *respectively).*

Remark 3.1.4. *The compactness of K is essential in the Weierstrass Theorem because otherwise the conclusion does not hold. As an example, let us consider the continuous function* $f : (0, 1] \to \mathbb{R}$, $f(x) = x^{-1} \sin(x^{-1})$. *Clearly,* $\inf_{(0,1]} f(x) = -\infty$, *and* $\sup_{(0,1]} f(x) = +\infty$.

We now present Fermat Theorem. The particular case of a function of one real variable was presented in Section 1.3. The proof of the theorem will be given later in this chapter.

Theorem 3.1.5 (Fermat Theorem). *Let $S \subset \mathbb{R}^p$ be a set and $a \in$ int S. If $f : S \to \mathbb{R}$ is of class C^1 in a neighborhood of a, and a is a local minimum or maximum point of f, then a is also a stationary (or critical) point of f, i.e., $\nabla f(a) = 0$.*

Some remarks are in order to understand the applications of Fermat Theorem.

Remark 3.1.6. *1. The converse of Fermat Theorem is not true: for instance, the derivative of $f : \mathbb{R} \to \mathbb{R}$, $f(x) = x^3$ vanishes at 0, but this point is neither minimum nor maximum of f.*

2. The interiority condition for a is essential since, without this assumption, the conclusion does not hold: take $f : [0, 1] \to [0, 1]$, $f(x) = x$ which has at $\overline{x} = 0$ a minimum point where the derivative is not 0. Therefore, in view of Remark 3.1.2, one can say that Fermat Theorem applies only to unconstrained problems.

3. If S is compact and f is continuous on S and differentiable on int S, it is possible to have $\nabla f(x) \neq 0$ for every $x \in$ int S, and in such a case the extreme points of f on S, which surely exist from Weierstrass Theorem, lie on the boundary of S.

We now start our discussion on the existence conditions for the minimum points.

Theorem 3.1.7. *Let $f : \mathbb{R}^p \to \mathbb{R}$ be a lower semicontinuous function and $M \subset \mathbb{R}^p$ be a nonempty, closed set. If there exists $v > \inf_{x \in M} f(x)$ such that the level set of f relative to M, i.e., $M \cap N_v f = \{x \in M \mid f(x) \leq v\}$, is bounded, then f attains its global minimum on M.*

Proof It is obvious that if f has a global minimum on M. Similarly, it also has a global minimum on $M \cap N_v f$. Since this set is compact and f is lower semicontinuous, from Weierstrass Theorem 3.1.3, we infer that f is lower bounded and attains its global minimum on $M \cap N_v f$, whence on M, too. □

Obviously, in the above result, if M is bounded, then the hypothesis is automatically fulfilled. The interesting case is where M is unbounded and in this situation we ensure the boundedness assumption by imposing a certain condition on f.

Proposition 3.1.8. *Let $f : \mathbb{R}^p \to \mathbb{R}$ be a function and $M \subset \mathbb{R}^p$ be a closed, unbounded set. If $\lim_{x \in M, \|x\| \to \infty} f(x) = \infty$ (i.e., for every sequence $(x_k) \subset M$ with $\|x_k\| \to \infty$, one has $f(x_k) \to \infty$), then the set $N_v f \cap M$ is bounded for every $v > \inf_{x \in M} f(x)$.*

Proof Let $v > \inf_{x \in M} f$. If the set $N_v f \cap M$ would be unbounded, then there exists $(x_k) \subset N_v f \cap M$ with $\|x_k\| \to \infty$. On one hand, by our assumption, $\lim f(x_k) = \infty$ and, on the other hand, $f(x_k) \leq v$ for every $k \in \mathbb{N}$, which is absurd. So $N_v f \cap M$ is bounded. □

If the condition $\lim_{x \in M, \|x\| \to \infty} f(x) = \infty$ holds, we say that f is coercive relative to M. If $\lim_{\|x\| \to \infty} f(x) = \infty$, the we say that f is coercive.

Some special conditions to ensure the existence and the attainment of the minimum on a bounded, not closed set can be given as well. We present here such a result which we are going to use in Chapter 6.

Proposition 3.1.9. *Let $D, C \subset \mathbb{R}^p$ be two sets such that C is compact and $D \cap C \neq \emptyset$. Let $\varphi : D \cap C \to \mathbb{R}$ be a continuous function. Suppose that the following condition holds: for every sequence $(x_k) \subset D \cap C$, $x_k \to \overline{x} \in \mathrm{bd}\, D \cap C$, the sequence $(\varphi(x_n))$ is unbounded above. Then φ is lower bounded and attains its minimum on $D \cap C$.*

Proof Firstly, let us observe that φ cannot be constant. Let $x_0 \in D \cap C$ such that $\varphi(x_0) > \inf_{x \in D \cap C} \varphi(x)$. It is enough to show that the level set $A := \{x \in D \cap C \mid \varphi(x) \le \varphi(x_0)\}$ is compact (see the proof of Theorem 3.1.7). Obviously, A is bounded (as a subset of C). It remains to show that A is closed. Let $(x_k) \subset A$, $x_k \to \overline{x}$. Suppose, by contradiction, that $\overline{x} \notin A$. Then, from the closedness of C,

$$\overline{x} \in (\mathrm{cl}\, D \setminus D) \cap C \subset \mathrm{bd}\, D \cap C.$$

By assumption, $(\varphi(x_k))$ is unbounded above, which contradicts the definition of the set A. Hence, A is compact and the conclusion follows. □

In general, a global minimum is local minimum, but, of course, the converse is false. We shall derive a condition for the fulfillment of this converse. We now give a necessary and sufficient condition for a point to be a global minimum.

Theorem 3.1.10. *Let $f : \mathbb{R}^p \to \mathbb{R}$ be a continuous function and $\overline{x} \in \mathbb{R}^p$. Then the following assertions are equivalent:*
(i) $\overline{x}$ is a global minimum point of f;
(ii) every $x \in \mathbb{R}^p$ with $f(x) = f(\overline{x})$ is a local minimum point of f.

Proof The implication $(i) \Rightarrow (ii)$ is obvious. Suppose that (ii) holds, but $\overline{x}$ is not a global minimum point. Then, there exists $u \in \mathbb{R}^p$ with $f(u) < f(\overline{x})$. We define $\varphi : [0, 1] \to \mathbb{R}$ by $\varphi(t) := f(t\overline{x} + (1 - t)u)$. The set $S := \{t \in [0, 1] \mid \varphi(t) = f(\overline{x})\}$ is nonempty ($1 \in S$), closed (from the continuity of f) and bounded. Then there is $t_0 = \min S$. Clearly, $t_0 \in (0, 1]$. From the hypothesis, $f(t_0\overline{x} + (1 - t_0)u) = f(\overline{x})$ tells us that the point $t_0\overline{x} + (1 - t_0)u$ is a local minimum of f. Consequently, there is $\varepsilon > 0$ such that for every $t \in [0, 1] \cap (t_0 - \varepsilon, t_0 + \varepsilon)$, $\varphi(t) \ge \varphi(t_0)$. Since $t_0 = \min S$, if one takes $t_1 \in [0, 1] \cap (t_0 - \varepsilon, t_0)$, the strict inequality $\varphi(t_1) > \varphi(t_0) > \varphi(0)$ holds. The function f being continuous, it has the Darboux property (or intermediate value property), whence there exists $t_2 \in (0, t_1)$ with $\varphi(t_2) = \varphi(t_0) = f(\overline{x})$, and this contradicts the minimality of t_0. Therefore, the assumption made was false, hence the conclusion holds. □

Notice that, if f is not continuous, the result does not hold. For this, it is sufficient to analyze the following function $f : \mathbb{R} \to \mathbb{R}$,

$$f(x) = \begin{cases} -x - 1, x \in (-\infty, -1] \\ x + 1, x \in (-1, 0) \\ -1, x = 0 \\ -x + 1, x \in (0, 1) \\ x - 1, x \in [1, \infty). \end{cases}$$

Let us observe that, by using the function $f : \mathbb{R} \to \mathbb{R}$, $f(x) = x^3$ and $\overline{x} = 0$, that one cannot replace in the item (ii) above the local minimality by the stationarity.

The above results ensure sufficient conditions for the existence of minimum points under compactness assumptions for the level sets. Conversely, it is clear that the lower boundedness of the function is a necessary condition for the existence of minimum points, but the boundedness of the level sets is not. For instance, the function $f : (0, \infty) \to \mathbb{R}$, $f(x) = (x - 1)^2 e^{-x}$ attains its minimum at $\overline{x} = 1$ and the minimal value is 0, but $N_v f$ is not bounded for every value of $v > 0 = \inf\{f(x) \mid x \in (0, \infty)\}$.

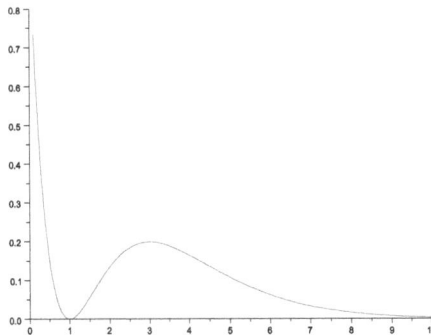

Figure 3.1: The graph of $(x - 1)^2 e^{-x}$.

One may like to have a notion of approximate minimum which is advantageous to always exist if f is lower bounded. We work here with unconstrained problems in order to better illustrate the main ideas.

Definition 3.1.11. *Let $f : \mathbb{R}^p \to \mathbb{R}$ be a lower bounded function and take $\varepsilon > 0$. A point x_ε is called ε–minimum of f if*

$$f(x_\varepsilon) \leq \inf_{x \in \mathbb{R}^p} f(x) + \varepsilon.$$

Since $\inf_{x \in \mathbb{R}^p} f(x) \in \mathbb{R}$, the existence of ε−minima for every positive ε is ensured. We use the generic term of approximate minima for ε−minima.

We now present a very important result, the Ekeland Variational Principle, which states that close to an approximate minimum point one can find a genuine minimum point for some perturbation of the initial function. This results was proved by the French mathematician Ivar Ekeland in 1974.

Theorem 3.1.12 (Ekeland Variational Principle). *Let $f : \mathbb{R}^p \to \mathbb{R}$ be a lower semicontinuous and lower bounded function. Let $\varepsilon > 0$ and let x_ε be an ε−minimum of f. Then, for every $\delta > 0$ there exists $\overline{x}_\varepsilon \in \mathbb{R}^p$ that have the following properties:*

$$f(\overline{x}_\varepsilon) \le f(x_\varepsilon),$$
$$\|\overline{x}_\varepsilon - x_\varepsilon\| \le \delta,$$
$$f(\overline{x}_\varepsilon) \le f(x) + \varepsilon \delta^{-1} \|x - \overline{x}_\varepsilon\|, \ \forall x \in \mathbb{R}^p.$$

Proof Let us consider the function $g : \mathbb{R}^p \to \mathbb{R}$

$$g(x) = f(x) + \varepsilon \delta^{-1} \|x - x_\varepsilon\|.$$

Using the assumptions on f, we infer that g is lower semicontinuous and lower bounded. Moreover, g is coercive, i.e.,

$$\lim_{\|x\| \to \infty} g(x) = +\infty.$$

From Theorem 3.1.7 and Proposition 3.1.8, the function g has a global minimum point, which we denote by $\overline{x}_\varepsilon$. Consequently,

$$f(\overline{x}_\varepsilon) + \varepsilon \delta^{-1} \|\overline{x}_\varepsilon - x_\varepsilon\| \le f(x) + \varepsilon \delta^{-1} \|x - x_\varepsilon\|, \ \forall x \in \mathbb{R}^p. \tag{3.1.1}$$

For $x = x_\varepsilon$, we get
$$f(\overline{x}_\varepsilon) \le f(x_\varepsilon).$$

That is the first relation in the conclusion. On the other hand, using this inequality

$$f(x_\varepsilon) \le \inf_{x \in \mathbb{R}^p} f(x) + \varepsilon,$$

with the relation (3.1.1), for $x = x_\varepsilon$, implies that:

$$\delta^{-1} \|\overline{x}_\varepsilon - x_\varepsilon\| \le 1.$$

This is the second part of the conclusion. Relation (3.1.1) allows us to write, successively, for any $x \in \mathbb{R}^p$,

$$f(\overline{x}_\varepsilon) \le f(x) + \varepsilon \delta^{-1}(\|x - x_\varepsilon\| - \|\overline{x}_\varepsilon - x_\varepsilon\|)$$
$$\le f(x) + \varepsilon \delta^{-1} \|x - \overline{x}_\varepsilon\|.$$

That is the last part of the conclusion. The proof is complete. $\qquad\square$

Remark 3.1.13. *Notice that the Ekeland Variational Principle holds (with minor changes in the proof) if instead of the whole space $\mathbb{R}^p$ one takes a closed subset of it.*

Clearly, the point $\overline{x}_\varepsilon$ is a global minimum point for the function

$$x \mapsto f(x) + \varepsilon \delta^{-1} \|x - \overline{x}_\varepsilon\| .$$

On the other hand, if we want $\overline{x}_\varepsilon$ to be close to x_ε (i.e., δ to be small), then the perturbation term, $\varepsilon \delta^{-1} \|\cdot - \overline{x}_\varepsilon\|$ is big. A compromise would be to choose $\delta := \sqrt{\varepsilon}$, and in this case one gets the next consequence.

Corollary 3.1.14. *Let $f : \mathbb{R}^p \to \mathbb{R}$ be a lower semicontinuous and lower bounded function. Take $\varepsilon > 0$ and let x_ε be an ε-minimum of f. Then there exists $\overline{x}_\varepsilon \in \mathbb{R}^p$ having the properties:*

$$f(\overline{x}_\varepsilon) \leq f(x_\varepsilon),$$
$$\|\overline{x}_\varepsilon - x_\varepsilon\| \leq \sqrt{\varepsilon},$$
$$f(\overline{x}_\varepsilon) \leq f(x) + \sqrt{\varepsilon} \|x - \overline{x}_\varepsilon\| , \ \forall x \in \mathbb{R}^p.$$

The Ekeland Variational Principle has many applications. Some of these refer to the same issues of extreme points. The next example of such an application asserts that every differentiable function has approximate critical points (for which the norm of the differential is arbitrarily small).

Theorem 3.1.15. *Let $f : \mathbb{R}^p \to \mathbb{R}$ be a differentiable, lower bounded function. Then for every $\varepsilon, \delta > 0$, there exists $\overline{x}_\varepsilon \in \mathbb{R}^p$ with $f(\overline{x}_\varepsilon) \leq \inf_{x \in \mathbb{R}^p} f(x) + \varepsilon$ and $\|\nabla f(\overline{x}_\varepsilon)\| \leq \varepsilon \delta^{-1}$. In particular, there exists $(x_n) \subset \mathbb{R}^p$ with*

$$f(x_n) \to \inf_{x \in \mathbb{R}^p} f(x), \ \nabla f(x_n) \to 0.$$

Proof Let x_ε be an ε-minimum of f. According to Theorem 3.1.12, there exists $\overline{x}_\varepsilon \in \mathbb{R}^p$ with the three mentioned properties. Since $f(\overline{x}_\varepsilon) \leq f(x_\varepsilon)$, we infer that $f(\overline{x}_\varepsilon) \leq \inf_{x \in \mathbb{R}^p} f(x) + \varepsilon$. Let $x := \overline{x}_\varepsilon + tu$ with $u \in \mathbb{R}^p$ and $t > 0$. The relation $f(\overline{x}_\varepsilon) \leq f(x) + \varepsilon \delta^{-1} \|x - \overline{x}_\varepsilon\|$ holds for any $x \in \mathbb{R}^p$, so we have

$$\frac{f(\overline{x}_\varepsilon + tu) - f(\overline{x}_\varepsilon)}{t} \geq -\varepsilon \delta^{-1} \|u\| .$$

Passing to the limit with $t \to 0$, we deduce

$$\nabla f(\overline{x}_\varepsilon)(u) \geq -\varepsilon \delta^{-1} \|u\| , \ \forall u \in \mathbb{R}^p,$$

that is,

$$-\nabla f(\overline{x}_\varepsilon)(u) \leq \varepsilon \delta^{-1} \|u\| , \ \forall u \in \mathbb{R}^p.$$

Changing u into $-u$, we get

$$\nabla f(\overline{x}_\varepsilon)(u) \leq \varepsilon \delta^{-1} \|u\| , \ \forall u \in \mathbb{R}^p,$$

whence

$$\left|\nabla f(\overline{x}_\varepsilon)(u)\right| \le \varepsilon\delta^{-1}\left\|u\right\|, \quad \forall u \in \mathbb{R}^p,$$

and this implies $\left\|\nabla f(\overline{x}_\varepsilon)\right\| \le \varepsilon\delta^{-1}$. For $\varepsilon := n^{-1}$, $\delta = \sqrt{n^{-1}}$, $n \in \mathbb{N}^*$ we obtain the second part of the conclusion. $\qquad\square$

The Ekeland Variational Principle also allows us to prove the equivalence of several existence conditions for minimum points.

Theorem 3.1.16. *Let $f : \mathbb{R}^p \to \mathbb{R}$ be a differentiable, lower bounded function. The following assertions are equivalent:*

(i) $\lim_{\|x\|\to\infty} f(x) = \infty$;

(ii) $N_v f$ is bounded for any $v > \inf_{x\in\mathbb{R}^p} f(x)$;

(iii) every sequence (x_n) for which $(f(x_n))$ is convergent and $\nabla f(x_n) \to 0$ has a convergent subsequence.

Proof The implication $(i) \Rightarrow (ii)$ was already proved in Proposition 3.1.8, while $(ii) \Rightarrow (iii)$ is a consequence of the fact that every bounded sequence has a convergent subsequence. Let us show that $(iii) \Rightarrow (i)$. Suppose, by way of contradiction, that there exist $c \in \mathbb{R}$ and a sequence $(x_n) \subset \mathbb{R}^n$ with $\|x_n\| \to \infty$ and $f(x_n) \le c$, for every $n \in \mathbb{N}$. Clearly, $c \ge \inf_{x\in\mathbb{R}^p} f(x)$. For every $n \in \mathbb{N}^*$ we choose

$$\varepsilon_n := c + n^{-1} - \inf_{x\in\mathbb{R}^p} f(x) > 0,$$

so,

$$f(x_n) < \inf_{x\in\mathbb{R}^p} f(x) + \varepsilon_n.$$

Let $\delta_n := 2^{-1}\|x_n\| > 0$. Like in Theorem 3.1.15 (and its proof) there exists $\overline{x}_n$ with

$$f(\overline{x}_n) \le f(x_n) \le \inf_{x\in\mathbb{R}^p} f(x) + \varepsilon_n,$$
$$\|\overline{x}_n - x_n\| \le \delta_n,$$
$$\|\nabla f(\overline{x}_n)\| \le \varepsilon_n \delta_n^{-1}.$$

But,

$$\|\overline{x}_n\| \ge \|x_n\| - \|\overline{x}_n - x_n\| \ge \|x_n\| - 2^{-1}\|x_n\| = 2^{-1}\|x_n\|,$$

whence $\|\overline{x}_n\| \to \infty$. On the other hand,

$$\|\nabla f(\overline{x}_n)\| \le \frac{2}{\|x_n\|}(c + n^{-1} - \inf_{x\in\mathbb{R}^p} f(x)) \to 0.$$

Since $(f(\overline{x}_n))$ is bounded, it has a convergent subsequence. From (iii), one deduces that $(\overline{x}_n)$ should also have such a subsequence, but this is not possible. Consequently, (i) holds. $\qquad\square$

The condition (iii) in the above result is called Palais-Smale condition.

On the basis of Ekeland Variational Principle one obtains a new condition for the existence of the minimum points.

Theorem 3.1.17. *Let $\alpha > 0$ and $f : \mathbb{R}^p \to \mathbb{R}$ be a lower semicontinuous and lower bounded function. Suppose that for every $x \in \mathbb{R}^p$ with $\inf_{x\in\mathbb{R}^p} f(x) < f(x)$ there exists $z \in \mathbb{R}^p \setminus \{x\}$ such that*

$$f(z) < f(x) - \alpha \|z - x\|.$$

Then f has a minimum global point.

Proof Suppose, to obtain a contradiction, that the conclusion does not hold. Then, for every $x \in \mathbb{R}^p$, $\inf_{x\in\mathbb{R}^p} f(x) < f(x)$, so, by the assumptions made, there exists $z_x \in \mathbb{R}^p \setminus \{x\}$ such that

$$f(z_x) < f(x) - \alpha \|z_x - x\|.$$

By the Ekeland Variational Principle for $\varepsilon > 0$, $\delta > 0$ with $\varepsilon\delta^{-1} = \alpha$, there is an element $u \in \mathbb{R}^p$ with

$$f(u) \leq f(v) + \alpha \|v - u\|, \ \forall v \in \mathbb{R}^p.$$

Then

$$f(u) \leq f(z_u) + \alpha \|z_u - u\| < f(u),$$

which is absurd. Hence the conclusion hold. $\qquad\square$

A straightforward example of a function which satisfies the conditions of the above result is $f : \mathbb{R} \to \mathbb{R}$, $f(x) = \beta |x|$ for $\beta > \alpha$.

In the second part of this section, we present necessary optimality conditions and sufficient optimality conditions. At first, we deduce necessary optimality conditions that use the ideas developed around the construction and the study of the Bouligand tangent cone.

Theorem 3.1.18 (First-order necessary optimality condition). *If $\overline{x}$ is a local solution of (P) and f is differentiable at $\overline{x}$, then $\nabla f(\overline{x})(u) \geq 0$ for every $u \in T_B(M, \overline{x})$.*

Proof Let V be a neighborhood of $\overline{x}$ where $f(\overline{x}) \leq f(x)$ for every $x \in V \cap M$. Let $u \in T_B(M, \overline{x})$. Then there exists $(t_n) \subset (0, \infty)$ with $t_n \to 0$ and $(u_n) \to u$ such that for every n,

$$\overline{x} + t_n u_n \in M.$$

Obviously, the sequence $(t_n u_n)$ converges towards $0 \in \mathbb{R}^p$ and, for n large enough, $\overline{x} + t_n u_n$ belongs to V. Taking into account the differentiability of f at $\overline{x}$, there exists $(\alpha_n) \subset \mathbb{R}$, $\alpha_n \to 0$ such that for every $n \in \mathbb{N}$,

$$f(\overline{x} + t_n u_n) = f(\overline{x}) + t_n \nabla f(\overline{x})(u_n) + t_n \|u_n\| \alpha_n,$$

whence,

$$\nabla f(\overline{x})(u_n) + \|u_n\| \alpha_n \geq 0,$$

for all n large enough. Passing to the limit for $n \to \infty$, we get the conclusion. $\qquad\square$

Remark 3.1.19. *The conclusion of Theorem 3.1.18 could equivalently be written as*

$$-\nabla f(\overline{x}) \in N_B(M, \overline{x}).$$

Remark 3.1.20. *Taking into account Proposition 2.1.12, if $\overline{x} \in$ int M (inactive restriction), Theorem 3.1.18 gives $\nabla f(\overline{x})(u) \geq 0$ for every $u \in \mathbb{R}^p$. The linearity of $\nabla f(\overline{x})$ implies $\nabla f(\overline{x}) = 0$, i.e., the Fermat Theorem on stationary points.*

We present now a second-order necessary optimality condition for the problem without restrictions.

Theorem 3.1.21 (Second-order necessary optimality condition). *Let $U \subset \mathbb{R}^p$ be an open set and $\overline{x} \in U$. If $f : U \to \mathbb{R}$ is of class C^2 on a neighborhood of $\overline{x}$, and $\overline{x}$ is a local minimum point of f, then $\nabla f(\overline{x}) = 0$ and $\nabla^2 f(\overline{x})$ is positive semidefinite (that is, $\nabla^2 f(\overline{x})(u, u) \geq 0$ for every $u \in \mathbb{R}^p$).*

Proof Let $V \subset U$ be a neighborhood of $\overline{x}$ such that $f(\overline{x}) \leq f(x)$ for every $x \in V$ and f is of class C^2 on V. The fact that $\nabla f(\overline{x}) = 0$ follows from Fermat Theorem. As before, take $u \in \mathbb{R}^p$ and $(t_n) \subset (0, \infty)$ with $t_n \to 0$. Taylor Theorem 1.3.4 says that for every $n \in \mathbb{N}$ there exists $c_n \in (\overline{x}, \overline{x} + t_n u)$ such that

$$f(\overline{x} + t_n u) - f(\overline{x}) = t_n \nabla f(\overline{x})(u) + \frac{1}{2} t_n^2 \nabla^2 f(c_n)(u, u) = \frac{1}{2} t_n^2 \nabla^2 f(c_n)(u, u).$$

For n sufficiently large, $f(\overline{x} + t_n u) - f(\overline{x}) \geq 0$, whence

$$\nabla^2 f(c_n)(u, u) \geq 0,$$

and passing to the limit as $n \to \infty$ we get $c_n \to \overline{x}$. Since f is of class C^2, we infer

$$\nabla^2 f(\overline{x})(u, u) \geq 0,$$

whence $\nabla^2 f(\overline{x})$ is positive semidefinite. □

Obviously, if $\overline{x} \in$ int U is a local maximum of f, then $\nabla f(\overline{x}) = 0$ and $\nabla^2 f(\overline{x})$ is negative semidefinite (i.e., $\nabla^2 f(\overline{x})(u, u) \leq 0$ for every $u \in \mathbb{R}^p$). Further, if $\nabla^2 f(\overline{x})$ is neither positive semidefinite, or negative semidefinite, then $\overline{x}$ is not an extreme point of f.

In fact, many results in this book are generalizations, refinements of these results, or answer to different issues which naturally arise from their analysis.

One may consider whether the converses of these results are true. The answer is negative in both cases: it is sufficient to consider the function $f : \mathbb{R} \to \mathbb{R}$, $f(x) = x^3$ and $\overline{x} = 0$.

One can however impose supplementary conditions in order to get some equivalences, and the most important of these relates to convex functions.

Theorem 3.1.22. *Let $U \subset \mathbb{R}^p$ be an open convex set and let $f : U \to \mathbb{R}$ be a convex differentiable function. The next assertions are equivalent:*
  (i) $\overline{x}$ is a global minimum point of f (on U);
  (ii) $\overline{x}$ is a local minimum point of f;
  (iii) $\overline{x}$ is a critical point of f (i.e., $\nabla f(\overline{x}) = 0$).

Proof The implication $(i) \Rightarrow (ii)$ is obvious for every function, and $(ii) \Rightarrow (iii)$ follows from Fermat Theorem. Finally, the implication $(iii) \Rightarrow (i)$ relies on the convexity of f and follows from Theorem 2.2.10. □

Therefore, for convex functions, the first-order necessary optimality condition (in the unconstrained case) is also sufficient. In this situation, the second order condition is automatically satisfied (according to Theorem 2.2.10).

Concerning the nature of the extreme points for convex functions, we record here some important aspects.

Proposition 3.1.23. *Let $M \subset \mathbb{R}^p$ be a convex set and let $f : M \to \mathbb{R}$ be a convex function. If $\overline{x} \in M$ is a local minimum point of f on M, then $\overline{x}$ is in fact a global minimum point of f on M. If $u \in \mathrm{int}\, M$ is a local maximum point of f, then u is a global minimum point of f.*

Proof Let $\overline{x}$ be a local minimum point of f on M. Then there exists a convex neighborhood V of $\overline{x}$ such that for every $x \in V \cap M$, $f(\overline{x}) \le f(x)$. Let $x \in M$. There exists $\lambda \in (0, 1)$ such that $y := (1 - \lambda)\overline{x} + \lambda x \in M \cap V$. Then,

$$f(\overline{x}) \le f(y) = f((1 - \lambda)\overline{x} + \lambda x) \le (1 - \lambda)f(\overline{x}) + \lambda f(x),$$

that is,

$$\lambda f(\overline{x}) \le \lambda f(x),$$

and the conclusion of the first part follows.

For the second part, there is a convex symmetric neighborhood V of 0 (a ball with the center 0, for instance) such that for every $v \in V$, $f(u + v) \le f(u)$ and $f(u - v) \le f(u)$. Then

$$f(u) = f\left(\frac{1}{2}(u + v) + \frac{1}{2}(u - v)\right) \le \frac{1}{2}f(u + v) + \frac{1}{2}f(u - v) \le f(u),$$

for all $v \in V$. Consequently, $f(u + v) = f(u)$ for every $v \in V$. Therefore, u is a local (hence global) minimum point of f. □

Proposition 3.1.24. *Let $M \subset \mathbb{R}^p$ be a convex set and let $f : M \to \mathbb{R}$ be a convex function. If nonempty, the set of minimum points of f on M is convex. If, moreover, f is strictly convex, then this set has at most one element.*

Proof From the preceding result, if $x_1, x_2 \in M$ are (global) minima of f on M, then $f(x_1) = f(x_2)$. The convexity implies $f(x) = f(x_1)$ for every $x \in [x_1, x_2]$. Therefore, the first part is proved. Suppose now that f is strictly convex. If this is so, then we would have two different global minima, then $f(x) < f(x_1)$ for every $x \in (x_1, x_2)$, which is not possible. □

For constrained problems involving convex functions, the first-order necessary optimality condition is, again, a sufficient optimality condition.

Proposition 3.1.25. *Let $U \subset \mathbb{R}^p$ be a convex open set and let $f : U \to \mathbb{R}$ be a convex, differentiable function. Let $M \subset U$ be convex. The element $\overline{x} \in M$ is a minimum point of f on M if and only if*

$$-\nabla f(\overline{x}) \in N(M, \overline{x}).$$

Proof Let $\overline{x} \in M$ be a minimum point of f on M. Then, according to Theorem 3.1.18, $\nabla f(\overline{x})(u) \geq 0$ for every $u \in T(M, \overline{x})$, that is

$$-\nabla f(\overline{x}) \in T(M, \overline{x})^- = N(M, \overline{x}).$$

Conversely, we know from the convexity of f (Theorem 2.2.10), that

$$f(x) \geq f(\overline{x}) + \nabla f(\overline{x})(x - \overline{x}), \ \forall x \in U.$$

But, using the hypothesis and the convexity of M (Proposition 2.1.15),

$$-\nabla f(\overline{x}) \in N(M, \overline{x}) = \{u \in \mathbb{R}^p \mid \langle u, x - \overline{x} \rangle \leq 0, \ \forall x \in M\},$$

whence $\nabla f(\overline{x})(x - \overline{x}) \geq 0$ for every $x \in M$. From these relations we know $f(x) \geq f(\overline{x})$ for every $x \in M$. □

Coming back to Theorems 3.1.18 and 3.1.21, in order to formulate sufficient optimality conditions, we strengthen the conclusion of these results. The good point is that we get stronger minimality concepts.

Definition 3.1.26. *Let $\alpha > 0$. One says that $\overline{x} \in M$ is a strict local solution of order α for (P), or a strict local minimum point of order α for f on M if there exist two constants $r, l > 0$ such that for every $x \in M \cap B(\overline{x}, r)$,*

$$f(x) \geq f(\overline{x}) + l \, \|x - \overline{x}\|^{\alpha}.$$

The announced results are as follows.

Theorem 3.1.27. *Suppose that f is differentiable at $\overline{x} \in M$ and*

$$\nabla f(\overline{x})(u) > 0, \ \forall u \in T_B(M, \overline{x}) \setminus \{0\}.$$

Then $\overline{x}$ is a strict local solution of order $\alpha = 1$ for (P).

Proof Suppose, by way of contradiction, that $\overline{x}$ is not a strictly local solution of order 1. Then, there exists a sequence $(x_n) \to \overline{x}$, $(x_n) \subset M$ such that for every $n \in \mathbb{N}^*$,

$$f(x_n) < f(\overline{x}) + n^{-1} \|x_n - \overline{x}\|.$$

By virtue of this inequality,

$$x_n \neq \overline{x}, \; \forall n \in \mathbb{N}^*.$$

Since f is differentiable, there exists a sequence of real numbers $(y_n) \to 0$ such that for every $n \in \mathbb{N}$,

$$f(x_n) = f(\overline{x}) + \nabla f(\overline{x})(x_n - \overline{x}) + y_n \|x_n - \overline{x}\|.$$

The combination of these two relation yields

$$n^{-1} \|x_n - \overline{x}\| > \nabla f(\overline{x})(x_n - \overline{x}) + y_n \|x_n - \overline{x}\|,$$

whence, by division with $\|x_n - \overline{x}\|$, we deduce

$$n^{-1} > \nabla f(\overline{x}) \left(\frac{x_n - \overline{x}}{\|x_n - \overline{x}\|} \right) + y_n, \; \forall n \in \mathbb{N}^*. \tag{3.1.2}$$

Since the sequence $\left(\frac{x_n - \overline{x}}{\|x_n - \overline{x}\|} \right)$ is bounded, there exists a convergent subsequence of it. The limit, denoted by u, of this subsequence is not zero (being of norm 1) and, furthermore, from $\|x_n - \overline{x}\| \to 0$, we infer that $u \in T_B(M, \overline{x})$. Consequently, $u \in T_B(M, \overline{x}) \setminus \{0\}$, and passing to the limit in the relation (3.1.2) we have

$$0 \geq \nabla f(\overline{x})(u),$$

which is in contradiction with the hypothesis. $\qquad \square$

Notice that for differentiable functions, the concept of a local strict solution of order 1 is specific to the case of active restrictions (that is, $\overline{x} \in M \setminus \operatorname{int} M$): if f is differentiable at $\overline{x} \in \operatorname{int} M$, then $\overline{x}$ cannot be a local strict solution of order 1. Indeed, if $\overline{x} \in \operatorname{int} M$ would be local strict solution of order 1, then, on one hand, $\nabla f(\overline{x}) = 0$ (Fermat Theorem), and, on the other hand, $\nabla f(\overline{x}) \neq 0$ from the definition of strict solutions.

Concerning second-order optimality conditions, one has the following results.

Theorem 3.1.28. *Suppose that f is of class C^2, $\nabla f(\overline{x}) = 0$ and*

$$\nabla^2 f(\overline{x})(u, u) > 0, \; \forall u \in T_B(M, \overline{x}) \setminus \{0\}.$$

Then $\overline{x}$ is a local strict solution of order $\alpha = 2$ for problem (P).

Proof As before, one supposes, by contradiction, that the conclusion does not hold. Then there exists a sequence $(x_n) \to \overline{x}$, $(x_n) \subset M \setminus \{\overline{x}\}$ such that for every $n \in \mathbb{N}^*$,

$$f(x_n) < f(\overline{x}) + n^{-1} \|x_n - \overline{x}\|^2.$$

From Taylor Theorem 1.3.4, for every $n \in \mathbb{N}$ there exists c_n on the segment joining $\bar{x}$ and x_n such that

$$f(x_n) - f(\bar{x}) = \nabla f(\bar{x})(x_n - \bar{x}) + \frac{1}{2}\nabla^2 f(c_n)(x_n - \bar{x}, x_n - \bar{x})$$

$$= \frac{1}{2}\nabla^2 f(c_n)(x_n - \bar{x}, x_n - \bar{x}).$$

We get

$$n^{-1}\|x_n - \bar{x}\|^2 > \frac{1}{2}\nabla^2 f(c_n)(x_n - \bar{x}, x_n - \bar{x}),$$

whence, in order to finish the proof, we divide by $\|x_n - \bar{x}\|^2$ and we repeat the above arguments. □

In the unconstrained case, this result gives the following consequence.

Corollary 3.1.29. *Let $U \subset \mathbb{R}^p$ be a nonempty, open set and $f : U \to \mathbb{R}$ be a C^2 function. If $\bar{x} \in U$ is a critical point of f and $\nabla^2 f(\bar{x})$ is positive definite (i.e., $\nabla^2 f(\bar{x})(u, u) > 0$ for every $u \in \mathbb{R}^p \setminus \{0\}$), then $\bar{x}$ is a local strict solution of order $\alpha = 2$ for f.*

One can identify $\nabla^2 f(\bar{x})$ with the Hessian matrix of f at $\bar{x}$, $\left(\frac{\partial^2 f}{\partial x^i \partial x^j}(\bar{x})\right)_{i,j\in\overline{1,p}}$, and a sufficient condition for the positive definiteness of it is given by the next criterion, known from linear algebra (Sylvester criterion): all the determinants of the matrices $\left(\frac{\partial^2 f}{\partial x^i \partial x^j}(\bar{x})\right)_{i,j\in\overline{1,k}}$, $k \in \overline{1,p}$ are strictly positive. Analogously (for $-f$), if the determinants of the matrices $\left(\frac{\partial^2 f}{\partial x^i \partial x^j}(\bar{x})\right)_{i,j\in\overline{1,k}}$, $k \in \overline{1,p}$ are not zero and change the signs starting with minus, then $\nabla^2 f(\bar{x})$ is negative definite and $\bar{x}$ is a maximum point. Furthermore, if all these determinants are not zero, then any other distribution of their signs leads to the conclusion that the reference point is not an extreme point.

3.2 Functional Restrictions

The restriction of the problem (P) introduced in the previous section is $x \in M$. Many times, in practice this set M of feasible points is defined by means of functions. Let us consider $g : \mathbb{R}^p \to \mathbb{R}^n$ and $h : \mathbb{R}^p \to \mathbb{R}^m$ as C^1 functions. As usual, g and h can be thought of as $g = (g_1, g_2, ..., g_n)$, and $h = (h_1, h_2, ..., h_m)$, respectively, where $g_i : \mathbb{R}^p \to \mathbb{R}$ ($i \in \overline{1,n}$) and $h_j : \mathbb{R}^p \to \mathbb{R}$ ($j \in \overline{1,m}$) are C^1 real valued functions.

Let the set of feasible points be defined as:

$$M := \{x \in U \mid g(x) \leq 0, \ h(x) = 0\} \subset \mathbb{R}^p.$$

Let us observe that we have two types of constraints: equalities and inequalities. Let $x \in M$. If for an $i \in \overline{1,n}$, one has that $g_i(x) < 0$, then the continuity of g ensures the existence of a neighborhood V of x such that $g_i(y) < 0$ for all $y \in V$. Therefore, when one looks for a certificate that x is a local solution of (P), the restriction $g_i \leq 0$ does not effectively influence the set of points u where one should compare $f(x)$ and $f(u)$. For this reason, one says that the restriction $g_i \leq 0$ is inactive at x and these kind of restrictions should be eliminated from the discussion. In the opposite case, when $g_i(x) = 0$, we call this active (inequality) restriction. For $\overline{x} \in M$, we denote the set of indexes corresponding to active inequality type restrictions by

$$A(\overline{x}) = \{i \in \overline{1, n} \mid g_i(\overline{x}) = 0\}.$$

We are now going to present two types of optimality conditions for problem (P) with functional constraints as described above. These two types of conditions are formally very close, but their differences are important for the detection of extreme points. We start with the Fritz John necessary optimality conditions where the objective function does not play any special role with respect to the functions which define the restrictions. We shall consider the drawbacks of these conditions, and next we shall impose supplementary conditions in order to eliminate then. By this procedure, we get the famous Karush-Kuhn-Tucker necessary optimality conditions which will be extensively used for solving nonlinear optimization problems.

3.2.1 Fritz John Optimality Conditions

The result of this subsection refers to necessary optimality conditions for problem (P) with functional restrictions without any additional assumption to the general framework already described. These conditions were obtained in 1948 by the German mathematician Fritz John.

Theorem 3.2.1 (Fritz John). *Let $\overline{x} \in M$ be a solution of (P). Then there exist $\lambda_0 \in \mathbb{R}$, $\lambda_0 \geq 0$, $\lambda = (\lambda_1, \lambda_2, ..., \lambda_n) \in \mathbb{R}^n$, $\mu = (\mu_1, \mu_2, ..., \mu_m) \in \mathbb{R}^m$, with $\lambda_0 + \|\lambda\| + \|\mu\| \neq 0$ such that*

$$\lambda_0 \nabla f(\overline{x}) + \sum_{i=1}^{n} \lambda_i \nabla g_i(\overline{x}) + \sum_{j=1}^{m} \mu_j \nabla h_j(\overline{x}) = 0$$

and

$$\lambda_i \geq 0, \ \lambda_i g_i(\overline{x}) = 0, \ \textit{for every } i \in \overline{1, n}.$$

Proof Let us take $\delta > 0$ such that $D(\overline{x}, \delta) \subset U$ and for every $x \in M \cap D(\overline{x}, \delta)$, $f(\overline{x}) \leq f(x)$. For all $k \in \mathbb{N}^*$ we consider the function $\varphi_k : D(\overline{x}, \delta) \to \mathbb{R}$ given by

$$\varphi_k(x) = f(x) + \frac{k}{2} \sum_{i=1}^{n} \left(g_i^+(x)\right)^2 + \frac{k}{2} \sum_{j=1}^{m} \left(h_j(x)\right)^2 + \frac{1}{2} \|x - \overline{x}\|^2,$$

where $g_i^+(x) = \max\{g_i(x), 0\}$. Clearly, φ_k attains its minimum on $D(\bar{x}, \delta)$ and we denote by x_k such a minimum point. We also observe that

$$0 \le \varphi_k(x_k) = f(x_k) + \frac{k}{2} \sum_{i=1}^{n} \left(g_i^+(x_k)\right)^2 + \frac{k}{2} \sum_{j=1}^{m} \left(h_j(x_k)\right)^2 + \frac{1}{2} \|x_k - \bar{x}\|^2$$

$$\le \varphi_k(\bar{x}) = f(\bar{x}).$$

Since the sequence (x_k) is bounded and f in continuous on $D(\bar{x}, \delta)$, we infer that $(f(x_k))$ is also a bounded sequence. Letting $k \to \infty$ in the above relation, we get

$$\lim_{k \to \infty} \sum_{i=1}^{n} \left(g_i^+(x_k)\right)^2 = 0$$

$$\lim_{k \to \infty} \sum_{j=1}^{m} \left(h_j(x_k)\right)^2 = 0.$$

The boundedness of (x_k) ensures that one can extract a convergent subsequence of it. Without relabeling, we can write $x_k \to x^* \in D(\bar{x}, \delta)$, and the previous relations yield $x^* \in M$. Consequently, passing to the limit in the inequality above, we have

$$f(x^*) + \frac{1}{2} \|x^* - \bar{x}\|^2 \le f(\bar{x}).$$

On the other hand, $f(\bar{x}) \le f(x^*)$, so $\|x^* - \bar{x}\| = 0$, that is $x^* = \bar{x}$. Therefore $x_k \to \bar{x}$.

An essential remark here is that φ_k is differentiable since the (nondifferentiable) scalar functions $g_i^+(x)$ are squared, whence $\nabla \left(g_i^+(x)\right)^2 = g_i^+(x)\nabla g(x)$. Since x_k is a minimum for φ_k on $D(\bar{x}, \delta)$, we deduce that

$$-\nabla \varphi_k(x_k) \in N(D(\bar{x}, \delta), x_k).$$

For k sufficiently large, x_k belongs to the interior of the ball $D(\bar{x}, \delta)$ and we conclude that for these numbers k, one has $N(D(\bar{x}, \delta), x_k) = \{0\}$. The combination of these facts allow us to write

$$\nabla f(x_k) + k \sum_{i=1}^{n} g_i^+(x_k)\nabla g(x_k) + k \sum_{j=1}^{m} h_j(x_k)\nabla h_j(x_k) + x_k - \bar{x} = 0, \qquad (3.2.1)$$

for every k large enough. For $i \in \overline{1, n}$, $j \in \overline{1, m}$, we denote $\alpha_i^k := kg_i^+(x_k)$, $\beta_j^k := kh_j(x_k)$ and $y^k = \sqrt{1 + \sum_{i=1}^{n} \left(\alpha_i^k\right)^2 + \sum_{j=1}^{m} \left(\beta_j^k\right)^2}$. It is clear that $y^k > 1$ and we take $\lambda_0^k := \frac{1}{y^k}$, $\lambda_i^k := \frac{\alpha_i^k}{y^k}$, $\mu_j^k := \frac{\beta_j^k}{y^k}$. We observe that

$$\left(\lambda_0^k\right)^2 + \sum_{i=1}^{n} \left(\lambda_i^k\right)^2 + \sum_{j=1}^{m} \left(\mu_j^k\right)^2 = 1,$$

whence the sequences (λ_0^k), (λ_i^k), (μ_j^k) $(i \in \overline{1,n}, j \in \overline{1,m})$ are bounded. Then there exist subsequences (we keep the indexes) convergent to some real numbers, respectively denoted by

$$\lambda_0, \lambda_1, \lambda_2, ..., \lambda_n, \mu_1, \mu_2, ..., \mu_m.$$

These numbers cannot be zero simultaneously. The positivity of the terms of the sequences (λ_0^k), (λ_i^k) $(i \in \overline{1,n})$ implies the positivity of their limits $\lambda_0, \lambda_1, \lambda_2, ..., \lambda_n$. Now, we divide relation (3.2.1) by y^k, and we get

$$\lambda_0^k \nabla f(x_k) + \sum_{i=1}^{n} \lambda_i^k \nabla g(x_k) + \sum_{j=1}^{m} \mu_j^k \nabla h_j(x_k) + \frac{1}{y^k}(x_k - \overline{x}) = 0.$$

Letting $k \to \infty$ we have the first relation in the conclusion. Now we show the second one. Let $i \in \overline{1,n}$. If $\lambda_i = 0$, there is nothing to prove. Otherwise, if $\lambda_i > 0$, from the definition of λ_i we infer that for k sufficiently large, $g_i^+(x_k) > 0$, whence $g_i^+(x_k) = g_i(x_k)$. The relation

$$0 < g_i(x_k) \to g_i(\overline{x}) \leq 0$$

leads us to the conclusion $g_i(\overline{x}) = 0$. So, the second part of the conclusion holds and the theorem is completely proved. □

The relations in the conclusion of Theorem 3.2.1 are called Fritz John necessary optimality conditions. The major drawback of this result is that it does not eliminate the possibility that the real number associated to the objective function (i.e., λ_0) can be zero. This means that it would be possible to have too many points where the conditions in the conclusion are satisfied and therefore, in such a case, the result would not give important practical hints on the solutions. For instance, if a feasible point x satisfies $\nabla g_i(x) = 0$ for a certain $i \in A(x)$ or $\nabla h_j(x) = 0$ for an $j \in \overline{1,m}$, then it satisfies Fritz John conditions (with $\lambda_0 = 0$), the objective function being then completely eliminated. In the next subsection we shall impose a condition in order to avoid $\lambda_0 = 0$.

Let us first illustrate the possibilities created by Theorem 3.2.1 through two concrete examples.

Example 3.2.2. *Let us consider the problem of minimization of* $f : \mathbb{R}^2 \to \mathbb{R}$,

$$f(x_1, x_2) = (x_1 - 3)^2 + (x_2 - 2)^2$$

under the restriction $g(x) \leq 0$, *where* $g : \mathbb{R}^2 \to \mathbb{R}^4$,

$$g(x_1, x_2) = (x_1^2 + x_2^2 - 5, x_1 + 2x_2 - 4, -x_1, -x_2).$$

One can observe graphically that $\overline{x} = (2,1)$ *is solution of the problem and* $A(\overline{x}) = \{1,2\}$. *We want to verify Fritz John condition at this point. From the second condition, since* $3, 4 \notin A(\overline{x})$, *we get* $\lambda_3 = \lambda_4 = 0$. *Since* $\nabla f(\overline{x}) = (-2,-2)$, $\nabla g_1(\overline{x}) = (4,2)$, $\nabla g_2(\overline{x}) =$

$(1, 2)$, *we have to find positive real numbers* $\lambda_0, \lambda_1, \lambda_2 \geq 0$, *not simultaneously zero, such that*

$$\lambda_0(-2, -2) + \lambda_1(4, 2) + \lambda_2(1, 2) = (0, 0).$$

We get $\lambda_1 = \frac{1}{3}\lambda_0$ *and* $\lambda_2 = \frac{2}{3}\lambda_0$, *whence, by taking* $\lambda_0 > 0$, *the first Fritz John condition is fulfilled.*

Let us now have a look to the point $x = (0, 0)$. *This time* $A(x) = \{3, 4\}$, *whence* $\lambda_1 = \lambda_2 = 0$. *We have that* $\nabla f(\overline{x}) = (-6, -4)$, $\nabla g_3(\overline{x}) = (-1, 0)$, $\nabla g_4(\overline{x}) = (0, -1)$. *A computation shows that the equation*

$$\lambda_0(-6, -4) + \lambda_3(-1, 0) + \lambda_4(0, -1) = (0, 0)$$

has no solution $(\lambda_0, \lambda_3, \lambda_4)$ *different to zero with positive components. Then x does not fulfill the Fritz John conditions, hence it is not a minimum point for the given problem.*

Example 3.2.3. *Let us consider the problem of minimization of* $f : (0, \infty) \times (0, \infty) \to \mathbb{R}$, $f(x_1, x_2) = -2x_2$ *under the restriction* $g(x) \leq 0$, *where* $g : \mathbb{R}^2 \to \mathbb{R}^3$, $g(x_1, x_2) = (x_1 - x_2 - 2, -x_1 + x_2 + 2, x_1 + x_2 - 6)$. *The set* M *of feasible points is* $[(2, 0), (4, 2)] \setminus \{(2, 0)\}$, *and the minimum point is* $\overline{x} = (4, 2)$. *It is easy to observe that any feasible point satisfies the Fritz John conditions, but the solution is the only point where one can choose* $\lambda_0 \neq 0$. *Indeed, if x is a feasible point different to* $\overline{x}$, *then* $\lambda_3 = 0$, $\lambda_1 = \lambda_2$ *and* $\lambda_0 = 0$.

3.2.2 Karush-Kuhn-Tucker Conditions

As seen before, it is desirable to have a Fritz John type result, but with $\lambda_0 \neq 0$. We could directly impose an extra condition in Theorem 3.2.1 in order to ensure this, but we prefer a direct approach because we aim at working with weak assumptions.

Let consider the sets

$$G(\overline{x}) = \left\{ \sum_{i \in A(\overline{x})} \lambda_i \nabla g_i(\overline{x}) + \sum_{j=1}^{m} \mu_j \nabla h_j(\overline{x}) \mid \lambda_i \geq 0, \ \forall i \in A(\overline{x}), \mu_j \in \mathbb{R}, \forall j \in \overline{1, m} \right\} \subset \mathbb{R}^p.$$

(where, as usual, we used the identification between $L(\mathbb{R}^p, \mathbb{R})$ and $\mathbb{R}^p$) and

$$D(\overline{x}) = \left\{ u \in \mathbb{R}^p \mid \nabla g_i(\overline{x})(u) \leq 0, \ \forall i \in A(\overline{x}) \text{ and } \nabla h_j(\overline{x})(u) = 0, \ \forall j \in \overline{1, m} \right\}.$$

Before the main result, we need to shed some light on some important relations for these sets.

Proposition 3.2.4. *For every* $\overline{x} \in M$ *we have:*
  (i) $G(\overline{x}) = D(\overline{x})^-$;
  (ii) $T_B(M, \overline{x}) \subset D(\overline{x})$.

Proof (i) The inclusion $G(\overline{x}) \subset D(\overline{x})^-$ is obvious, while the reverse one is a direct consequence of Farkas Lemma (Theorem 2.1.8).

(ii) Clearly, $0 \in D(\overline{x})$. Let $u \in T_B(M, \overline{x}) \setminus \{0\}$. By the definition of tangent vectors, there exist $(t_n) \subset (0, \infty)$, $t_n \to 0$ and $(u_n) \to u$ such that for every n,

$$\overline{x} + t_n u_n \in M.$$

The sequence $(t_n u_n)$ converges towards 0 in $\mathbb{R}^p$. Taking into account the differentiability of h at $\overline{x}$, there exists $(\alpha_n) \subset \mathbb{R}^p$, $\alpha_n \to 0$ such that for every $n \in \mathbb{N}$,

$$h(\overline{x} + t_n u_n) = h(\overline{x}) + t_n \nabla h(\overline{x})(u_n) + t_n \|u_n\| \alpha_n.$$

Since $h(\overline{x} + t_n u_n) = h(\overline{x}) = 0$, dividing by t_n and passing to the limit as $n \to \infty$, we get $\nabla h(\overline{x})(u) = 0$. Now, for every $i \in A(\overline{x})$ there exist $(\alpha_n^i) \subset \mathbb{R}$, $\alpha_n^i \to 0$ such that for every $n \in \mathbb{N}$,

$$g_i(\overline{x} + t_n u_n) = g_i(\overline{x}) + t_n \nabla g_i(\overline{x})(u_n) + t_n \|u_n\| \alpha_n^i.$$

As before, since $g_i(\overline{x} + t_n u_n) \le 0$ and $g_i(\overline{x}) = 0$, we have $\nabla g_i(\overline{x})(u) \le 0$, and the proposition is proved. $\qquad\square$

The next example shows that the reverse inclusion in the item (*ii*) above is false.

Example 3.2.5. *Let* $g : \mathbb{R}^2 \to \mathbb{R}$, $g(x_1, x_2) = -x_1 - x_2$, $h : \mathbb{R}^2 \to \mathbb{R}$, $h(x_1, x_2) = x_1 x_2$ *and the feasible point* $\overline{x} = (0, 0)$. *Then:*

$$D(\overline{x}) = \{(u_1, u_2) \mid -u_1 - u_2 \le 0\},$$
$$T_B(M, \overline{x}) = \{(u_1, u_2) \mid u_1 \ge 0, u_2 \ge 0, u_1 u_2 = 0\}.$$

We establish now a generalized form of a classical result known under the name of Karush-Kuhn-Tucker Theorem, since it was obtained (with stronger assumptions) by the American mathematicians William Karush, Harold William Kuhn and Albert William Tucker. It is interesting to note that William Karush obtained the result in 1939, but the mathematical community become aware of its importance when Harold William Kuhn and Albert William Tucker got the result, in a different way, in 1950.

Theorem 3.2.6 (Karush-Kuhn-Tucker). *Let* $\overline{x} \in M$ *be a solution of the problem* (P). *Suppose that* $T_B(M, \overline{x})^- = D(\overline{x})^-$. *Then there exist* $\lambda = (\lambda_1, \lambda_2, ..., \lambda_n) \in \mathbb{R}^n$, $\mu = (\mu_1, \mu_2, ..., \mu_m) \in \mathbb{R}^m$, *such that*

$$\nabla f(\overline{x}) + \sum_{i=1}^{n} \lambda_i \nabla g_i(\overline{x}) + \sum_{j=1}^{m} \mu_j \nabla h_j(\overline{x}) = 0 \tag{3.2.2}$$

and

$$\lambda_i \ge 0, \ \lambda_i g_i(\overline{x}) = 0, \text{ for every } i \in \overline{1, n}. \tag{3.2.3}$$

Proof From Theorem 3.1.18, $\nabla f(\overline{x})(u) \geq 0$ for every $u \in T_B(M, \overline{x})$, whence $-\nabla f(\overline{x}) \in T_B(M, \overline{x})^-$. We use now the assumption $T_B(M, \overline{x})^- = D(\overline{x})^-$ to infer that $-\nabla f(\overline{x}) \in D(\overline{x})^-$. From Proposition 3.2.4 (*i*), we get $-\nabla f(\overline{x}) \in G(\overline{x})$. Consequently, there exist $\lambda_i \geq 0$, $i \in A(\overline{x})$, $\mu_j \in \mathbb{R}$, $j \in \overline{1, m}$ such that $-\nabla f(\overline{x}) = \sum_{i \in A(\overline{x})} \lambda_i \nabla g_i(\overline{x}) + \sum_{j=1}^{m} \mu_j \nabla h_j(\overline{x})$. Now, for indexes $i \in \overline{1, n} \setminus A(\overline{x})$ we take $\lambda_i = 0$, and we obtain the conclusion. □

If one compares Theorem 3.2.6 and Theorem 3.2.1, one notices the announced difference concerning the real number associated to the objective function.

The function $L : U \times \mathbb{R}^{n+m} \to \mathbb{R}$,

$$L(x, (\lambda, \mu)) := f(x) + \sum_{i=1}^{n} \lambda_i g_i(x) + \sum_{j=1}^{m} \mu_j h_j(x)$$

is called the Lagrangian of (*P*). Therefore, the conclusion given by relation (3.2.2) can be written as

$$\nabla_x L(\overline{x}, (\lambda, \mu)) = 0,$$

and the elements $(\lambda, \mu) \in \mathbb{R}_+^n \times \mathbb{R}^m$ are called Lagrange multipliers. This name is due to the fact that the first time this method was used to investigate constrained optimization problems was given in some of Lagrange's works on calculus of variations problems.

The preceding theorem does not ensure the uniqueness of these multipliers. We denote by $M(\overline{x})$ the set of Lagrange multipliers at $\overline{x}$, i.e.,

$$M(\overline{x}) := \{(\lambda, \mu) \in \mathbb{R}_+^n \times \mathbb{R}^m \mid \nabla_x L(\overline{x}, (\lambda, \mu)) = 0\},$$

where $\mathbb{R}_+^n := [0, \infty)^n$.

On the other hand, $L(x, (\lambda, \mu))$ is an affine function with respect to the variables (λ, μ). We can observe the following fact which will appear later in the discussion: if $\overline{x} \in M$ and $(\overline{\lambda}, \overline{\mu})$ is a maximum on $\mathbb{R}_+^n \times \mathbb{R}^m$ for $(\lambda, \mu) \mapsto L(\overline{x}, (\lambda, \mu))$, then $\overline{\lambda}_i g_i(\overline{x}) = 0$ for every $i \in \overline{1, n}$.

Theorem 3.2.6 gives necessary optimality conditions for (*P*). If, instead of minimization, we are looking for maximization of the objective function f under the same constraints, then, from $\max f = -\min(-f)$, the necessary condition (3.2.2) can be written as

$$-\nabla f(\overline{x}) + \sum_{i=1}^{n} \lambda_i \nabla g_i(\overline{x}) + \sum_{j=1}^{m} \mu_j \nabla h_j(\overline{x}) = 0.$$

Furthermore, let us notice that if one has only equalities as constraints, taking into account that $h(x) = 0$ is equivalent to $-h(x) = 0$, the necessary optimality condition can by written, for both maxima and minima, as

$$\nabla f(\overline{x}) + \sum_{j=1}^{m} \mu_j \nabla h_j(\overline{x}) = 0.$$

Coming back to the main results, let us observe two more things. Firstly, if the problem has no restrictions (for instance, $U = M = \mathbb{R}^p$), then relation (3.2.2) reduces to the first-order necessary optimality condition (Fermat Theorem): $\nabla f(\overline{x}) = 0$. Secondly, the key relation (3.2.2) does not hold without supplementary conditions (here, $T_B(M, \overline{x})^- = D(\overline{x})^-$). To illustrate this consider the following example.

Example 3.2.7. *Let $f : \mathbb{R}^2 \to \mathbb{R}$ and $g : \mathbb{R}^2 \to \mathbb{R}^2$ given by $f(x_1, x_2) = x_1$ and $g(x_1, x_2) = (-x_2 + (1 - x_1)^3, x_2)$. It is easy to see that $\overline{x} = (1, 0)$ is a minimum point of the associated problem, but (3.2.2) does not hold. Clearly, Fritz John conditions are fulfilled for $\lambda_0 = 0$.*

So, in the next section, every condition which ensures the validity of the Karush-Kuhn-Tucker Theorem is called a qualification condition, and in view of the decisive importance of such requirements, we shall discuss it into detail in the next section.

Before that, let us observe that under certain assumptions, Karush-Kuhn-Tucker conditions (3.2.2) and (3.2.3) are also sufficient for minimality.

Theorem 3.2.8. *Suppose that U is convex, f is convex on U, h is affine and $g_i, i \in \overline{1, n}$ are convex. Let $\overline{x} \in M$. If there exists $(\lambda, \mu) \in \mathbb{R}^n \times \mathbb{R}^m$ such that (3.2.2) and (3.2.3) hold, then $\overline{x}$ is a minimum point for (P) (or minimum of f on M).*

Proof The condition (3.2.2) expresses the fact that

$$\nabla_x L(\overline{x}, (\lambda, \mu)) = 0.$$

Under our assumptions, L is a convex function in x, so according to Theorem 3.1.22, $\overline{x}$ is a minimum (without constraints) of the map $x \mapsto L(x, (\lambda, \mu))$. Therefore, for every $x \in U$,

$$L(x, (\lambda, \mu)) = f(x) + \sum_{i=1}^{n} \lambda_i g_i(x) + \sum_{j=1}^{m} \mu_j h_j(x) \geq L(\overline{x}, (\lambda, \mu)) = f(\overline{x}).$$

But, for any $x \in M$,

$$\sum_{i=1}^{n} \lambda_i g_i(x) + \sum_{j=1}^{m} \mu_j h_j(x) \leq 0,$$

whence $f(x) \geq f(\overline{x})$. The proof is complete. □

Concerning the structure of the set of Lagrange multipliers, we have the following result.

Proposition 3.2.9. *For data with the structure mentioned in the above theorem, the set $M(\overline{x})$ of the Lagrange multipliers is the same for all minimum points of f an M.*

Proof Clearly, M is a convex set. Let $x_1, x_2 \in M$ be two minimum points of (P). According to Proposition 3.1.23, one has $f(x_1) = f(x_2)$. Let $(\lambda, \mu) \in M(x_1)$. Then

$$\nabla f(x_1) + \sum_{i=1}^{n} \lambda_i \nabla g_i(x_1) + \sum_{j=1}^{m} \mu_j \nabla h_j(x_1) = 0$$

and

$$\lambda_i \geq 0, \ \lambda_i g_i(x_1) = 0, \ \text{for every } i \in \overline{1, n}.$$

As before,

$$f(x_2) + \sum_{i=1}^{n} \lambda_i g_i(x_2) \geq f(x_1) = f(x_2).$$

Taking into account the information on the numbers λ_i and $g_i(x_2)$, we infer that $\lambda_i g_i(x_2) = 0$ for every $i \in \overline{1, n}$. From

$$L(x_2, (\lambda, \mu)) = f(x_2) = f(x_1) = L(x_1, (\lambda, \mu)),$$

we get that x_2 is a minimum point for the convex function $L(\cdot, (\lambda, \mu))$ on U. Hence

$$\nabla f(x_2) + \sum_{i=1}^{n} \lambda_i \nabla g_i(x_2) + \sum_{j=1}^{m} \mu_j \nabla h_j(x_2) = 0.$$

We have that $(\lambda, \mu) \in M(x_2)$. The other inclusion follows by exchanging x_1 and x_2 in the above proof. $\qquad\square$

We now interpret Theorem 3.2.6 by using the concept of saddle point applied to the Lagrangian function. Firstly, we define the concept.

Definition 3.2.10. *Let X, Y be two sets and $F : X \times Y \to \mathbb{R}$. A saddle point of F is a pair $(\overline{x}, \overline{y}) \in X \times Y$ with the property that*

$$\max_{y \in Y} F(\overline{x}, y) = F(\overline{x}, \overline{y}) = \min_{x \in X} F(x, \overline{y}). \qquad (3.2.4)$$

It is clear that the relation (3.2.4) is equivalent to

$$F(\overline{x}, y) \leq F(\overline{x}, \overline{y}) \leq F(x, \overline{y}), \ \forall (x, y) \in X \times Y$$

and to

$$F(\overline{x}, y) \leq F(x, \overline{y}), \ \forall (x, y) \in X \times Y.$$

For instance, the point $(0, 0)$ is a saddle point of $F : \mathbb{R} \times \mathbb{R} \to \mathbb{R}$, $F(x, y) = x^2 - y^2$ (the figure below).

The following general result is in order.

Proposition 3.2.11. *For all saddle points $(\overline{x}, \overline{y})$ of F, the value $F(\overline{x}, \overline{y})$ is constant. If (x_1, y_1) and (x_2, y_2) are saddle points, then (x_1, y_2) and (x_2, y_1) are saddle points as well.*

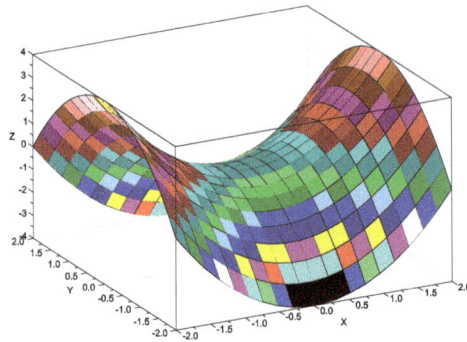

Figure 3.2: A saddle point.

Proof The following relations hold:

$$F(x_1, y) \le F(x_1, y_1) \le F(x, y_1), \ \forall (x, y) \in X \times Y$$
$$F(x_2, y) \le F(x_2, y_2) \le F(x, y_2), \ \forall (x, y) \in X \times Y.$$

If, in the first one, we take $x = x_2$ and $y = y_2$, and in the second one we put $x = x_1$ and $y = y_1$, we get $F(x_1, y_1) = F(x_2, y_2) = F(x_2, y_1) = F(x_1, y_2)$. Moreover, we can write for every $(x, y) \in X \times Y$,

$$F(x_1, y) \le F(x_1, y_2) \le F(x, y_2),$$

whence (x_1, y_2) is a saddle point. For (x_2, y_1), the proof is similar. $\qquad \square$

For the general form of problem (P), we consider again the Lagrangian function $L : U \times (\mathbb{R}_+^n \times \mathbb{R}^m) \to \mathbb{R}$,

$$L(x, (\lambda, \mu)) = f(x) + \sum_{i=1}^n \lambda_i g_i(x) + \sum_{j=1}^m \mu_j h_j(x).$$

Theorem 3.2.12. *An element* $(\overline{x}, (\overline{\lambda}, \overline{\mu})) \in U \times (\mathbb{R}_+^n \times \mathbb{R}^m)$ *is a saddle point for the Lagrangian function L if and only if the following relations hold:*
  (i) $\overline{x}$ is a minimum point for $L(\cdot, (\overline{\lambda}, \overline{\mu}))$ on the open set U;
  (ii) $\overline{x} \in M$;
  (iii) $\lambda_i g_i(\overline{x}) = 0$, for every $i \in \overline{1, n}$.

Proof Let $(\overline{x}, (\overline{\lambda}, \overline{\mu})) \in U \times (\mathbb{R}_+^n \times \mathbb{R}^m)$ be a saddle point for L. Then, according to the definition,

$$\max_{(\lambda, \mu) \in \mathbb{R}_+^n \times \mathbb{R}^m} L(\overline{x}, (\lambda, \mu)) = L(\overline{x}, (\overline{\lambda}, \overline{\mu})) = \min_{x \in U} L(x, (\overline{\lambda}, \overline{\mu})).$$

The second part of this relation is equivalent to (i). It remains to be shown that the first equality is equivalent to the combination of (ii) and (iii), and this is based on the fact that L is affine with respect to (λ, μ) and, moreover, the particular form of $\mathbb{R}^n_+ \times \mathbb{R}^m$ allows us to easily compute the polar of its Bouligand tangent cone. According to Proposition 3.1.25, $(\overline{\lambda}, \overline{\mu})$ with the property

$$\max_{(\lambda,\mu) \in \mathbb{R}^n_+ \times \mathbb{R}^m} L(\overline{x}, (\lambda, \mu)) = L(\overline{x}, (\overline{\lambda}, \overline{\mu}))$$

is characterized by the relation

$$-\nabla_{(\lambda,\mu)} L(\overline{x}, (\overline{\lambda}, \overline{\mu})) \in N(\mathbb{R}^n_+ \times \mathbb{R}^m, (\overline{\lambda}, \overline{\mu})).$$

It is not difficult to see that

$$N(\mathbb{R}^n_+ \times \mathbb{R}^m, (\overline{\lambda}, \overline{\mu})) = \{u \in \mathbb{R}^n \mid u_i = 0 \text{ if } \overline{\lambda}_i > 0, u_i \leq 0 \text{ if } \overline{\lambda}_i = 0\} \times \{0\}_{\mathbb{R}^m}.$$

Then

$$\frac{\partial L}{\partial \lambda_i}(\overline{x}, (\overline{\lambda}, \overline{\mu})) = g_i(\overline{x}) : \begin{cases} = 0, & \text{if } \overline{\lambda}_i > 0 \\ \leq 0, & \text{if } \overline{\lambda}_i = 0 \end{cases}, \ \forall i = \overline{1, n},$$

and

$$\frac{\partial L}{\partial \mu_j}(\overline{x}, (\overline{\lambda}, \overline{\mu})) = h_j(\overline{x}) = 0, \ \forall j = \overline{1, m}.$$

The proof is complete. $\qquad\qquad\qquad\qquad\qquad\qquad\qquad\qquad\qquad\qquad\quad$ □

Corollary 3.2.13. *If $(\overline{x}, (\overline{\lambda}, \overline{\mu})) \in U \times (\mathbb{R}^n_+ \times \mathbb{R}^m)$ is a saddle point for the Lagrangian function L, then $\overline{x}$ is a solution of (P).*

Proof The preceding result shows that $\overline{x} \in M$ and

$$f(\overline{x}) = L(\overline{x}, (\overline{\lambda}, \overline{\mu})) \leq L(x, (\overline{\lambda}, \overline{\mu})), \ \forall x \in U.$$

Since for $x \in M$,

$$L(x, (\overline{\lambda}, \overline{\mu})) \leq f(x),$$

we get $f(\overline{x}) \leq f(x)$ for every $x \in M$. $\qquad\qquad\qquad\qquad\qquad\qquad\qquad\quad$ □

As usual, for convex data the converse holds as well.

Theorem 3.2.14. *Suppose that U is convex, f is convex on U, h is affine and g_i, $i \in \overline{1, n}$ are convex. The next relations are equivalent:*
  (i) $(\overline{x}, (\overline{\lambda}, \overline{\mu})) \in U \times (\mathbb{R}^n_+ \times \mathbb{R}^m)$ is a saddle point for the Lagrangian function L;
  (ii) $\overline{x}$ is a minimum point for (P) and $(\overline{\lambda}, \overline{\mu})$ is a Lagrange multiplier.

Proof According to Theorem 3.2.12, relation (i) above is equivalent to all three relations in that result. One applies now Theorem 3.1.22 and the conclusion follows. $\qquad$ □

3.2.3 Qualification Conditions

The qualification condition $T_B(M, \bar{x})^- = D(\bar{x})^-$ imposed in Theorem 3.2.6 is called the Guignard condition at $\bar{x}$ (after the name of the French mathematician Monique Guignard who proposed it back in 1969) and it is one of the weakest qualification conditions. The difficulty with this condition is that the effective calculations of the involved objects can be tricky in certain situation, and for this reason we want to investigate and to compare it with other qualification conditions as well. Clearly, relation $T_B(M, \bar{x}) = D(\bar{x})$ is in turn a qualification condition (called the quasiregularity condition), since implies Guignard condition. As expected, the two conditions are not equivalent, as one can see from the next example (see also Example 2.1.7).

Example 3.2.15. *Let $g : \mathbb{R}^2 \to \mathbb{R}^2$, $g(x_1, x_2) = (-x_1, x_2)$ and $h : \mathbb{R}^2 \to \mathbb{R}$, $h(x_1, x_2) = x_1 x_2$. Let us consider the feasible point $\bar{x} = (0, 0)$. Then:*

$$D(\bar{x}) = \{(u_1, u_2) \mid u_1 \geq 0, u_2 \leq 0\},$$
$$T_B(M, \bar{x}) = \{(u_1, u_2) \mid u_1 \geq 0, u_2 \leq 0, u_1 u_2 = 0\}$$

and

$$T_B(M, \bar{x})^- = D(\bar{x})^- = \{(u_1, u_2) \mid u_1 \leq 0, u_2 \geq 0\}.$$

The qualification conditions are linked to the reference point ($\bar{x}$ in our notation). Every time when no confusion concerning the reference point could appear, we avoid, for simplicity, writing it explicitly.

Two of the most important (from a practical point of view) qualification conditions are listed below. The first one is called the linear independence qualification condition (at $\bar{x}$) and is as follows:

the set $\{\nabla g_i(\bar{x}) \mid i \in A(\bar{x})\} \cup \{\nabla h_j(\bar{x}) \mid j \in \overline{1, m}\}$ is linearly independent.

The second one is called Mangasarian-Fromovitz qualification condition (at $\bar{x}$):

the set $\{\nabla h_j(\bar{x}) \mid j \in \overline{1, m}\}$ is linearly independent and
$\exists u \in \mathbb{R}^p : \nabla h(\bar{x})(u) = 0$ and $\nabla g_i(\bar{x})(u) < 0, \forall i \in A(\bar{x}).$

(The American mathematicians Olvi Leon Mangasarian and Stanley Fromovitz published this condition in 1967.)

We will now establish the relations between these conditions and then show that they are indeed qualification conditions.

Theorem 3.2.16. *If the linear independence qualification condition at $\bar{x} \in M$ holds, then the Mangasarian-Fromovitz qualification condition at $\bar{x}$ is satisfied.*

Proof Without loss of generality, we suppose that $A(\bar{x}) = \{1, ..., q\}$. Let T be the matrix of dimensions $(q + m) \times p$ with the lines $\nabla g_i(\bar{x})$, $i \in \overline{1, q}$, $\nabla h_j(\bar{x})$, $j \in \overline{1, m}$ and let b be the column vector with $b_i = -1$, $i \in \overline{1, q}$, $b_j = 0$, $j \in \overline{q+1, q+m}$. Since the lines of T are linearly independent, the system $Td = b$ has a solution. If one denotes by u such a solution, then

$$\nabla g_i(\bar{x})(u^t) = -1, \ \forall i \in \overline{1, q} \text{ and } \nabla h_j(\bar{x})(u^t) = 0, \ \forall j \in \overline{1, m},$$

hence the Mangasarian-Fromovitz condition at $\bar{x}$ is satisfied. □

The two conditions are not, however, equivalent.

Example 3.2.17. *Let $g_i : \mathbb{R}^2 \to \mathbb{R}$, $i \in \overline{1, 3}$ defined by:*

$$g_1(x) = (x_1 - 1)^2 + (x_2 - 1)^2 - 2$$
$$g_2(x) = (x_1 - 1)^2 + (x_2 + 1)^2 - 2$$
$$g_3(x) = -x_1$$

and the feasible point $\bar{x} = (0, 0)$. Surely, the set

$$\{\nabla g_1(\bar{x}), \nabla g_2(\bar{x}), \nabla g_3(\bar{x}), \ i \in \overline{1, 3}\}$$

is not linearly independent since it consists of three elements in the two dimensional space $\mathbb{R}^2$. On the other hand, for $u = (1, 0)$, $\nabla g_i(\bar{x})(u) < 0$ for every $i \in \overline{1, 3}$.

Theorem 3.2.16 tells us that in order to show that the two conditions above are qualifications conditions, it is enough to show this only for Mangasarian-Fromovitz condition. This becomes obvious if one applies Theorem 3.2.1 and argues by contradiction. Suppose that $\lambda_0 = 0$. Then

$$\sum_{i \in A(\bar{x})} \lambda_i \nabla g_i(\bar{x}) + \sum_{j=1}^{m} \mu_j \nabla h_j(\bar{x}) = 0.$$

We multiply by the vector u from Mangasarian-Fromovitz condition and deduce that

$$\sum_{i \in A(\bar{x})} \lambda_i \langle \nabla g_i(\bar{x}), u \rangle = 0,$$

whence $\lambda_i = 0$ for every $i \in A(\bar{x})$. Therefore,

$$\sum_{j=1}^{m} \mu_j \nabla h_j(\bar{x}) = 0,$$

and the linear independence of the gradients $\{\nabla h_j(\bar{x}) \mid j \in \overline{1, m}\}$ implies that $\mu_j = 0$ for every $j \in \overline{1, m}$. Putting together these remarks, we get the contradiction to $|\lambda_0| + \|\lambda\| + \|\mu\| \neq 0$. Consequently, $\lambda_0 \neq 0$.

In order to more precisely classify the qualification conditions introduced so far, we show that the Mangasarian-Fromovitz condition implies the quasiregularity condition.

We need an auxiliary result.

Lemma 3.2.18. *Let $\varepsilon > 0$ and $y : (-\varepsilon, \varepsilon) \to \mathbb{R}^p$ be a differentiable function such that $y(0) = \overline{x}, y'(0) = u \neq 0$. Then there exists a sequence $(x_k) \subset \mathrm{Im}\, y \setminus \{\overline{x}\}, (x_k) \to \overline{x}$ such that*

$$\frac{x_k - \overline{x}}{\|x_k - \overline{x}\|} \to \frac{u}{\|u\|}.$$

Proof We have

$$\lim_{t \to 0} \frac{y(t) - \overline{x}}{t} = \lim_{t \to 0} \frac{y(t) - y(0)}{t} = y'(0) = u \neq 0.$$

In particular, for $t \neq 0$ sufficiently small one has $y(t) \neq \overline{x}$. We consider a sequence $(t_k) \to 0$ of positive numbers and we define $x_k = y(t_k)$. Then,

$$\frac{x_k - \overline{x}}{\|x_k - \overline{x}\|} = \frac{x_k - \overline{x}}{t_k} \frac{t_k}{\|x_k - \overline{x}\|} \to \frac{u}{\|u\|}.$$

This ends the proof. □

Theorem 3.2.19. *If the Mangasarian-Fromovitz condition is satisfied at $\overline{x} \in M$, then $T_B(M, \overline{x}) = D(\overline{x})$.*

Proof As already observed, one inclusion is always true. We show only the opposite one, so we start with an element $u \in D(\overline{x})$. Denote by $\overline{u} \in \mathbb{R}^p$ the vector given by the Mangasarian-Fromovitz condition. Let $\lambda \in (0, 1)$ and $d_\lambda := (1 - \lambda)u + \lambda\overline{u}$. We show that $d_\lambda \in T_B(M, \overline{x})$ for every $\lambda \in (0, 1)$, and then, taking $\lambda \to 0$ and using the closedness of $T_B(M, \overline{x})$ the conclusion will follow. Suppose that $d_\lambda \neq 0$, since otherwise, there is nothing to prove.

Let P be the operator defined by the matrix (of dimensions $m \times p$) which has on the lines the vectors $\nabla h_j(\overline{x}), j \in \overline{1, m}$ of $\mathbb{R}^p$. These vectors are linearly independent and form a basis in the linear space $\mathrm{Im}(P)$. Clearly, from the linear independence of $\nabla h_j(\overline{x}), j \in \overline{1, m}$ one deduces that $m \leq p$. But $p = \dim(\mathrm{Im}(P)) + \dim(\mathrm{Ker}(P))$, and we complete the above linear independent set up to a base of $\mathbb{R}^p$ with a set of vectors $\{v_1, v_2, \ldots, v_{p-m}\}$, and we denote by Z the matrix (of dimensions $(p - m) \times p$) which has on the lines these vectors (which give a base in $\mathrm{Ker}(P)$). Then the square matrix $\begin{pmatrix} P \\ Z \end{pmatrix}$ is nonsingular. We define $\varphi : \mathbb{R}^{p+1} \to \mathbb{R}^p$ by

$$\varphi(x, \tau) = (h(x), (Z(x - \overline{x} - \tau d_\lambda)^t)^t).$$

Then $\nabla_x \varphi(\overline{x}, 0) = \begin{pmatrix} P \\ Z \end{pmatrix}$ is a nonsingular matrix, whence, from Implicit Functions Theorem (Theorem 1.3.5), there exist $\varepsilon > 0$ and a differentiable function $y : (-\varepsilon, \varepsilon) \to$

$\mathbb{R}^p$ such that

$$\varphi\left(y(\tau), \tau\right) = 0,$$

for every $\tau \in (-\varepsilon, \varepsilon)$. Then

$$h(y(\tau)) = 0 \text{ and } Z(y(\tau) - \overline{x} - \tau d_\lambda)^t = 0. \tag{3.2.5}$$

At the same time, for every $\tau \in (-\varepsilon, \varepsilon)$ and every x close enough to $\overline{x}$ we have

$$\varphi(x, \tau) = 0 \Rightarrow x = y(\tau).$$

Since $\varphi(\overline{x}, 0) = 0$, we infer that $y(0) = \overline{x}$. According to the relations (3.2.5), we get, on one hand (by differentiation),

$$Py'(0) = 0,$$

and, on the other hand (by dividing with $\tau \neq 0$ and passing to the limit),

$$Zy'(0)^t = Zd_\lambda^t.$$

Since $u, \overline{u} \in D(\overline{x})$, we get $P(d_\lambda) = 0$. We obtain

$$\begin{pmatrix} P \\ Z \end{pmatrix} (y'(0)) = \begin{pmatrix} P \\ Z \end{pmatrix} (d_\lambda),$$

that is $d_\lambda = y'(0)$. Using the above lemma, there exists a sequence $(x_k) \subset \operatorname{Im} y \setminus \{\overline{x}\}$, $(x_k) \to \overline{x}$ with

$$\frac{x_k - \overline{x}}{\|x_k - \overline{x}\|} \to \frac{d_\lambda}{\|d_\lambda\|}.$$

Then $h(x_k) = 0$. In order to deduce that $d_\lambda \in T_B(M, \overline{x})$, it is sufficient to prove that, for k large enough, $g(x_k) \leq 0$. If $i \notin A(\overline{x})$, then $g_i(\overline{x}) < 0$, and the continuity of g_i implies that $g_i(x_k) < 0$ for large k. If $i \in A(\overline{x})$, $\langle \nabla g_i(\overline{x}), u \rangle \leq 0$ and $\langle \nabla g_i(\overline{x}), \overline{u} \rangle < 0$, hence $\langle \nabla g_i(\overline{x}), d_\lambda \rangle < 0$. Since g_i is smooth (of class C^1), there exists a sequence $(\alpha_k) \to 0$ such that for every $k \in \mathbb{N}$,

$$g_i(x_k) = g_i(\overline{x}) + \nabla g_i(\overline{x})(x_k - \overline{x}) + \alpha_k \|x_k - \overline{x}\|.$$

Therefore,

$$\frac{g_i(x_k)}{\|x_k - \overline{x}\|} = \frac{\nabla g_i(\overline{x})(x_k - \overline{x})}{\|x_k - \overline{x}\|} + \alpha_k \overset{k \to \infty}{\to} \nabla g_i(\overline{x}) \left(\frac{d_\lambda}{\|d_\lambda\|} \right) < 0.$$

Then $g_i(x_k) < 0$ for sufficiently large k. Since there are a finite number of indexes i, we obtain the conclusion. $\square$

In order to show that all four qualification conditions introduced are different, it remains to prove that the quasiregularity condition does not imply the Mangasarian-Fromovitz condition.

Example 3.2.20. *Let* $g : \mathbb{R}^2 \to \mathbb{R}^2$, $g(x_1, x_2) = (-x_1^2 + x_2, -x_1^2 - x_2)$ *and the feasible point* $\bar{x} = (0, 0)$. *Then,* $D(\bar{x}) = \{(u_1, 0) \mid u_1 \in \mathbb{R}\}$. *On the other hand, is it easy to check that* $T_B(M, \bar{x}) \supset D(\bar{x})$ *(whence the equality holds), but there is no* $\bar{u} \in \mathbb{R}^2$ *with* $\nabla g(\bar{x})(\bar{u}) < 0$.

We have shown the following implications :

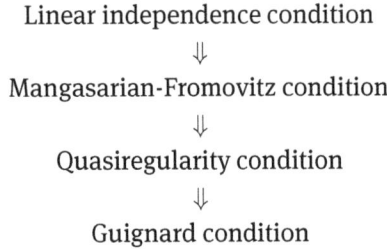

Linear independence condition

$\Downarrow$

Mangasarian-Fromovitz condition

$\Downarrow$

Quasiregularity condition

$\Downarrow$

Guignard condition

and none of the converses hold.

Remark 3.2.21. *Let us notice that, in particular, Theorem 3.2.19 shows as well that if* $h : \mathbb{R}^p \to \mathbb{R}$ *is a* C^1 *function, and* $\bar{x} \in \mathbb{R}^p$ *has the property that* $\nabla h(\bar{x}) \neq 0$, *then the Bouligand tangent cone to the level curve* $\{x \in \mathbb{R}^p \mid h(x) = h(\bar{x})\}$ *at* $\bar{x}$ *is the hyperplane* $\{u \in \mathbb{R}^p \mid \nabla h(\bar{x})(u) = 0\}$ *(or* $\mathrm{Ker}\, \nabla h(\bar{x})$*). Therefore,* $\nabla h(\bar{x})$ *is a normal vector to this hyperplane. We recall here that the affine subspace (of* $\mathbb{R}^{p+1}$*) tangent to the graph of* h *at* $(\bar{x}, h(\bar{x}))$ *has the equation*

$$y = h(\bar{x}) + \nabla h(\bar{x})(x - \bar{x}),$$

and a normal vector to it is $(\nabla h(\bar{x}), -1)$.

Therefore, the Mangasarian-Fromovitz condition ensures that the set $M(\bar{x})$ is nonempty at $\bar{x}$, which is local minimum of the problem (P). Moreover, we will now show that this condition implies special properties of the set of Lagrange multipliers.

Proposition 3.2.22. *If the Mangasarian-Fromovitz condition at* $\bar{x}$ *holds, then* $M(\bar{x})$ *is convex and compact (in* $\mathbb{R}^{n+m}$*).*

Proof According to the definition of $M(\bar{x})$, an element $(\lambda, \mu) \in M(\bar{x})$ satisfies

$$\nabla f(\bar{x}) + \sum_{i=1}^{n} \lambda_i \nabla g_i(\bar{x}) + \sum_{j=1}^{m} \mu_j \nabla h_j(\bar{x}) = 0$$

and

$$\lambda_i \geq 0, \ \lambda_i g_i(\bar{x}) = 0, \ \text{for every } i \in \overline{1, n}.$$

Therefore, checking the convexity and the closedness of $M(\bar{x})$ is straightforward. We will now show that $M(\bar{x})$ is bounded. Let, from the Mangasarian-Fromovitz condition,

$u \in \mathbb{R}^p$ such that

$$\nabla h(\overline{x})(u) = 0 \text{ and } \nabla g_i(\overline{x})(u) < 0, \forall i \in A(\overline{x}).$$

Then, for every $(\lambda, \mu) \in M(\overline{x})$,

$$\nabla f(\overline{x})(u) + \sum_{i \in A(\overline{x})} \lambda_i \nabla g_i(\overline{x})(u) + \sum_{j=1}^{m} \mu_j \nabla h_j(\overline{x})(u) = 0,$$

whence

$$\sum_{i \in A(\overline{x})} \lambda_i(-\nabla g_i(\overline{x})(u)) = \nabla f(\overline{x})(u),$$

from where we deduce

$$\sum_{i \in A(\overline{x})} \lambda_i \min_{i \in A(\overline{x})} \left(-\nabla g_i(\overline{x})(u)\right) \le \nabla f(\overline{x})(u),$$

so

$$\sum_{i \in A(\overline{x})} \lambda_i \le \frac{\nabla f(\overline{x})(u)}{\min_{i \in A(\overline{x})} \left(-\nabla g_i(\overline{x})(u)\right)}.$$

Since the right-hand side is constant, we deduce that the set of multipliers associated to inequalities constraints is bounded. Suppose, by contradiction, that there exists a sequence $(\mu_k)_{k \in \mathbb{N}} \subset \mathbb{R}^m$ unbounded (without loss of generality, we can suppose that $\|\mu_k\| \to \infty$) and a sequence $(\lambda_k)_{k \in \mathbb{N}} \subset \mathbb{R}_+^n$ such that $(\lambda_k, \mu_k) \in M(\overline{x})$. Then, for every $k \in \mathbb{N}$,

$$\nabla f(\overline{x}) + \sum_{i \in A(\overline{x})} (\lambda_i)_k \nabla g_i(\overline{x}) + \sum_{j=1}^{m} (\mu_j)_k \nabla h_j(\overline{x}) = 0.$$

We divide by $\|\mu_k\|$ and we infer that

$$\|\mu_k\|^{-1} \nabla f(\overline{x}) + \sum_{i \in A(\overline{x})} \|\mu_k\|^{-1} (\lambda_i)_k \nabla g_i(\overline{x}) + \sum_{j=1}^{m} \|\mu_k\|^{-1} (\mu_j)_k \nabla h_j(\overline{x}) = 0. \qquad (3.2.6)$$

From the previous step of the proof we have

$$\sum_{i \in A(\overline{x})} \|\mu_k\|^{-1} (\lambda_i)_k \nabla g_i(\overline{x}) \overset{k \to \infty}{\to} 0$$

and,

$$\|\mu_k\|^{-1} \nabla f(\overline{x}) \overset{k \to \infty}{\to} 0.$$

On the other hand, the sequence $(\|\mu_k\|^{-1} \mu_k)$ is bounded (in $\mathbb{R}^m$), whence, without relabeling, we can suppose that $(\|\mu_k\|^{-1} \mu_k)$ is convergent towards a limit denoted by $\overline{\mu} \in \mathbb{R}^m \setminus \{0\}$. Passing to the limit in (3.2.6), we get

$$\sum_{j=1}^{m} \overline{\mu}_j \nabla h_j(\overline{x}) = 0.$$

Since $\overline{\mu} \neq 0$, this is in contradiction to the linear independence assumed in the Mangasarian-Fromovitz condition. Then $M(\overline{x})$ is bounded. □

Let us discuss now two special cases of the problem data.

Firstly, we consider the situation where the inequality restrictions are convex functions, while the equality restriction is affine. The Slater condition takes place if h is affine, g_i, $i \in \overline{1, n}$ are convex and there exists $u \in \mathbb{R}^p$ such that $h(u) = 0$ and $g(u) < 0$. This condition was introduced in 1950 by the American mathematician Morton Slater.

Theorem 3.2.23. *The Slater condition implies $T(M, x) = D(x)$ for every $x \in M$ whence, in particular, is a qualification condition.*

Proof Let $\overline{x} \in M$. The inclusion $T(M, \overline{x}) \subset D(\overline{x})$ is always true. Let $v \in D(\overline{x})$. By the Slater condition (using the convexity of g_i) we deduce (by virtue of Theorem 2.2.10) that

$$0 > g_i(u) \geq g_i(\overline{x}) + \nabla g_i(\overline{x})(u - \overline{x}),$$

whence, for $i \in A(\overline{x})$, $\nabla g_i(\overline{x})(u - \overline{x}) < 0$. We denote $w := u - \overline{x}$, and for $\lambda \in (0, 1)$, we define

$$w_\lambda := (1 - \lambda)v + \lambda w.$$

We show that $w_\lambda \in T(M, \overline{x})$ for every $\lambda \in (0, 1)$. For $i \in A(\overline{x})$,

$$\nabla g_i(\overline{x})(v) \leq 0, \ \nabla g_i(\overline{x})(w) < 0,$$

hence $\nabla g_i(\overline{x})(w_\lambda) < 0$. By Taylor's Formula, there exists $t > 0$ such that $g_i(\overline{x} + tw_\lambda) < g_i(\overline{x}) = 0$ for every $i \in A(\overline{x})$. Let $(t_k) \subset (0, \infty)$, $t_k \to 0$. Then

$$x_k := (1 - t_k)\overline{x} + t_k(\overline{x} + tw_\lambda) = \overline{x} + t_k t w_\lambda \overset{k \to \infty}{\to} \overline{x}.$$

In order for the conclusion to follow, we need to show that for k sufficiently large all (x_k) are in M. As usual, for $i \notin A(\overline{x})$, the continuity of g ensures this, while for $i \in A(\overline{x})$, we have

$$g_i(x_k) \leq (1 - t_k)g_i(\overline{x}) + t_k g_i(\overline{x} + tw_\lambda) < 0.$$

Since h is affine and $h(\overline{x}) = 0$, we get

$$h(x_k) = h(\overline{x} + t_k tw_\lambda) = t_k t \nabla h(\overline{x})(w_\lambda).$$

But $v \in D(\overline{x})$, $\nabla h(\overline{x})(v) = 0$, so

$$\nabla h(\overline{x})(w_\lambda) = \lambda \nabla h(\overline{x})(w) = \lambda \nabla h(\overline{x})(u - \overline{x}) = \lambda h(u) = 0.$$

Therefore, $h(x_k) = 0$ for any k, so, finally, $(x_k)_{k \geq k_0} \subset M$, and this means that $w_\lambda \in T(M, \overline{x})$. We let now $\lambda \to 0$; since $T(M, \overline{x})$ is closed, we get $v \in T(M, \overline{x})$, and the proof is complete. □

We consider now the case of affine restrictions. Take a matrix A of dimensions $n \times p$, a matrix B of dimensions $m \times p$ and $b \in \mathbb{R}^n$, $c \in \mathbb{R}^m$. Therefore the set M become $M = \{x \in \mathbb{R}^p \mid Ax^t \leq b^t, Bx^t = c^t\}$, where the relationship "$\leq$" is understood in the componentwise sense. Hence $g(x) = (Ax^t - b^t)^t$, $h(x) = (Bx^t - c^t)^t$.

Theorem 3.2.24. *In the above conditions and notation, the quasiregularity condition is automatically fulfilled.*

Proof As before, it is enough to prove that $D(\overline{x}) \subset T(M, \overline{x})$. Without loss of generality, one can suppose that $A(\overline{x}) = \overline{1, n}$. Let $v \in D(\overline{x})$. Then $Av^t \leq 0$, $Bv^t = 0$. If $v = 0$, there is nothing to prove. Otherwise, we define

$$x_k := \overline{x} + \frac{1}{k}v, \ \forall k \in \mathbb{N}^*.$$

The relations

$$Ax_k^t \leq b^t, \ Bx_k^t = c^t, \ x_k \to \overline{x}$$

show that $v \in T(M, \overline{x})$. $\qquad\square$

For affine restrictions, every minimum point of (P) satisfies the conclusions of Theorem 3.2.6.

3.3 Second-order Conditions

In this section we obtain second-order optimality conditions for the optimization problem with functional constraints and to this end, we assume that the data are C^2 functions. Let $\overline{x} \in M$ and $(\overline{\lambda}, \overline{\mu}) \in \mathbb{R}^{n+m}$ be a vector which satisfies Karush-Kuhn-Tucker conditions (i.e., the conclusion of Theorem 3.2.6). We define the set of critical directions $C(\overline{x}, (\overline{\lambda}, \overline{\mu}))$ as the set of vectors $u \in \mathbb{R}^p$ for which

$$\begin{cases} \nabla g_i(\overline{x})(u) = 0, \ \text{if } i \in A(\overline{x}) \text{ and } \overline{\lambda}_i > 0, \\ \nabla g_i(\overline{x})(u) \leq 0, \ \text{if } i \in A(\overline{x}) \text{ and } \overline{\lambda}_i = 0, \\ \nabla h_j(\overline{x})(u) = 0, \ \text{for every } j \in \overline{1, m}. \end{cases}$$

Clearly, $C(\overline{x}, (\overline{\lambda}, \overline{\mu}))$ is a cone.

Remark 3.3.1. *Obviously, $C(\overline{x}, (\overline{\lambda}, \overline{\mu})) \subset D(\overline{x})$. In particular, under quasiregularity qualification condition, i.e., $T_B(M, \overline{x}) = D(\overline{x})$, one has the inclusion $C(\overline{x}, (\overline{\lambda}, \overline{\mu})) \subset T_B(M, \overline{x})$. Moreover, if one has only equalities constraints, then one has the equality, since $\overline{\lambda}$ does not intervene in such a case.*

Theorem 3.3.2. *Let $\overline{x} \in M$ be a solution of the problem (P) and $(\overline{\lambda}, \overline{\mu}) \in \mathbb{R}^{n+m}$ a vector which satisfies Karush-Kuhn-Tucker conditions. If the linear independence condition* *holds at $\overline{x}$, then*

$$\nabla^2_{xx} L(\overline{x}, (\overline{\lambda}, \overline{\mu}))(u, u) \geq 0$$

for every $u \in C(\overline{x}, (\overline{\lambda}, \overline{\mu}))$.

Proof Without loss of generality, we suppose that all the inequality constraints are active. We split the proof into several steps.

At the first step, we repeat, with some modifications, several arguments from the proof of Theorem 3.2.19 in order to get a sequence of feasible points with special properties. Let $d \in D(\overline{x})$, and let P be the operator defined by the matrix (of dimensions $(n+m) \times p$) with the lines consisting of vectors $\nabla g_i(\overline{x})$, $i \in \overline{1, n}$, $\nabla h_j(\overline{x})$, $j \in \overline{1, m}$ in $\mathbb{R}^p$. These vectors are linearly independent and form a basis in the linear subspace $\text{Im}(P)$. Let us denote by Z the matrix (of dimensions $(p - (n + m)) \times p$) whose lines are some vectors that form a basis in $\text{Ker}(P)$. The the square matrix $\begin{pmatrix} P \\ Z \end{pmatrix}$ is nonsingular. We define $\varphi : \mathbb{R}^{p+1} \to \mathbb{R}^p$ by

$$\varphi(x, \tau) = ((g(x), h(x)) - \tau(Pd^t)^t, (Z(x - \overline{x} - \tau d)^t)^t).$$

Then $\nabla_x \varphi(\overline{x}, 0) = \begin{pmatrix} P \\ Z \end{pmatrix}$ is nonsingular and, from Implicit Function Theorem (i.e., Theorem 1.3.5), there exists $\varepsilon > 0$ and a differentiable function $y : (-\varepsilon, \varepsilon) \to \mathbb{R}^p$ such that

$$\varphi(y(\tau), \tau) = 0,$$

for every $\tau \in (-\varepsilon, \varepsilon)$. Moreover, for every $\tau \in (-\varepsilon, \varepsilon)$ and every x close enough to $\overline{x}$,

$$\varphi(x, \tau) = 0 \Rightarrow x = y(\tau).$$

Let $(t_k) \subset (0, \infty)$, $(t_k) \to 0$. Then, using the fact that $\varphi(y(t_k), t_k) = 0$, there exists, for every k large enough, $z_k = y(t_k)$ such that

$$g_i(z_k) = t_k \nabla g_i(\overline{x})(d) \leq 0, \ \forall i \in \overline{1, n} \tag{3.3.1}$$
$$h_j(z_k) = t_k \nabla h_j(\overline{x})(d) = 0, \ \forall j \in \overline{1, m}.$$

Therefore, $(z_k) \subset M$ and the sequence $(t_k^{-1}(z_k - \overline{x}))$ is convergent. We show that $t_k^{-1}(z_k - \overline{x}) \to d$, which would imply $d \in T_B(M, \overline{x})$. According to the Taylor Theorem, from $\varphi(z_k, t_k) = 0$, there exists $(\mu_k) \subset \mathbb{R}^{n+m}$, $\mu_k \to 0$ such that

$$0 = \left((P(z_k - \overline{x} - t_k d)^t)^t, (Z(z_k - \overline{x} - t_k d)^t)^t\right) + \|z_k - \overline{x}\| \mu_k,$$

whence

$$\left(\frac{z_k - \overline{x}}{t_k} - d\right)^t = \begin{pmatrix} P \\ Z \end{pmatrix}^{-1} (-t_k^{-1} \|z_k - \overline{x}\| \mu_k),$$

from where, after passing to the limit, one gets the announced relation.

Let now $u \in C(\bar{x}, (\bar{\lambda}, \bar{\mu})) \subset D(\bar{x})$. We use now the above construction of the sequence $(z_k) \to \bar{x}$ corresponding to u. We have

$$L(z_k, (\bar{\lambda}, \bar{\mu})) = f(z_k) + \sum_{i=1}^{n} \bar{\lambda}_i g_i(z_k) + \sum_{j=1}^{m} \bar{\mu}_j h_j(z_k) = f(z_k) - t_k \sum_{i \in A(\bar{x})} \bar{\lambda}_i \nabla g_i(\bar{x})(u) = f(z_k).$$

From Taylor second-order condition, there exists $(y_k) \to 0$ such that for every k,

$$L(z_k, (\bar{\lambda}, \bar{\mu})) = L(\bar{x}, (\bar{\lambda}, \bar{\mu})) + \nabla_x L(\bar{x}, (\bar{\lambda}, \bar{\mu}))(z_k - \bar{x})$$
$$+ \frac{1}{2} \nabla_{xx} L(\bar{x}, (\bar{\lambda}, \bar{\mu}))(z_k - \bar{x}, z_k - \bar{x}) + y_k \|z_k - \bar{x}\|^2.$$

But, from the Karush-Kuhn-Tucker conditions, $L(\bar{x}, (\bar{\lambda}, \bar{\mu})) = f(\bar{x})$ and $\nabla_x L(\bar{x}, (\bar{\lambda}, \bar{\mu})) = 0$, whence

$$f(z_k) = f(\bar{x}) + \frac{1}{2} \nabla_{xx} L(\bar{x}, (\bar{\lambda}, \bar{\mu}))(z_k - \bar{x}, z_k - \bar{x}) + y_k \|z_k - \bar{x}\|^2.$$

Since $z_k \to \bar{x}$ and $\bar{x}$ is a solution for the problem (P), we obtain $f(z_k) - f(\bar{x}) \geq 0$ for every sufficiently large k. Then

$$\frac{1}{2} \nabla_{xx} L(\bar{x}, (\bar{\lambda}, \bar{\mu}))(z_k - \bar{x}, z_k - \bar{x}) + y_k \|z_k - \bar{x}\|^2 \geq 0.$$

We divide by t_k^2 and we pass to the limit in order to get

$$\nabla_{xx} L(\bar{x}, (\bar{\lambda}, \bar{\mu}))(u, u) \geq 0.$$

The proof is complete. $\qquad\qquad\qquad\qquad\qquad\qquad\qquad\qquad\qquad\qquad\qquad\qquad\qquad\qquad\square$

We formulate now a converse of the previous result. As shown before, the sufficient optimality condition returns a stronger type of solution (i.e., strict solution).

Theorem 3.3.3. *Let $\bar{x} \in M$ and $(\bar{\lambda}, \bar{\mu}) \in \mathbb{R}^{n+m}$ a vector which satisfies Karush-Kuhn-Tucker conditions. Suppose that*

$$\nabla_{xx}^2 L(\bar{x}, (\bar{\lambda}, \bar{\mu}))(u, u) > 0$$

for every $u \in C(\bar{x}, (\bar{\lambda}, \bar{\mu})) \setminus \{0\}$. Then $\bar{x}$ is a local strict solution of second order of (P).

Proof Since the set $C(\bar{x}, (\bar{\lambda}, \bar{\mu})) \cap \{u \in \mathbb{R}^p \mid \|u\| = 1\}$ is compact, and $C(\bar{x}, (\bar{\lambda}, \bar{\mu}))$ is a cone, the relation $\nabla_{xx}^2 L(\bar{x}, (\bar{\lambda}, \bar{\mu}))(u, u) > 0$ for every $u \in C(\bar{x}, (\bar{\lambda}, \bar{\mu})) \setminus \{0\}$ is equivalent to the existence of a strictly positive number ρ with the property

$$\nabla_{xx}^2 L(\bar{x}, (\bar{\lambda}, \bar{\mu}))(u, u) \geq \rho \|u\|^2, \forall u \in C(\bar{x}, (\bar{\lambda}, \bar{\mu})).$$

Suppose that there exists $(z_k) \to \bar{x}$, $(z_k) \subset M$ such that

$$f(z_k) < f(\bar{x}) + k^{-1} \|z_k - \bar{x}\|^2,$$

for every k sufficiently large. Then, without loss of generality, we suppose that $\|z_k - \overline{x}\|^{-1}(z_k - \overline{x}) \to d \in T_B(M, \overline{x}) \setminus \{0\} \subset D(\overline{x})$. On the other hand,

$$L(z_k, (\overline{\lambda}, \overline{\mu})) = f(z_k) + \sum_{i \in A(\overline{x})} \overline{\lambda}_i g_i(z_k) \leq f(z_k),$$

and, as before, there exists $(y_k) \to 0$ such that for every k,

$$L(z_k, (\overline{\lambda}, \overline{\mu})) = f(\overline{x}) + \frac{1}{2} \nabla_{xx} L(\overline{x}, (\overline{\lambda}, \overline{\mu}))(z_k - \overline{x}, z_k - \overline{x}) + y_k \|z_k - \overline{x}\|^2 . \tag{3.3.2}$$

Suppose that $d \notin C(\overline{x}, (\overline{\lambda}, \overline{\mu}))$. Then there exists $i_0 \in A(\overline{x})$ with $\overline{\lambda}_{i_0} \nabla g_{i_0}(\overline{x})(d) < 0$. For the other indices $i \in A(\overline{x})$ we have $\overline{\lambda}_i \nabla g_i(\overline{x})(d) \leq 0$. Then, it exists $(\tau_k) \to 0$ such that for every k,

$$\overline{\lambda}_{i_0} g_{i_0}(z_k) = \overline{\lambda}_{i_0} g_{i_0}(\overline{x}) + \overline{\lambda}_{i_0} \nabla g_{i_0}(\overline{x})(z_k - \overline{x}) + \tau_k \overline{\lambda}_{i_0} \|z_k - \overline{x}\|$$

$$= \|z_k - \overline{x}\| \overline{\lambda}_{i_0} \nabla g_{i_0}(\overline{x}) \left(\frac{z_k - \overline{x}}{\|z_k - \overline{x}\|} \right) + \tau_k \overline{\lambda}_{i_0} \|z_k - \overline{x}\| .$$

Hence

$$L(z_k, (\overline{\lambda}, \overline{\mu})) = f(z_k) + \sum_{i \in A(\overline{x})} \overline{\lambda}_i g_i(z_k) \leq f(z_k) + \overline{\lambda}_{i_0} g_{i_0}(z_k)$$

$$= f(z_k) + \|z_k - \overline{x}\| \overline{\lambda}_{i_0} \nabla g_{i_0}(\overline{x}) \left(\frac{z_k - \overline{x}}{\|z_k - \overline{x}\|} \right) + \tau_k \overline{\lambda}_{i_0} \|z_k - \overline{x}\| .$$

From (3.3.2), we get

$$f(\overline{x}) + \frac{1}{2} \nabla_{xx} L(\overline{x}, (\overline{\lambda}, \overline{\mu}))(z_k - \overline{x}, z_k - \overline{x}) + y_k \|z_k - \overline{x}\|^2$$

$$\leq f(z_k) + \|z_k - \overline{x}\| \overline{\lambda}_{i_0} \nabla g_{i_0}(\overline{x}) \left(\frac{z_k - \overline{x}}{\|z_k - \overline{x}\|} \right) + \tau_k \overline{\lambda}_{i_0} \|z_k - \overline{x}\| .$$

Furthermore,

$$\lim_k \|z_k - \overline{x}\|^{-1} \nabla_{xx} L(\overline{x}, (\overline{\lambda}, \overline{\mu}))(z_k - \overline{x}, z_k - \overline{x})$$

$$= \lim_k \|z_k - \overline{x}\| \nabla_{xx} L(\overline{x}, (\overline{\lambda}, \overline{\mu})) \left(\frac{z_k - \overline{x}}{\|z_k - \overline{x}\|}, \frac{z_k - \overline{x}}{\|z_k - \overline{x}\|} \right) = 0.$$

After relabeling, one can see that there exists $(v_k) \to 0$ such that

$$f(z_k) \geq f(\overline{x}) - \|z_k - \overline{x}\| \overline{\lambda}_{i_0} \nabla g_{i_0}(\overline{x}) \left(\frac{z_k - \overline{x}}{\|z_k - \overline{x}\|} \right) + v_k \|z_k - \overline{x}\| .$$

From the assumption made, $f(z_k) < f(\overline{x}) + k^{-1} \|z_k - \overline{x}\|^2$, whence

$$f(\overline{x}) + k^{-1} \|z_k - \overline{x}\|^2 \geq f(\overline{x}) - \|z_k - \overline{x}\| \overline{\lambda}_{i_0} \nabla g_{i_0}(\overline{x}) \left(\frac{z_k - \overline{x}}{\|z_k - \overline{x}\|} \right) + v_k \|z_k - \overline{x}\| ,$$

that is

$$k^{-1} \|z_k - \overline{x}\| \geq -\overline{\lambda}_{i_0} \nabla g_{i_0}(\overline{x}) \left(\frac{z_k - \overline{x}}{\|z_k - \overline{x}\|} \right) + v_k.$$

Passing to the limit, we arrive at a contradiction to the relation $\overline{\lambda}_{i_0} \nabla g_{i_0}(\overline{x})(d) < 0$. Consequently, $d \in C(\overline{x}, (\overline{\lambda}, \overline{\mu})) \setminus \{0\}$, whence $\nabla_{xx}^2 L(\overline{x}, (\overline{\lambda}, \overline{\mu}))(d, d) \geq \rho$. Since $L(z_k, (\overline{\lambda}, \overline{\mu})) \leq f(z_k)$, coming back to (3.3.2), we can write

$$f(z_k) \geq f(\overline{x}) + \frac{1}{2} \nabla_{xx} L(\overline{x}, (\overline{\lambda}, \overline{\mu}))(z_k - \overline{x}, z_k - \overline{x}) + y_k \|z_k - \overline{x}\|^2 .$$

But $\nabla_{xx}^2 L(\overline{x}, (\overline{\lambda}, \overline{\mu}))$ is continuous, whence for k large enough,

$$\nabla_{xx} L(\overline{x}, (\overline{\lambda}, \overline{\mu}))(z_k - \overline{x}, z_k - \overline{x}) > 2^{-1}\rho \|z_k - \overline{x}\|^2 .$$

Finally,

$$f(\overline{x}) + k^{-1} \|z_k - \overline{x}\|^2 \geq f(\overline{x}) + \frac{\rho}{4} \|z_k - \overline{x}\|^2 + y_k \|z_k - \overline{x}\|^2 ,$$

and a new contradiction occurs. □

3.4 Motivations for Scientific Computations

In this section, we examine the computational limits of the theoretical results from the previous sections. In some cases, it is not possible to get the exact solution of the optimization problems. The theory leads us to solve some nonlinear (systems of) equations, which do not admit analytical expressions for the solutions. This motivates us to subsequently study numerical algorithms for solving such equations, and this will be done in Chapter 6.

   1. (Least squares method) We discuss now a special case of an optimization problem without restrictions. This problem belongs to the general approach called the method of least squares which is designed for the interpretation of numerical data issued from experiments in physics, biology, astronomy, chemistry. From historical perspective, this kind of problem arose from the study of the movements of planets and in questions linked to navigation techniques. The mathematician who founded this method is considered to be Gauss, but the method was published for the first time by Legendre.

   In few words, the method of least squares refers to the following situation: we dispose of a data set $v_1, v_2, \ldots, v_N$ obtained after some measurements made at the (different) moments $t_1, t_2, \ldots, t_N$. The objective is to determine the best model function of the form $t \mapsto \varphi(t, x)$ (where $x = (x_1, x_2, \ldots, x_k)$ are parameters to optimize) which fits with the measurements. Therefore, for every $i \in \overline{1, N}$, one defines the residual at the moment t_i as the absolute difference between the measurement v_i and the value of the model at same time:

$$r_i := |v_i - \varphi(t_i, x)| ,$$

and now the problem is to minimize the function

$$f(x) = \sum_{i=1}^{N} r_i^2 = \sum_{i=1}^{N} \left[v_i - \varphi(t_i, x) \right]^2.$$

It should be said that another possible objective function (even more natural to be considered) would be

$$\sum_{i=1}^{N} \left| v_i - \varphi(t_i, x) \right|$$

but this construction does not preserve the differentiability. For this reason, one prefers the sum of the squares of the residuals, whence the name of the method.

Let us now consider the simplest case of a linear dependence. Let us suppose that one has made N measurements at the different moments of time $t_1, t_2, \ldots, t_N > 0$ and, correspondingly, one has the values $v_1, v_2, \ldots, v_N$. We know that the dependence between these two sets of data is linear, and we are interested in obtaining a line which better fits the collection of observations. Let be a line $v = at + b$. As above, the residual at the moment t_i is $\left| v_i - (at_i + b) \right|$ and in order to "measure" the sum of these residuals, we consider the function $f : \mathbb{R}^2 \to \mathbb{R}$,

$$f(a, b) = \sum_{i=1}^{N} \left[v_i - (at_i + b) \right]^2.$$

The line with respect to which this sum of residuals will be the smallest, will be that which we seek. Then, we arrive at the problem of minimization (without restrictions) of the function f. We compute the partial derivatives of f:

$$\frac{\partial f}{\partial a}(a, b) = \sum_{i=1}^{N} 2(-t_i) \left[v_i - (at_i + b) \right]$$

$$\frac{\partial f}{\partial b}(a, b) = \sum_{i=1}^{N} -2 \left[v_i - (at_i + b) \right],$$

and the calculus of critical points is reduced to computation of the solutions of the system:

$$\begin{cases} \left(\sum_{i=1}^{N} t_i^2 \right) a + \left(\sum_{i=1}^{N} t_i \right) b = \sum_{i=1}^{N} t_i v_i \\ \left(\sum_{i=1}^{N} t_i \right) a + Nb = \sum_{i=1}^{N} v_i. \end{cases}$$

The determinant of this system is

$$\Delta := N \left(\sum_{i=1}^{N} t_i^2 \right) - \left(\sum_{i=1}^{N} t_i \right)^2 = \left(\sum_{i=1}^{N} 1^2 \right) \left(\sum_{i=1}^{N} t_i^2 \right) - \left(\sum_{i=1}^{N} t_i \right)^2.$$

From the Hölder inequality, this number is positive (equality would be possible only if all the values t_i are equal, but this is not possible). Then the system admits a unique solution:

$$\begin{pmatrix} a \\ b \end{pmatrix} = \begin{pmatrix} \sum_{i=1}^{N} t_i^2 & \sum_{i=1}^{N} t_i \\ \sum_{i=1}^{N} t_i & N \end{pmatrix}^{-1} \begin{pmatrix} \sum_{i=1}^{N} t_i v_i \\ \sum_{i=1}^{N} v_i \end{pmatrix}.$$

Let us observe that, furthermore, the function is coercive since $\lim_{\|(a,b)\| \to \infty} f(a, b) = \infty$, hence, according to Theorem 3.1.7, it admits a minimum point which is necessarily the critical point determined above. Therefore this pair (a, b) is the solution of the problem.

Another remark is that an important part of the above calculations can be repeated with obvious changes if one supposes a dependence of the type $v = a \cdot p(t) + b \cdot q(t)$, where $p, q : \mathbb{R} \to \mathbb{R}$. One obtains the system

$$\begin{pmatrix} \sum_{i=1}^{N} p^2(t_i) & \sum_{i=1}^{N} p(t_i)q(t_i) \\ \sum_{i=1}^{N} p(t_i)q(t_i) & \sum_{i=1}^{N} q^2(t_i) \end{pmatrix} \begin{pmatrix} a \\ b \end{pmatrix} = \begin{pmatrix} \sum_{i=1}^{N} p(t_i)v_i \\ \sum_{i=1}^{N} q(t_i)v_i \end{pmatrix}.$$

Again, the Hölder inequality ensures that the associated matrix is invertible if and only if $(p^2(t_i))_{i=\overline{1,N}}$ and $(q^2(t_i))_{i=\overline{1,N}}$ are not proportional.

In general, for more complicated models (nonlinear in x) the method of least squares does not have an easily computable solution and this will be an impetus for us to study several algorithm in order to get good approximation of solution in fast computational time.

2. (The projection on a closed convex set) Let $a_1, a_2, \ldots, a_p \in (0, \infty)$ and the set (generalized ellipsoid)

$$M := \left\{ x \in \mathbb{R}^p \mid \sum_{i=1}^{p} \left(\frac{x_i}{a_i} \right)^2 \leq 1 \right\}.$$

Obviously, this is a convex and compact set. Let $v \notin M$. From Theorem 2.1.5, there exists $\bar{v} \in M$, the projection of v on M. Again, we want to find an expression of this element.

As before, $\bar{v}$ is the solution of the minimization problem of $f(x) = \|x - v\|^2$ under the restriction $x \in M$. If it would exist a solution $\bar{x} \in \text{int } M$, then $\nabla f(\bar{x}) = 0$, whence $\bar{x} - v = 0$, that is $v \in M$, which is false. Therefore, the restriction is active in $\bar{v}$, that is $\sum_{i=1}^{p} \left(\frac{\bar{v}_i}{a_i} \right)^2 = 1$. Moreover, the function which defines (with inequality) the constraint $x \in M$, i.e., $g : \mathbb{R}^p \to \mathbb{R}$,

$$g(x) := \sum_{i=1}^{p} \left(\frac{x_i}{a_i} \right)^2 - 1,$$

is convex, and the Slater condition holds. Moreover, f is convex as well, so we can conclude that $\bar{v}$ is a solution of the problem if and only of there exists $\lambda \geq 0$ such that for all indexes i,

$$\bar{v}_i - v_i + \lambda \frac{\bar{v}_i}{a_i^2} = 0.$$

Since $v \neq \bar{v}$, $\lambda > 0$ and

$$\bar{v}_i = \frac{a_i^2 v_i}{a_i^2 + \lambda}.$$

On the other hand, by $\sum_{i=1}^{p} \left(\frac{\bar{v}_i}{a_i} \right)^2 = 1$,

$$\sum_{i=1}^{p} \frac{a_i^2 v_i^2}{\left(a_i^2 + \lambda \right)^2} = 1,$$

so finding λ (and then $\bar{v}$) requires solving the above equation. Let us remark that the equation has a unique solution, since the mapping

$$0 \leq \lambda \mapsto \sum_{i=1}^{p} \frac{a_i^2 v_i^2}{\left(a_i^2 + \lambda \right)^2}$$

is strictly decreasing, its value at 0 is strictly greater than 1 (notice that $v \notin M$), while its limit at $+\infty$ is 0. So, to get $\bar{v}$ one must to solve an algebraic equation of degree $2p$, and, in general, this is impossible. We will be interested in approximation methods of the solutions of nonlinear equations, and this will be one of the subjects of Chapter 6.

4 Convex Nonsmooth Optimization

In this chapter we study optimization problems involving convex functions that are not necessarily differentiable. Naturally, several new tools are needed in order to compensate for the lack of differentiability. On one hand, we need to study convex sets which were briefly defined and studied in Section 2.1. A new object to replace the differential is introduced and studied. With all these tools in hand, we will be able to derive a generalized Karush-Kuhn-Tucker theorem in the case of convex nonsmooth optimization.

4.1 Further Properties and Separation of Convex Sets

We start with some results concerning the fundamental topological properties of convex sets in $\mathbb{R}^p$.

Theorem 4.1.1. *Let $C \subset \mathbb{R}^p$ be a convex set. Then*
  (i) cl C *is convex;*
  (ii) if $x \in$ int C and $y \in$ cl C, then $[x, y) \subset$ int C;
  (iii) int C *is convex;*
  (iv) if int $C \neq \emptyset$, *then* cl $C =$ cl(int C) *and* int $C =$ int(cl C).

Proof (i) Let us take $x, y \in$ cl C and $\alpha \in (0, 1)$, and a neighborhood V of $0 \in \mathbb{R}^p$. It is well known that there is a neighborhood U of 0 such that $\alpha U + (1 - \alpha)U \subset V$. Since $x, y \in$ cl C, there are $x_U, y_U \in C$ such that $x_U \in (x + U) \cap C$ and $y_U \in (y + U) \cap C$. Consequently, from the convexity of C, one has

$$C \ni \alpha x_U + (1 - \alpha)y_U \in \alpha(x + U) + (1 - \alpha)(y + U)$$
$$= \alpha x + (1 - \alpha)y + \alpha U + (1 - \alpha)U \subset \alpha x + (1 - \alpha)y + V,$$

whence $C \cap (\alpha x + (1 - \alpha)y + V) \neq \emptyset$. Since V is an arbitrary neighborhood of 0, this shows that $\alpha x + (1 - \alpha)y \in$ cl C. Of course, an argument based on a characterization of cl C that used sequences is also possible.

(ii) Take $\alpha \in (0, 1)$. It is enough to show that $\alpha x + (1 - \alpha)y \in$ int C. Since $x \in$ int C, there is a neighborhood V of $0 \in \mathbb{R}^p$ with $x + V + V \subset C$. On the other hand, $y \in$ cl C implies that $C \cap (y - \alpha(1 - \alpha)^{-1}V) \neq \emptyset$, whence y can be written as $c + \alpha(1 - \alpha)^{-1}v$, with $c \in C$ and $v \in V$. We get that

$$\alpha x + (1 - \alpha)y + \alpha V = \alpha x + (1 - \alpha)c + \alpha v + \alpha V = (1 - \alpha)c + \alpha(x + v + V)$$
$$\subset (1 - \alpha)c + \alpha C \subset C.$$

Since αV is still a neighborhood of 0, we conclude that $\alpha x + (1 - \alpha)y \in$ int C.

(iii) If $x, y \in \text{int } C$, then the above implication means that $[x, y] \subset \text{int } C$, whence int C is convex.

(iv) Clearly, $\text{cl}(\text{int } C) \subset \text{cl } C$. Consider $x \in \text{cl } C$. By (ii), for any $y \in \text{int } C$, $(x, y] \subset \text{int } C$, which means that x can be approached by a sequence of points in int C, that is $x \in \text{cl}(\text{int } C)$. For the second part, one inclusion always holds: $\text{int } C \subset \text{int}(\text{cl } C)$. Consider $x \in \text{int}(\text{cl } C)$, which means that there exists a neighborhood V of $0 \in \mathbb{R}^p$ such that $x + V \subset \text{cl } C$, whence, once again from (ii), for any $y \in \text{int } C$, $\alpha \in (0, 1)$ and $v \in V$,

$$\alpha(x + v) + (1 - \alpha)y \in \text{int } C.$$

But V is absorbing (that is, for every $x \in \mathbb{R}^p$, there is $\lambda > 0$ such that $\lambda x \in V$), so for α sufficiently close to 1,

$$\overline{v} := \frac{(1 - \alpha)(x - y)}{\alpha} \in V$$

and for such an α,

$$x = \alpha \left(x + \frac{(1 - \alpha)(x - y)}{\alpha} \right) + (1 - \alpha)y$$

$$= \alpha(x + \overline{v}) + (1 - \alpha)y \in \text{int } C.$$

The proof is complete. $\qquad\square$

Now we need a supplementary investigation into the projection of a point on a closed convex set (see Theorem 2.1.5 (iii)).

Proposition 4.1.2. *Let $C \subset \mathbb{R}^p$ be a nonempty closed and convex set. Then the application $\mathbb{R}^p \ni x \mapsto \text{pr}_C x$ is 1–Lipschitz. In particular, it is continuous.*

Proof As proved in Theorem 2.1.5 (iii), the projection $\text{pr}_C x$ of a point x on C is characterized by the properties

$$\begin{cases} \text{pr}_C x \in C \\ \langle x - \text{pr}_C x, u - \text{pr}_C x \rangle \le 0, \ \forall u \in C. \end{cases}$$

Take $x_1, x_2 \in \mathbb{R}^p$. Then the above system gives

$$\langle x_1 - \text{pr}_C x_1, \text{pr}_C x_2 - \text{pr}_C x_1 \rangle \le 0$$
$$\langle x_2 - \text{pr}_C x_2, \text{pr}_C x_1 - \text{pr}_C x_2 \rangle \le 0.$$

This means that $\langle x_1 - \text{pr}_C x_1 - x_2 + \text{pr}_C x_2, \text{pr}_C x_2 - \text{pr}_C x_1 \rangle \le 0$, whence

$$\langle x_1 - x_2 + \text{pr}_C x_2 - \text{pr}_C x_1, \text{pr}_C x_2 - \text{pr}_C x_1 \rangle \le 0.$$

We get that

$$\|\text{pr}_C x_2 - \text{pr}_C x_1\|^2 \le \langle x_2 - x_1, \text{pr}_C x_2 - \text{pr}_C x_1 \rangle \le \|x_2 - x_1\| \cdot \|\text{pr}_C x_2 - \text{pr}_C x_1\|,$$

so the desired inequality holds. Since every Lipschitz function is continuous, so the second part follows as well. $\qquad\square$

Now we are able to prove a separation result for convex sets.

Theorem 4.1.3. *Let $C \subset \mathbb{R}^p$ be a nonempty convex set and $\overline{x} \notin C$. Then there is an $a \in \mathbb{R}^p \setminus \{0\}$ such that for all $c \in C$ one has*

$$\langle a, \overline{x} \rangle \le \langle a, c \rangle .$$

Proof First, suppose that $\overline{x} \notin \text{cl } C$. In fact, in this case we can prove a stronger inequality. Notice that Theorem 4.1.1 (*i*) ensures that cl C is a closed convex set, so there exists $\text{pr}_{\text{cl } C} \overline{x} \in \text{cl } C$, which is not equal to $\overline{x}$. The properties of projection ensure that

$$\langle \text{pr}_{\text{cl } C} \overline{x} - \overline{x}, x - \text{pr}_{\text{cl } C} \overline{x} \rangle \ge 0, \ \forall x \in \text{cl } C.$$

Define $a := \text{pr}_{\text{cl } C} \overline{x} - \overline{x} \in \mathbb{R}^p \setminus \{0\}$ and rewrite the above inequality as

$$\langle a, x - \overline{x} - a \rangle \ge 0, \ \forall x \in \text{cl } C.$$

This means that

$$\langle a, c - \overline{x} \rangle \ge \|a\|^2 > 0, \ \forall c \in C,$$

and the desired inequality follows.

Take now that case when $\overline{x} \in \text{cl } C \setminus C$. Observe (by the use of Theorem 4.1.1 (*iv*)) that

$$\text{cl}(\mathbb{R}^p \setminus \text{cl } C) = \mathbb{R}^p \setminus \text{int cl } C = \mathbb{R}^p \setminus \text{int } C \supset \text{cl } C \setminus C,$$

so $\overline{x} \in \text{cl}(\mathbb{R}^p \setminus \text{cl } C)$, which means that there exists a sequence $(x_k)_{k \in \mathbb{N}}$ of points outside cl C with $x_k \to \overline{x}$ as $k \to \infty$. Then, for all $k \in \mathbb{N}$, $\text{pr}_{\text{cl } C} x_k \ne x_k$ and we can define

$$a_k := \frac{\text{pr}_{\text{cl } C} x_k - x_k}{\|\text{pr}_{\text{cl } C} x_k - x_k\|}$$

with the property that for every $k \in \mathbb{N}$ and $x \in \text{cl } C$,

$$\langle a_k, x - \text{pr}_{\text{cl } C} x_k \rangle \ge 0.$$

The sequence (a_k) is bounded, whence it has a convergent subsequence. As usual, without relabeling, we assume that (a_k) converges to a limit $a \in \mathbb{R}^p$. Since $\|a_k\| = 1$ for all k, we deduce that $a \ne 0$. By means of Proposition 4.1.2,

$$\text{pr}_{\text{cl } C} x_k \to \text{pr}_{\text{cl } C} \overline{x} = \overline{x},$$

and passing to the limit in the previous inequality, one gets

$$\langle a, c - \overline{x} \rangle \ge 0, \ \forall c \in C.$$

The result is proved. $\qquad\square$

Theorem 4.1.4. *Let A and B be two nonempty convex sets in $\mathbb{R}^p$. If $A \cap B = \emptyset$, then there exists $a \in \mathbb{R}^p \setminus \{0\}$ such that*

$$\langle a, x \rangle \le \langle a, y \rangle, \ \forall x \in A, \ \forall y \in B.$$

Proof Consider the convex set

$$C := A - B.$$

Since $A \cap B = \emptyset$, we deduce that $\overline{x} := 0 \notin C$, and from Theorem 4.1.3 we get the existence of an element $a \in \mathbb{R}^p \setminus \{0\}$ such that for all $c \in C$ one has

$$\langle a, \overline{x} \rangle \le \langle a, c \rangle.$$

This proves the result. $\qquad\qquad\qquad\qquad\qquad\qquad\qquad\qquad\qquad\qquad \square$

4.2 The Subdifferential of a Convex Function

We have seen (Theorem 2.2.10) that if f is a differentiable function on an open convex set $D \subset \mathbb{R}^p$, then f is convex if and only if for every $x, y \in D$,

$$f(y) \ge f(x) + \nabla f(x)(y - x).$$

As explained before, the differential $\nabla f(x)$ is a linear functional from $\mathbb{R}^p$ to $\mathbb{R}$ and it can actually be identified to an element of $\mathbb{R}^p$. By this identification, the expression $\nabla f(x)(y - x)$ can be written as well as $\langle \nabla f(x), y - x \rangle$.

Consider now the case of a convex function defined on a convex subset D of $\mathbb{R}^p$, i.e., $f : D \to \mathbb{R}$ which is not necessarily differentiable. Fix $x \in D$. Then it is natural to consider those elements $u \in \mathbb{R}^p$ having the property that for all $y \in D$,

$$\langle u, y - x \rangle \le f(y) - f(x). \tag{4.2.1}$$

In this way, it is possible to replace in many results the missing differential at x with the set consisting of such elements. More precisely, an element satisfying (4.2.1) is called a subgradient of f at x, and the set of all such elements is denoted by $\partial f(x)$ and is called the subdifferential of f at x.

We will observe that, in the case of a differentiable convex function, $\partial f(x)$ reduces to $\{\nabla f(x)\}$ for any $x \in \text{int } D$. Before that, we investigate some generalized differentiation properties of a convex function.

Proposition 4.2.1. *Let $D \subset \mathbb{R}^p$ be a convex set, $f : D \to \mathbb{R}$ be a convex function, and let $x \in \text{int } D$. Then for every $v \in \mathbb{R}^p$, then a directional derivative of f at x in the direction v exists and is defined as*

$$f'(x, v) := \lim_{t \to 0+} \frac{f(x + tv) - f(x)}{t},$$

which can be written as

$$f'(x, v) = \inf_{t>0} \frac{f(x + tv) - f(x)}{t}.$$

Proof Fix $x \in \text{int } D$. For any $v \in \mathbb{R}^p$ there exists $t > 0$ such that $x + tv \in D$, so by the convexity of D, $x + sv \in D$ for all $s \in [0, t]$. It is enough to prove that the map

$$t \mapsto \frac{f(x + tv) - f(x)}{t}$$

is increasing (since every monotone function admits lateral limits, the conclusions follow). Indeed, if $0 < s < t$, then

$$\frac{f(x + sv) - f(x)}{s} \le \frac{f(x + tv) - f(x)}{t}$$

means that

$$f(x + sv) \le \frac{s}{t} f(x + tv) + \left(1 - \frac{s}{t}\right) f(x),$$

that is

$$f\left(\frac{s}{t}(x + tv) + \left(1 - \frac{s}{t}\right) x\right) \le \frac{s}{t} f(x + tv) + \left(1 - \frac{s}{t}\right) f(x),$$

an inequality which follows from the convexity of f. □

Proposition 4.2.2. *Let $D \subset \mathbb{R}^p$ be a convex set, $f : D \to \mathbb{R}$ be a convex function, and let $x \in \text{int } D$. Then*

$$\partial f(x) = \{u \in \mathbb{R}^p \mid f'(x, v) \ge \langle u, v \rangle , \ \forall v \in \mathbb{R}^p\}.$$

Proof Take $u \in \partial f(u)$. Then

$$f(x + tv) - f(x) \ge \langle u, tv \rangle ,$$

for all $v \in \mathbb{R}^p$ and $t > 0$ with $x + tv \in D$. This implies that

$$f'(x, v) \ge \langle u, v \rangle , \ \forall v \in \mathbb{R}^p.$$

Conversely, suppose that the above inequality holds. Take $y \in D$. Then $v := y - x$ satisfies the inclusion $x + v \in D$. The monotonicity of

$$t \mapsto \frac{f(x + tv) - f(x)}{t}$$

(see the proof of Proposition 4.2.1) allows us to write

$$f(y) - f(x) = f(x + v) - f(x) \ge \inf_{t>0} \frac{f(x + tv) - f(x)}{t} \ge \langle u, v \rangle = \langle u, y - x \rangle ,$$

which confirms that $u \in \partial f(x)$. □

Proposition 4.2.3. *Let $D \subset \mathbb{R}^p$ be a convex set, $f : D \to \mathbb{R}$ be a convex function, and let $\overline{x} \in \text{int } D$. If f is differentiable at $\overline{x}$, then $\partial f(\overline{x}) = \{\nabla f(\overline{x})\}$.*

Proof Theorem 2.2.10 yields the inclusion $\nabla f(\overline{x}) \in \partial f(\overline{x})$. Now, if $u \in \partial f(\overline{x})$ then, by Proposition 4.2.2,

$$f'(x, v) \geq \langle u, v \rangle, \ \forall v \in \mathbb{R}^p.$$

But, since f is differentiable it is easy to observe that $f'(x, v) = \langle \nabla f(x), v \rangle$, so we have that

$$\langle \nabla f(x), v \rangle \geq \langle u, v \rangle, \ \forall v \in \mathbb{R}^p.$$

This implies that $u = \nabla f(x)$ and the proof is complete. $\qquad\square$

It is important to see that at a point of nondifferentiability, the subdifferential is not, in general, a singleton set, as the next example illustrates.

Example 4.2.4. *Consider $f : \mathbb{R} \to \mathbb{R}$, $f(x) = |x|$. It is easy to see that this function is convex and it is differentiable on $\mathbb{R} \setminus \{0\}$. An easy computation reveals that the subdifferential of this convex function is*

$$\partial f(x) = \begin{cases} -1, & \textit{if } x < 0 \\ [-1, 1], & \textit{if } x = 0 \\ 1, & \textit{if } x > 0. \end{cases}$$

Another important fact about convex functions refers to their local Lipschitz behaviour, which induces important properties of the subdifferential. One can say that a function is locally Lipschitz if for every point in its domain there exists a neighborhood where f is Lipschitz. More precisely, $f : D \subset \mathbb{R}^p \to \mathbb{R}$ is locally Lipschitz on D if for every point $x \in D$, there exist $\varepsilon > 0$ and $L > 0$ such that

$$\left| f(x) - f(y) \right| \leq L \, \|x - y\|$$

for every $x, y \in B(x, \varepsilon) \cap D$.

Theorem 4.2.5. *Let $D \subset \mathbb{R}^p$ be a convex set and $f : D \to \mathbb{R}$ be a convex function. Then f it is locally Lipschitz on int D.*

Proof As shown in the first part of the proof of Theorem 2.2.4, for every point in int D, there is a neighborhood on which f is bounded. Take arbitrary $\overline{x} \in$ int D and $r > 0$, $M > 0$ such that for every $x \in D(\overline{x}, r) \subset$ int D, $f(x) \leq f(\overline{x}) + M$. We show that for any $r' \in (0, r)$ and any $x, y \in D(\overline{x}, r')$,

$$\left| f(x) - f(y) \right| \leq \frac{M}{r} \cdot \frac{r + r'}{r - r'} \cdot \|x - y\|,$$

which implies that f is Lipschitz around $\overline{x}$. We can suppose, without loss of generality, that $\overline{x} = 0$ and $f(\overline{x}) = 0$ (otherwise, we consider instead of f the function $g(x) := f(x + \overline{x}) - f(\overline{x})$). Consider $r' \in (0, r)$ and $x, y \in D(\overline{x}, r')$, $x \neq y$. There exist z with $\|z\| = r$

and $\alpha \in (0, 1)$ such that $y = (1 - \alpha)x + \alpha z$. Indeed, taking $\varphi(t) := ty + (1 - t)x$, the map $t \mapsto \|\varphi(t)\|$ is continuous on $[1, \infty)$, $\|\varphi(1)\| = \|y\| < r'$ and $\lim_{t \to \infty} \|\varphi(t)\| \to \infty$, since $\|\varphi(t)\| \geq t \|x - y\| - \|x\|$. Therefore, there exists $\bar{t} > 1$ such that $\|\varphi(\bar{t})\| = r \in (r', \infty)$. Take $z := \varphi(\bar{t})$, i.e.,

$$z = \bar{t}y + (1 - \bar{t})x.$$

Taking $\alpha := \bar{t}^{-1}$, we obtain that

$$y = (1 - \alpha)x + \alpha z.$$

Moreover, $\alpha = \|z - x\|^{-1} \|y - x\|$. Since

$$f(y) \leq (1 - \alpha)f(x) + \alpha f(z),$$

we get

$$f(y) - f(x) \leq \frac{\|y - x\|}{\|z - x\|}(f(z) - f(x)).$$

From

$$0 = f(0) = f\left(2^{-1}x + 2^{-1}(-x)\right) \leq 2^{-1}f(x) + 2^{-1}f(-x),$$

we obtain $-f(x) \leq f(-x)$. But, if $x \neq 0$, $u := r\|x\|^{-1}(-x) \in D(0, r)$, so

$$f(-x) = f\left(\frac{\|x\|}{r}u\right) \leq \frac{\|x\|}{r}f(u) \leq M\frac{\|x\|}{r}.$$

Clearly, this is true also for $x = 0$. Since $\|z - x\| \geq r - r'$ and $r + \|x\| \leq r + r'$, we finally obtain

$$f(y) - f(x) \leq \frac{\|y - x\|}{\|z - x\|}\left(M + M\frac{\|x\|}{r}\right) \leq \frac{M}{r} \cdot \frac{r + r'}{r - r'} \cdot \|x - y\|.$$

Interchanging x and y, the theorem is proved. $\qquad\square$

A cornerstone for the development of a subdifferential calculus (i.e., calculus with subdifferentials) is to observe that, in general, the subdifferential at a given point is a nonempty set.

Theorem 4.2.6. *Let $D \subset \mathbb{R}^p$ be a convex set and $f : D \to \mathbb{R}$ be a convex function. Then for every $x \in \text{int } D$, the subdifferential $\partial f(x)$ is a nonempty compact set.*

Proof Fix $x \in \text{int } D$. First we show that $\partial f(x)$ is nonempty. From Theorem 2.2.4, the function f is continuous at x. In particular, this shows that the convex set $\text{int epi } f$ is nonempty (one applies here both Proposition 2.2.3 (*ii*) and Theorem 4.1.1 (*iii*)). Clearly, $(x, f(x)) \notin \text{int epi } f$, so, from Theorem 4.1.3, there exists $(u, \alpha) \in \mathbb{R}^p \times \mathbb{R} \setminus \{(0, 0)\}$ such that for every $(y, t) \in \text{int epi } f$,

$$\langle u, y \rangle + \alpha t \leq \langle u, x \rangle + \alpha f(x).$$

Moreover, this is true for every $(y, t) \in \text{cl(int epi} f) = \text{cl epi} f = \text{epi} f$. If we suppose that $\alpha = 0$, then the inequality $\langle u, x - y \rangle \geq 0$ holds for all $y \in D$, and this means that $u = 0$, which contradicts the fact that the pair (u, α) is not zero. Consequently, $\alpha \neq 0$. But, letting $t \to \infty$ in the above inequality, one deduces that $\alpha < 0$. Therefore, we can take $\alpha = -1$, and then we get

$$\langle u, y \rangle - t \leq \langle u, x \rangle - f(x), \ \forall (y, t) \in \text{epi} f.$$

In particular, for every $y \in D$,

$$\langle u, y \rangle - f(y) \leq \langle u, x \rangle - f(x),$$

that is

$$\langle u, y - x \rangle \leq f(y) - f(x),$$

i.e., $u \in \partial f(x)$.

Let us notice now that the fact that $\partial f(x)$ is always a closed set is an easy remark. It remains to show that $\partial f(x)$ is a bounded set. By virtue of Theorem 4.2.5, f is Lipschitz on a neighborhood of x. Denote by L the corresponding Lipschitz constant on the given neighborhood $V := D(x, \varepsilon) \subset D$ of x. Take $u \in \partial f(x)$. Then for any $y \in V$,

$$\langle u, y - x \rangle \leq f(y) - f(x) \leq |f(y) - f(x)| \leq L \|y - x\|.$$

Since $y = x + \varepsilon v \in V$ for any v in the unit ball, this implies that

$$\langle u, v \rangle \leq L \|v\|, \ \forall v \in D(0, 1),$$

which means that $\|u\| \leq L$. Then $\partial f(x)$ is bounded. $\qquad \square$

The next calculus rule is fundamental.

Theorem 4.2.7 (sum rule). *$D \subset \mathbb{R}^p$ be a convex set with nonempty interior and let $f, g : D \to \mathbb{R}$ be convex functions. Then for any $x \in D$,*

$$\partial(f + g)(x) = \partial f(x) + \partial g(x).$$

Proof The inclusion $\partial f(x) + \partial g(x) \subset \partial(f + g)(x)$ easily follows from the definition of the subdifferential. Consider $u \in \partial(f + g)(x)$. We can consider the case $x = 0$, $f(0) = g(0) = 0$ directly, simply by taking $f_1(y) := f(y + x) - f(x)$ and $g_1(y) := g(y + x) - g(x)$. Denote

$$A := \text{int epi} f \text{ and } B := \{(v, t) \in D \times \mathbb{R} \mid t \leq \langle u, v \rangle - g(v)\}$$

Both these sets are nonempty and convex. If $(v, t) \in A \cap B$ did exist, then, on one hand, $f(v) < t$, and, on the other hand, since $u \in \partial(f + g)(0)$,

$$f(v) < t \leq \langle u, v \rangle - g(v) \leq f(v),$$

which is not possible. Therefore, $A \cap B = \emptyset$. Theorem 4.1.4 ensures the existence of an element $(a, \alpha) \in \mathbb{R}^p \times \mathbb{R} \setminus \{(0, 0)\}$ such that for all $(u, s) \in A$ and $(v, t) \in B$,

$$\langle a, v \rangle + \alpha t \leq \langle a, u \rangle + \alpha s.$$

It follows that this inequality holds for any $(u, s) \in \mathrm{cl}\, A = \mathrm{epi}\, f$ and any $(v, t) \in B$. If $\alpha = 0$, then $\langle a, v \rangle \leq \langle a, u \rangle$ for every $u, v \in D$, which means that $a = 0$ (otherwise, a would have positive scalar product with all the elements of a ball centered at 0, because D has nonempty interior), an impossible situation. Therefore, $\alpha \neq 0$. Letting $s \to \infty$, we infer that $\alpha > 0$ and therefore we can take $\alpha = 1$, so

$$\langle a, v \rangle + t \leq \langle a, u \rangle + s, \ \forall (u, s) \in \mathrm{epi}\, f, \ \forall (v, t) \in B.$$

For $(v, t) = (0, 0)$ and $s = f(u)$, we get

$$\langle -a, u \rangle \leq f(u), \ \forall u \in \mathrm{epi}\, f,$$

i.e., $-a \in \partial f(0)$. On the other hand, for $(u, s) = (0, 0)$, we obtain

$$\langle a, v \rangle + t \leq 0, \ \forall (v, t) \in B,$$

which implies that

$$\langle a + u, v \rangle \leq g(v), \ \forall v \in D,$$

i.e., $a + u \in \partial g(0)$, so $u = -a + a + u \in \partial f(0) + \partial g(0)$. $\qquad\square$

4.3 Optimality Conditions

We give now optimality conditions for convex nondifferentiable data for the basic optimization problems already studied in the smooth case. We start with the case of optimization without constraints.

Theorem 4.3.1. *Let $D \subset \mathbb{R}^p$ be a convex set, and $f : D \to \mathbb{R}$ be a convex function. An element $\overline{x} \in D$ is a (local) minimum point of f if and only if $0 \in \partial f(\overline{x})$.*

Proof If $\overline{x}$ is a local minimum point, then it is a global one (see Proposition 3.1.23), and therefore

$$0 = \langle 0, y - \overline{x} \rangle \leq f(y) - f(\overline{x})$$

for all $y \in D$. Then, $0 \in \partial f(\overline{x})$.

Conversely, suppose that $0 \in \partial f(\overline{x})$. From the definition of the subgradients, this means that $\langle 0, y - \overline{x} \rangle \leq f(y) - f(\overline{x})$ for all $y \in D$, whence $\overline{x}$ is a minimum point for f. $\square$

We now consider the optimization problems with geometric constraints. In order to tackle this case, we need a general penalization result. Generally speaking, to penalize an optimization problem means to transform it in such a way that a (local) minimum point of the initial constrained problem becomes a (local) minimum point of the unconstrained problem. This can be done by adding to the objective function a penalty term which somehow contains the constraints of the initial problem. We present here a method to penalize a given optimization problem (not necessarily convex) with geometric constraints due to Canadian–French mathematician Frank H. Clarke, who proposed it in 1983.

Theorem 4.3.2. *Let $f : \mathbb{R}^p \to \mathbb{R}$ be a function and $M \subset \mathbb{R}^p$ a closed set. Consider the problem*

$$\min f(x), \; x \in M.$$

Suppose that $\overline{x}$ is a (local) solution of this problem and f is locally Lipschitz of constant $L \geq 0$ around $\overline{x}$. Then $\overline{x}$ is a (local) minimum without constraints of the mapping $x \mapsto f(x) + L d_M(x)$.

Proof Let U be the neighborhood of $\overline{x}$ such that f is L–Lipschitz on U and $f(\overline{x}) \leq f(x)$ for every $x \in U \cap M$. Then there exists $\alpha > 0$ such that $B(\overline{x}, \alpha) \subset U$. Consider $V = B(\overline{x}, 3^{-1}\alpha)$ and take $x \in V$ arbitrarily. First, if $x \in V \cap M \subset U \cap M$, it is clear that

$$f(\overline{x}) + L d_M(\overline{x}) \leq f(x) + L d_M(x).$$

Then, if $x \in V \setminus M$, for every $\varepsilon \in (0, 3^{-1}\alpha)$, there exists $x_\varepsilon \in M$ such that

$$\begin{aligned}
\|x - x_\varepsilon\| &< d_M(x) + \varepsilon \\
&\leq \|x - \overline{x}\| + \varepsilon \\
&\leq 3^{-1}\alpha + \varepsilon < 2 \cdot 3^{-1}\alpha.
\end{aligned}$$

Then,

$$\begin{aligned}
\|x_\varepsilon - \overline{x}\| &\leq \|x_\varepsilon - x\| + \|x - \overline{x}\| \\
&< 2 \cdot 3^{-1}\alpha + 3^{-1}\alpha = \alpha.
\end{aligned}$$

Consequently, $x_\varepsilon \in U \cap M$, whence,

$$\begin{aligned}
f(\overline{x}) + L d_M(\overline{x}) \leq f(x_\varepsilon) &\leq f(x) + L\|x - x_\varepsilon\| \\
&\leq f(x) + L(d_M(x) + \varepsilon) \\
&= f(x) + L d_M(x) + L\varepsilon.
\end{aligned}$$

Letting $\varepsilon \to 0$, we get the conclusion. $\qquad\square$

We now connect the normal cone to a convex set to the subdifferential of the (non-differentiable) distance function: see Propositions 2.1.15 and 2.2.9. We shall see that similar relations hold in more general settings, as shown in Chapter 5.

Proposition 4.3.3. *Let $C \subset \mathbb{R}^p$ be a nonempty closed convex set and $\overline{x} \in C$. Then*

$$[0, \infty)\partial d_C(\overline{x}) = N(C, \overline{x}).$$

Proof Take $u \in \partial d_C(\overline{x})$. Then

$$\langle u, y - \overline{x} \rangle \le d_C(y) - d_C(\overline{x}), \ \forall y \in \mathbb{R}^p.$$

In particular, for $y \in C$,

$$\langle u, y - \overline{x} \rangle \le 0,$$

which proves that $u \in N(C, \overline{x})$. Consequently, $\partial d_C(\overline{x}) \subset N(C, \overline{x})$ and since the latter set is a cone, one infers that $[0, \infty)\partial d_C(\overline{x}) \subset N(C, \overline{x})$.

Conversely, take $u \in N(C, \overline{x})$ with $\|u\| \le 1$, i.e.,

$$\langle u, c - \overline{x} \rangle \le 0, \ \forall c \in C.$$

Then

$$\langle -u, c \rangle \ge \langle -u, \overline{x} \rangle, \ \forall c \in C,$$

which means that $\overline{x}$ is a minimum point of the mapping $f : \mathbb{R}^p \to \mathbb{R}$, $f(x) := \langle -u, x \rangle$ on the set C. It is clear that f is 1–Lipschitz (since $\|u\| \le 1$), and therefore, by virtue of Theorem 4.3.2, $\overline{x}$ is a minimum without restrictions of the convex function $f + d_C$. Therefore, Theorem 4.3.1 gives

$$0 \in \partial(f + d_C)(\overline{x}),$$

and, by means of Theorem 4.2.7, we get

$$0 \in \partial f(\overline{x}) + \partial d_C(\overline{x}).$$

Since f is differentiable and $\nabla f(\overline{x}) = -u$, from Proposition 4.2.3, $\partial f(\overline{x}) = \{-u\}$. Consequently, $u \in \partial d_C(\overline{x})$, so we have shown that

$$N(C, \overline{x}) \cap D(0, 1) \subset \partial d_C(\overline{x}),$$

and the conclusion follows. $\qquad\square$

Now we are able to present the optimality condition for convex geometric constraint problems.

Theorem 4.3.4 (Pshenichnyi-Rockafellar). *Let $D \subset \mathbb{R}^p$ be an open convex set, $f : D \to \mathbb{R}$ be a convex function, and $C \subset D$ be a nonempty closed convex set. A point $\overline{x} \in C$ is a local minimum point of f on C if and only if*

$$0 \in \partial f(\overline{x}) + N(C, \overline{x}).$$

Proof Suppose that $\overline{x} \in C$ is a local minimum point of f on C. Since f is locally Lipschitz (Theorem 4.2.5) on D (notice that we assume that int $D = D$), we can apply Theorem 4.3.2 to deduce that $\overline{x}$ is a local minimum (without constraints) of the convex function

$f + Ld_C$ defined on D, where $L > 0$ is the Lipschitz constant of f around $\bar{x}$. Using Theorem 4.3.1, we can write

$$0 \in \partial(f + Ld_C)(\bar{x}),$$

and by virtue of Theorem 4.2.7,

$$0 \in \partial f(\bar{x}) + \partial(Ld_C)(\bar{x}).$$

It is easy to see that $\partial(Ld_C)(\bar{x}) = L\partial(d_C)(\bar{x})$ and, finally, Proposition 4.3.3 allows us to write

$$0 \in \partial f(\bar{x}) + N(C, \bar{x}).$$

Conversely, suppose that the above condition holds. This means that there exists $a \in -N(C, \bar{x}) \cap \partial f(\bar{x})$, i.e., for every $y \in C$,

$$0 \le \langle a, y - \bar{x} \rangle \le f(y) - f(\bar{x}),$$

which confirms that $\bar{x}$ is a minimum point of f on C. $\qquad\square$

Let now us consider an open convex set $D \subset \mathbb{R}^p$. We study the case of convex optimization with functional constraints, that is the problem (P) of minimizing a convex objective function $f : D \to \mathbb{R}$ under the constraint $x \in C := \{x \in D \mid g(x) \le 0, h(x) = 0\}$ where $g : D \to \mathbb{R}^n$ has all the component functions $g_1, g_2, ..., g_n : D \to \mathbb{R}$ convex and $h : D \to \mathbb{R}^m$ is affine. Due to the assumptions on D, g and h, the set C is convex. We discussed this problem in Section 3.2 with smoothness assumptions on f, g and h. Now, this problem is investigated without the differentiability hypotheses, but using subdifferential calculus.

To begin with, we give a Fritz John type result. We use the notations $\mathbb{R}^p_+ := (\mathbb{R}_+)^p$ and $\mathbb{R}^p_- := (-\mathbb{R}_+)^p$.

Theorem 4.3.5. *Suppose that $\bar{x}$ is a solution of the convex problem (P). Then there exist $\lambda_0 \ge 0$, $\lambda = (\lambda_1, \lambda_2, ..., \lambda_n) \in \mathbb{R}^n$, $\mu = (\mu_1, \mu_2, ..., \mu_m) \in \mathbb{R}^m$, with $\lambda_0 + \|\lambda\| + \|\mu\| \ne 0$, such that*

$$0 \in \lambda_0 \partial f(\bar{x}) + \sum_{i=1}^{n} \lambda_i \partial g_i(\bar{x}) + \sum_{j=1}^{m} \mu_j \partial h_j(\bar{x}),$$

and

$$\lambda_i \ge 0, \ \lambda_i g_i(\bar{x}) = 0, \ \forall i \in \overline{1, n}.$$

Proof As discussed in the smooth case, we can suppose, without loss of generality, that $A(\bar{x}) = \overline{1, n}$. Take

$$A := \{(f(x) - f(\bar{x}) + t, g(x) + s, h(x)) \mid x \in D, t \in [0, \infty), s \in \mathbb{R}^p_+\},$$

and

$$B := (-\infty, 0) \times \mathbb{R}^p_- \times \{0\}^m.$$

It is easy to see that the properties of f, g and h (i.e., f convex, g_i convex for all $i \in \overline{1, n}$, and h affine) ensure the convexity of A, while the convexity of B is obvious. On the other hand, if a common element of A and B exists, then there also exists $x \in D$ with $f(x) - f(\overline{x}) < 0$, $g(x) \leq 0$, $h(x) = 0$, which would contradict the (global) minimality of $\overline{x}$. Consequently, $A \cap B = \emptyset$. We apply Theorem 4.1.4 in order to deduce the existence of some elements $\lambda_0 \in \mathbb{R}$, $\lambda = (\lambda_1, \lambda_2, ..., \lambda_n) \in \mathbb{R}^n$, $\mu = (\mu_1, \mu_2, ..., \mu_m) \in \mathbb{R}^m$, with $|\lambda_0| + \|\lambda\| + \|\mu\| \neq 0$ such that

$$\lambda_0 a + \langle \lambda, u \rangle \leq \lambda_0 (f(x) - f(\overline{x}) + t) + \langle \lambda, g(x) + s \rangle + \langle \mu, h(x) \rangle$$

for all $x \in D$, $t \in [0, \infty)$, $s \in \mathbb{R}_+^p$, $a \in (-\infty, 0)$, $u \in \mathbb{R}^p$. It is not possible to have $\lambda_0 < 0$. Indeed, if we suppose so, then for fixed x, t, s and u, letting $a \to -\infty$, we arrive at a contradiction, since the right-hand side is fixed, while the left-hand side goes to $+\infty$. A similar argument employed for $s_i \to \infty$ (for all $i \in \overline{1, n}$) allows us to conclude that $\lambda_i \geq 0$, for all $i \in \overline{1, n}$. Letting $a \to 0$, $t \to 0$, $u \to 0$ and $s \to 0$, we actually get that

$$0 \leq \lambda_0 (f(x) - f(\overline{x})) + \langle \lambda, g(x) \rangle + \langle \mu, h(x) \rangle$$

for all $x \in D$. Since $g(\overline{x}) = 0$, we deduce that $\lambda_i g_i(\overline{x}) = 0$, for all $i \in \overline{1, n}$. Finally, we can write

$$0 \leq (\lambda_0 f(x) - \lambda_0 f(\overline{x})) + \sum_{i=1}^{n} (\lambda_i g(x) - \lambda_i g_i(\overline{x})) + \sum_{j=1}^{m} (\mu_j h_j(x) - \mu_j h_j(\overline{x})), \ \forall x \in D,$$

which means that

$$0 \in \partial \left(\lambda_0 f + \sum_{i=1}^{n} \lambda_i g_i + \sum_{j=1}^{m} \mu_j h_j \right) (\overline{x}).$$

Theorem 4.2.7 allows us to write

$$0 \in \partial(\lambda_0 f)(\overline{x}) + \sum_{i=1}^{n} \partial(\lambda_i g_i)(\overline{x}) + \sum_{j=1}^{m} \partial(\mu_j h_j)(\overline{x}).$$

Since $\lambda_0 \geq 0$, one has that $\partial \lambda_0 f(\overline{x}) = \lambda_0 \partial f(\overline{x})$. The same argument is applicable to the equality $\partial \lambda_i g_i(\overline{x}) = \lambda_i \partial g_i(\overline{x})$ for all $i \in \overline{1, n}$. The equality $\partial \mu_j h_j(\overline{x}) = \mu_j \partial h_j(\overline{x})$ for all $j \in \overline{1, m}$ is true by the fact that h is affine. The proof is complete. $\square$

As we already saw several times, in order to pass from Fritz John type conditions to Karush-Kuhn-Tucker type condition (i.e., to ensure $\lambda_0 \neq 0$) one needs a supplementary (constraint qualification) assumption. Since we are now in the convex case, we use here a Slater type condition, which for the nondifferentiable case looks as follows: there exists $u \in D$ such that $g(u) < 0$, $h(u) = 0$, and the set $\{\nabla h_j(u) \mid j \in \overline{1, m}\}$ is linearly independent (for $u \in D$; notice that $\nabla h_j(u)$ is the same for any $u \in D$). Under this condition, we can deduce from Theorem 4.3.5 a Karush-Kuhn-Tucker theorem, as follows.

Theorem 4.3.6. *Suppose that $\overline{x}$ is a solution of the convex problem (P) and that the Slater condition holds. Then there exist $\lambda = (\lambda_1, \lambda_2, \ldots, \lambda_n) \in \mathbb{R}^n$, $\mu = (\mu_1, \mu_2, \ldots, \mu_m) \in \mathbb{R}^m$ such that*

$$0 \in \partial f(\overline{x}) + \sum_{i=1}^{n} \lambda_i \partial g_i(\overline{x}) + \sum_{j=1}^{m} \mu_j \partial h_j(\overline{x}),$$

and

$$\lambda_i \geq 0, \ \lambda_i g_i(\overline{x}) = 0, \ \forall i \in \overline{1, n}.$$

Proof It is sufficient to show that in Theorem 4.3.5 is not possible to have $\lambda_0 = 0$. Suppose, by way of contradiction, that $\lambda_0 = 0$. Then

$$0 \in \sum_{i=1}^{n} \lambda_i \partial g_i(\overline{x}) + \sum_{j=1}^{m} \mu_j \partial h_j(\overline{x}) = \partial \left(\sum_{i=1}^{n} \lambda_i g_i + \sum_{j=1}^{m} \mu_j h_j \right)(\overline{x})$$

that is, for every $x \in D$,

$$\sum_{i=1}^{n} \lambda_i (g_i(x) - g_i(\overline{x})) + \sum_{j=1}^{m} \mu_j (h_j(x) - h_j(\overline{x})) \geq 0,$$

i.e.,

$$\sum_{i=1}^{n} \lambda_i g_i(x) + \sum_{j=1}^{m} \mu_j h_j(x) \geq 0.$$

For $x = u$ (the element from Slater condition), this becomes

$$\sum_{i=1}^{n} \lambda_i g_i(u) \geq 0,$$

which is possible only if $\lambda_i = 0$ for all $i \in \overline{1, n}$. So, in fact,

$$\sum_{j=1}^{m} \mu_j h_j(x) \geq \sum_{j=1}^{m} \mu_j h_j(\overline{x}) = 0, \ \forall x \in D.$$

This means that $\overline{x}$ is a minimum point (without constraint) for the function $x \mapsto \sum_{j=1}^{m} \mu_j h_j(x)$, so, from the Fermat Theorem,

$$\sum_{j=1}^{m} \mu_j \nabla h_j(\overline{x}) = 0.$$

The linear independence condition says that the above relation holds only if $\mu_j = 0$ for all $j \in \overline{1, m}$. The facts obtained up to now, collectively imply that $\lambda_0 + \|\lambda\| + \|\mu\| = 0$, which is a contradiction. The proof is complete. $\qquad\square$

5 Lipschitz Nonsmooth Optimization

Smooth and convex functions, which we discussed in the previous two chapters, are also (locally) Lipschitz. This chapter aims to introduce a more general framework, which provides optimality conditions for problems with Lipschitz data. We first present the theory of generalized gradients for locally Lipschitz functions, developed by Frank H. Clarke in the 1970's, and we then pass to a more general framework, by giving some comprehensive elements of the modern theory developed by Boris S. Mordukhovich and his collaborators in the last three decades. Of course, we limit our exposition to the finite dimensional spaces framework, but both of these theories can be developed in much more general situations. It is beyond the scope of this book to discuss the case of multifunctions (i.e., mappings with values given by sets), and of the associated generalized differentiation objects (derivatives and coderivatives, associated by means of corresponding tangent and normal cones), although many interesting results, which link the whole machinery of calculus for functions, multifunctions and tangent cones exist in literature.

5.1 Clarke Generalized Calculus

Consider the following optimization problem:

$$\min f(x_1, x_2) \quad \text{such that } x_1^2 + x_2^2 \le 1, \tag{5.1.1}$$

where $f : \mathbb{R}^2 \to \mathbb{R}$ is the function $f(x_1, x_2) = \max \{\min \{2x_1, x_2\}, x_1 + 2x_2\}$. Remark that this function is neither convex nor smooth on the admissible set of points given by the unit ball. Therefore, the theory developed until now cannot apply. Remark that f, as well as the function which gives the restrictions (i.e., $g(x_1, x_2) := x_1^2 + x_2^2 - 1$), do exhibit nice regularity properties: they are both Lipschitz functions.

It is interesting to develop a theory which could solve such a problem. This is the main purpose of this chapter.

5.1.1 Clarke Subdifferential

Consider a function $f : \mathbb{R}^p \to \mathbb{R}$ and suppose that f is locally Lipschitz with modulus $L > 0$ around a point $\overline{x} \in \mathbb{R}^p$, which means that there is an $\varepsilon > 0$ such that f is L–Lipschitz on $D(\overline{x}, \varepsilon)$, i.e.,

$$\left| f(x) - f(u) \right| \le L \cdot \|x - u\|, \ \forall x, u \in D(\overline{x}, \varepsilon).$$

One can define the (Clarke) generalized directional derivative of f at $\bar{x}$ in the direction u, denoted $f^{\circ}(\bar{x}, u)$, as follows:

$$f^{\circ}(\bar{x}, u) := \limsup_{x \to \bar{x},\, t \downarrow 0} \frac{f(x + tu) - f(x)}{t}. \tag{5.1.2}$$

Some useful basic properties of this function are given bellow.

Proposition 5.1.1. *Let $f : \mathbb{R}^p \to \mathbb{R}$ be a locally Lipschitz function with modulus $L > 0$ around a point $\bar{x} \in \mathbb{R}^p$. Then*

(i) The function $u \mapsto f^{\circ}(\bar{x}, u)$ is finite, sublinear and L–Lipschitz on $\mathbb{R}^p$, and it satisfies

$$\left| f^{\circ}(\bar{x}, u) \right| \le L \, \|u\|, \ \forall u \in \mathbb{R}^p. \tag{5.1.3}$$

Moreover,

$$f^{\circ}(\bar{x}, -u) = (-f)^{\circ}(\bar{x}, u), \ \forall u \in \mathbb{R}^p. \tag{5.1.4}$$

(ii) For every $u \in \mathbb{R}^p$, the function $(x, v) \mapsto f^{\circ}(x, v)$ is upper semicontinuous at $(\bar{x}, u)$.

Proof (i) By the Lipschitz condition of f around $\bar{x}$, it follows that for every x and positive t sufficiently close to $\bar{x}$ and 0, respectively, one has

$$\left| \frac{f(x + tu) - f(x)}{t} \right| \le \frac{L \, \|tu\|}{t} = L \, \|u\|,$$

which shows (5.1.3) and the fact that $f^{\circ}(\bar{x}, \cdot)$ is everywhere finite.

Moreover, for every $\lambda \ge 0$, one has

$$\frac{f(x + t\lambda u) - f(x)}{t} = \lambda \frac{f(x + t\lambda u) - f(x)}{\lambda t},$$

which shows the positive homogeneity of $f^{\circ}(\bar{x}, \cdot)$, i.e.,

$$f^{\circ}(\bar{x}, \lambda u) = \lambda f^{\circ}(\bar{x}, u), \ \forall \lambda \ge 0.$$

Also, for every $u, v \in \mathbb{R}^p$,

$$f^{\circ}(\bar{x}, u + v) = \limsup_{x \to \bar{x},\, t \downarrow 0} \frac{f(x + tu + tv) - f(x)}{t}$$

$$\le \limsup_{x \to \bar{x},\, t \downarrow 0} \frac{f(x + tu + tv) - f(x + tu)}{t} + \limsup_{x \to \bar{x},\, t \downarrow 0} \frac{f(x + tu) - f(x)}{t}.$$

But, since for any $r, \delta > 0$,

$$\sup_{x \in B(\bar{x},r),\, t \in (0,\delta)} \frac{f(x + tu + tv) - f(x + tu)}{t} \le \sup_{y \in B(\bar{x},r+\delta\|u\|),\, t \in (0,\delta)} \frac{f(y + tv) - f(y)}{t},$$

it follows, by letting $r \to 0$, $\delta \to 0$, that

$$\limsup_{x \to \bar{x}, \, t \downarrow 0} \frac{f(x + tu + tv) - f(x + tu)}{t} \leq \limsup_{y \to \bar{x}, \, t \downarrow 0} \frac{f(y + tv) - f(y)}{t}$$

and, finally, that $f^\circ(\bar{x}, \cdot)$ is subadditive, i.e.,

$$f^\circ(\bar{x}, u + v) \leq f^\circ(\bar{x}, u) + f^\circ(\bar{x}, v), \quad \forall u, v \in \mathbb{R}^p.$$

Concerning relation (5.1.4), observe that for every $u \in \mathbb{R}^p$,

$$f^\circ(\bar{x}, -u) = \limsup_{x \to \bar{x}, \, t \downarrow 0} -\frac{f(x) - f(x - tu)}{t} = \limsup_{x \to \bar{x}, \, t \downarrow 0} \frac{(-f)(x) - (-f)(x - tu)}{t}$$

$$= \limsup_{x \to \bar{x}, \, t \downarrow 0} \frac{(-f)((x - tu) + tu) - (-f)(x - tu)}{t}$$

$$= \limsup_{y \to \bar{x}, \, t \downarrow 0} \frac{(-f)(y + tu) - (-f)(y)}{t} = (-f)^\circ(\bar{x}, u).$$

Now, by the subadditivity and relations (5.1.3), (5.1.4), one gets that

$$f^\circ(\bar{x}, u) - f^\circ(\bar{x}, v) \leq f^\circ(\bar{x}, u - v) \leq L \|u - v\|, \quad \forall u, v \in \mathbb{R}^p,$$

which implies the expected Lipschitz property of $f^\circ(\bar{x}, \cdot)$.

(ii) Concerning the upper semicontinuity of the function $(x, v) \mapsto f^\circ(x, v)$ at $(\bar{x}, u)$, take an arbitrary sequence (x_n, u_n) converging to $(\bar{x}, u)$. It follows that for each n,

$$f^\circ(x_n, u_n) - \frac{1}{n} < \limsup_{x \to x_n, \, t \downarrow 0} \frac{f(x + tu_n) - f(x)}{t}.$$

Using the definition of the upper limit, one can find $y_n \in \mathbb{R}^p$ with $\|y_n - x_n\| < \frac{1}{n}$ and $t_n \in \left(0, \frac{1}{n}\right)$ such that

$$f^\circ(x_n, u_n) - \frac{1}{n} < \frac{f(y_n + t_n u_n) - f(y_n)}{t_n}$$

$$= \frac{f(y_n + t_n u) - f(y_n)}{t_n} + \frac{f(y_n + t_n u_n) - f(y_n + t_n u)}{t_n}.$$

By the Lipschitz property of f, we get that the final term is smaller than $L \|u_n - u\|$. Passing to lim sup for $n \to \infty$, it follows that

$$\limsup_{n \to \infty} f^\circ(x_n, u_n) \leq f^\circ(\bar{x}, u),$$

which shows the desired upper semicontinuity. $\qquad \square$

Given a nonempty subset $A \subset \mathbb{R}^p$, its support function is the function $h_A : \mathbb{R}^p \to (-\infty, \infty]$ given as

$$h_A(u) := \sup \{\langle \xi, u \rangle \mid \xi \in A\}.$$

Some properties of the support function, useful in the sequel, are given in the next result.

Proposition 5.1.2. *(i) Let A be a nonempty subset of $\mathbb{R}^p$. Then h_A is positively homogeneous, subadditive, hence convex and continuous.*

(ii) If A is convex and closed, then

$$\xi \in A \iff \langle \xi, u \rangle \leq h_A(u), \ \forall u \in \mathbb{R}^p.$$

(iii) If A and B are two nonempty, convex and closed subsets of $\mathbb{R}^p$, then

$$A \subset B \iff h_A(u) \leq h_B(u), \ \forall u \in \mathbb{R}^p. \tag{5.1.5}$$

(iv) If $\varphi : \mathbb{R}^p \to \mathbb{R}$ is positively homogeneous, subadditive and bounded on the unit ball, then there is a uniquely defined nonempty, convex and compact subset A of $\mathbb{R}^p$ such that $\varphi = h_A$, and the supremum is realized at every $u \in \mathbb{R}^p$.

Proof (i) It is straightforward from the definition that h_A is positively homogeneous and subadditive.

(ii) It is obvious that $\xi \in A$ implies $\langle \xi, u \rangle \leq h_A(u)$ for every $u \in \mathbb{R}^p$. Consider now $\xi \notin A$. Since $\{\xi\}$ is convex, and A is convex and closed, it follows by applying the first argument of the proof of the separation Theorem 4.1.3, that there exist $u \in \mathbb{R}^p$ and $\alpha \in \mathbb{R}$ such that $\langle \xi, u \rangle > \alpha > \langle \chi, u \rangle$, for any $\chi \in A$. This means $\langle \xi, u \rangle > h_A(u)$, hence the converse implication is proved.

(iii) Again, the direct implication is straightforward. For the converse, take $\xi \in A$ such that $\xi \notin B$. Using the proof of (ii), one gets that there is $u \in \mathbb{R}^p$ such that $\langle \xi, u \rangle > h_B(u)$, which means that $h_A(u) \geq \langle \xi, u \rangle > h_B(u)$.

(iv) Given φ, define

$$A := \left\{ \xi \in \mathbb{R}^p \mid \langle \xi, u \rangle \leq \varphi(u), \ \forall u \in \mathbb{R}^p \right\}.$$

Clearly, A is a convex set. Now, because A can be written as

$$A = \bigcap_{u \in \mathbb{R}^p} \left\{ \xi \in \mathbb{R}^p \mid \langle \xi, u \rangle \leq \varphi(u) \right\},$$

it follows that A is closed as intersection of closed sets. Moreover, denoting by $M > 0$ the boundedness constant of φ on the unit ball, one has, for any $\xi \in A$ and any $u \in D(0, 1)$, that $\langle \xi, u \rangle \leq \varphi(u) \leq M$, hence $\|\xi\| \leq M$. So A is bounded, therefore compact. The inequality $h_A \leq \varphi$ is obvious from the definition of A. Let us show the equality. Take arbitrary $u \in \mathbb{R}^p$ and consider the linear subspace $H := \{\lambda u \mid \lambda \in \mathbb{R}\}$. Pick $\zeta \in \mathbb{R}^p$ such that $\langle \zeta, u \rangle = \varphi(u)$. For instance, one can take $\zeta := 0$ if $u = 0$, and $\zeta := \|u\|^{-2} \varphi(u) u$ if $u \neq 0$. Then, if $\lambda \geq 0$, one has $\langle \zeta, \lambda u \rangle = \varphi(\lambda u)$, and if $\lambda < 0$,

$$\langle \zeta, \lambda u \rangle = -(-\lambda)\varphi(u) = -\varphi(-\lambda u) \leq \varphi(\lambda u).$$

In any case, the linear functional $\langle \zeta, \cdot \rangle$ is majorized on H by the sublinear functional φ. By applying the Hahn-Banach Theorem, one gets that there is $\xi \in \mathbb{R}^p$ such that $\langle \xi, v \rangle \leq \varphi(v)$ for any $v \in \mathbb{R}^p$ and $\langle \zeta, u \rangle = \langle \xi, u \rangle = \varphi(u)$. It means that $\xi \in A$ and $\varphi(u) \leq h_A(u)$. Since u was chosen arbitrarily, we have $\varphi = h_A$. The proof of the uniqueness follows by the use of (iii). $\qquad \square$

Now, consider a locally Lipschitz function f, and take instead of the function φ from the previous proposition the generalized directional derivative $f°(\overline{x}, \cdot)$, Clarke defined the generalized gradient as the nonempty compact set whose support function is $f°(\overline{x}, \cdot)$.

Definition 5.1.3. *Let $f : \mathbb{R}^p \to \mathbb{R}$ be a locally Lipschitz function with modulus $L > 0$ around a point $\overline{x} \in \mathbb{R}^p$. The Clarke generalized gradient, or the Clarke subdifferential of f at $\overline{x}$ is the set*

$$\partial_C f(\overline{x}) = \left\{ \xi \in \mathbb{R}^p \mid f°(\overline{x}, u) \geq \langle \xi, u \rangle, \ \forall u \in \mathbb{R}^p \right\}. \tag{5.1.6}$$

Some basic properties of the Clarke subdifferential are given next.

Proposition 5.1.4. *Let $f : \mathbb{R}^p \to \mathbb{R}$ be a locally Lipschitz function with modulus $L > 0$ around a point $\overline{x} \in \mathbb{R}^p$. Then*
 (i) $\partial_C f(\overline{x})$ is a nonempty, compact and convex set;
 (ii) One has

$$f°(\overline{x}, u) = \max \left\{ \langle \xi, u \rangle \mid \xi \in \partial_C f(\overline{x}) \right\}. \tag{5.1.7}$$

Proof Both (i) and (ii) follow from Proposition 5.1.2. $\qquad \square$

Example 5.1.5. *Consider the function $f : \mathbb{R}^p \to \mathbb{R}$, $f(x) = \|x\|$. Of course, this function is Lipschitz with modulus 1. We know by Proposition 5.1.1 that $f°(0, u) \leq \|u\|$ for any $u \in \mathbb{R}^p$. Also, since*

$$\frac{f(x + tu) - f(x)}{t} = \frac{\|x + tu\| - \|x\|}{t}, \ \forall x, u \in \mathbb{R}^p, \forall t > 0,$$

it follows, by passing to $\limsup$ for $x \to 0$, that $f°(0, u) \geq \|u\|$, for any $u \in \mathbb{R}^p$, hence $f°(0, u) = \|u\|$, for any $u \in \mathbb{R}^p$. Then

$$\partial_C f(0) = \left\{ \xi \in \mathbb{R}^p \mid \|u\| \geq \langle \xi, u \rangle, \ \forall u \in \mathbb{R}^p \right\} = D(0, 1).$$

The next proposition provides a convergence result, which is useful in the subsequent sections.

Proposition 5.1.6. *Let $f : \mathbb{R}^p \to \mathbb{R}$ be locally Lipschitz around $\overline{x}$, and take the sequences (x_n) and (ξ_n) such that*

$$x_n \to \overline{x} \text{ and } \xi_n \in \partial_C f(x_n) \text{ for every } n.$$

If ξ is a cluster point of (ξ_n) (in particular, if $\xi_n \to \xi$), then $\xi \in \partial_c f(\overline{x})$.

Proof Consider $u \in \mathbb{R}^p$. Then, for every n, one has $f^\circ(x_n, u) \geq \langle \xi_n, u \rangle$. Denote the subsequence of (ξ_n) which converges to ξ by (χ_n). It follows that $\langle \chi_n, u \rangle \to \langle \xi, u \rangle$ and, moreover, using the upper semicontinuity of f° proven in Proposition 5.1.1, we deduce that $f^\circ(x, u) \geq \langle \xi, u \rangle$. Since $u \in \mathbb{R}^p$ was arbitrarily chosen, we get the conclusion. $\square$

Usually, the generalized directional derivative will not be computed directly, but, rather by taking the upper limit of the difference quotients (keep in mind the usual case of the derivative, where the same applies). This is quite clear from the definition: one must compute the upper limit of a quotient which contains two variables changing simultaneously. Nevertheless, we emphasize two cases where the situation is simpler, and which also justify the name of generalized gradient: the cases of smooth and convex functions.

Theorem 5.1.7. *Let $f : \mathbb{R}^p \to \mathbb{R}$ be a function.*
    (i) If f is C^1, then $f^\circ(x, u) = \nabla f(x)(u)$ for every $x, u \in \mathbb{R}^p$, and $\partial_c f(x) = \{\nabla f(x)\}$.
    (ii) If f is convex, then $f^\circ(x, u) = f'(x, u)$ for every $x, u \in \mathbb{R}^p$, and $\partial_c f(x) = \partial f(x)$.

Proof (i) Remark, firstly, that because f is C^1, it is locally Lipschitz around any $x \in \mathbb{R}^p$ (by the mean value theorem). Consider $x, u \in \mathbb{R}^p$ and take $x_n \to x$ and $t_n \downarrow 0$ (that is, $(t_n) \subset (0, \infty)$ and $t_n \to 0$) two sequences for which lim sup in the definition of $f^\circ(x, u)$ is attained, that is

$$\lim_{n \to \infty} \frac{f(x_n + t_n u) - f(x_n)}{t_n} = f^\circ(x, u).$$

By Lagrange Theorem, one can find $y_n \in [x_n, x_n + t_n u]$ such that

$$f(x_n + t_n u) - f(x_n) = \nabla f(y_n)(t_n u) = t_n \nabla f(y_n)(u).$$

As f is C^1, $\nabla f(y_n) \to \nabla f(x)$ for $n \to \infty$, which finally gives

$$f^\circ(x, u) = \nabla f(x)(u), \quad \forall x, u \in \mathbb{R}^p.$$

Then
$$\partial_c f(x) = \{\xi \in \mathbb{R}^p \mid \langle \xi, u \rangle \leq \nabla f(x)(u), \ \forall u \in \mathbb{R}^p\},$$

which shows, due to the linearity of $\nabla f(x)(u)$ (by changing $u \to -u$), that in fact one has equality above, hence the conclusion.

    (ii) Consider now the case of a convex function f. We know in this case that f is locally Lipschitz (Theorem 4.2.5), and also its directional derivative $f'(x, u)$ always exists and is finite (Proposition 4.2.1). Because

$$f'(x, u) = \lim_{t \downarrow 0} \frac{f(x + tu) - f(x)}{t},$$

immediately follows that $f'(x, u) \leq f^\circ(x, u)$. On the other hand, since $t \to \frac{f(y + tu) - f(y)}{t}$ is always increasing (see the proof of Proposition 4.2.1), using the definition of $f^\circ(x, u)$, one has

$$f^\circ(x, u) = \inf_{r,\delta>0} \sup_{\|y-x\|<r} \sup_{t\in(0,\delta)} \frac{f(y + tu) - f(y)}{t}$$
$$= \inf_{r,\delta>0} \sup_{\|y-x\|<r} \frac{f(y + \delta u) - f(y)}{\delta}.$$

Due to the Lipschitz property of f (with modulus L), it follows that

$$f(y + \delta u) - f(y) = f(x + \delta u) - f(x) + [f(y + \delta u) - f(y) - f(x + \delta u) + f(x)]$$
$$\leq f(x + \delta u) - f(x) + 2L \|y - x\| \leq f(x + \delta u) - f(x) + 2Lr,$$

hence

$$f^\circ(x, u) \leq \inf_{r,\delta>0} \left(\frac{f(x + \delta u) - f(x)}{\delta} + 2Lr \right) = f'(x, u).$$

It means that $f'(x, u) = f^\circ(x, u)$. Then

$$\partial_c f(x) = \{\xi \in \mathbb{R}^p \mid f^\circ(x, u) \geq \langle \xi, u \rangle \ \forall u \in \mathbb{R}^p\}$$
$$= \{\xi \in \mathbb{R}^p \mid f'(x, u) \geq \langle \xi, u \rangle \ \forall u \in \mathbb{R}^p\}$$
$$= \partial f(x). \qquad \square$$

We now consider the general case, where the next theorem will provide a useful method to compute the generalized gradient of a locally Lipschitz function. We will use the celebrated theorem of Rademacher, which asserts that locally Lipschitz functions are differentiable almost everywhere (in the sense of the Lebesgue measure), i.e., the points where the function is not differentiable form a set whose Lebesgue measure is zero.

Theorem 5.1.8. *Let $f : \mathbb{R}^p \to \mathbb{R}$ be a locally Lipschitz function around a point $x \in \mathbb{R}^p$. Let $\Omega \subset \mathbb{R}^p$ be any set of zero measure in $\mathbb{R}^p$, and Ω_f be the set of points where f fails to be differentiable. Then*

$$\partial_c f(x) = \text{conv} \left\{ \lim_{n\to\infty} \nabla f(x_n) \mid x_n \to x, \ x_n \notin \Omega \cup \Omega_f \right\}. \tag{5.1.8}$$

Proof Remark that, by the Rademacher Theorem, the measure of $\Omega \cup \Omega_f$ is zero, hence there are sequences which satisfy the conditions from (5.1.8). Since f is locally Lipschitz around x (we denote its Lipschitz modulus by L), we deduce that its differential satisfies

$$\nabla f(y)(u) = \lim_{t\to 0} \frac{f(y + tu) - f(y)}{t} \leq L \|u\|$$

for every y near x where f is differentiable, hence

$$\|\nabla f(y)\| \le L,$$

for any such y. Consequently, if $x_n \to x$, $x_n \notin \Omega \cup \Omega_f$, one can extract a subsequence of (x_n), denoted also by (x_n) for simplicity, for which $(\nabla f(x_n))$ converges to a limit ξ. By the very definitions of the differential and of the generalized directional derivative, it follows that

$$\langle \nabla f(x_n), u \rangle = \nabla f(x_n)(u) = \lim_{t \to 0} \frac{f(x_n + tu) - f(x_n)}{t}$$

$$\le \limsup_{y \to x_n, t \downarrow 0} \frac{f(y + tu) - f(y)}{t} = f^\circ(x_n, u),$$

for any $n \in \mathbb{N}$ and $u \in \mathbb{R}^p$. Using the upper semicontinuity of $f^\circ(x_n, u)$ at (x, u), and passing to the limit for $n \to \infty$ above, it follows that

$$\langle \xi, u \rangle \le f^\circ(x, u)$$

for any $u \in \mathbb{R}^p$. Hence $\xi \in \partial_c f(x)$, and in fact

$$A := \left\{ \lim_{n \to \infty} \nabla f(x_n) \mid x_n \to x, \ x_n \notin \Omega \cup \Omega_f \right\} \subset \partial_c f(x), \tag{5.1.9}$$

since ξ is an arbitrary limit of the kind taken in A. Since $\partial_c f(x)$ is a convex set, it contains also the convex hull of A.

For the reverse inclusion, we will show that for any $u \ne 0$ in $\mathbb{R}^p$,

$$f^\circ(x, u) \le \limsup_{y \to x, \ y \notin \Omega \cup \Omega_f} \nabla f(y)(u).$$

For this, denote the lim sup in the right-hand side by ℓ, and take arbitrary $\varepsilon > 0$. Since

$$\limsup_{y \to x, \ y \notin \Omega \cup \Omega_f} \nabla f(y)(u) = \inf_{r > 0} \sup_{\substack{y \in B(x, r) \\ y \notin \Omega \cup \Omega_f}} \nabla f(y)(u) < \ell + \varepsilon,$$

it follows that there exists $r > 0$ such that

$$\nabla f(y)(u) < \ell + \varepsilon, \ \forall y \in B(x, r), y \notin \Omega \cup \Omega_f.$$

Without loss of generality, we may suppose that r is sufficiently small such that f is Lipschitz on $B(x, r)$. By the definition of the differential, we know that

$$\nabla f(y)(u) = \lim_{t \to 0} \frac{f(y + tu) - f(y)}{t} < \ell + \varepsilon, \ \forall y \in B(x, r), y \notin \Omega \cup \Omega_f.$$

Then there exists $\delta > 0$ such that, for any $t \in (0, \delta)$, and any $y \in B(x, r), y \notin \Omega \cup \Omega_f$, one has

$$f(y + tu) - f(y) < t(\ell + \varepsilon).$$

Since f is continuous, and the above inequality is true for any y in $B(x, r)$ with the exception of a set of zero measure, it follows that the same inequality, but nonstrict, is satisfied by the all points y in $B(x, r)$. Consequently,

$$f(y + tu) - f(y) \le t(\ell + \varepsilon), \; \forall t \in (0, \delta), y \in B(x, r),$$

hence

$$f^\circ(x, u) \le \ell + \varepsilon.$$

As ε was taken arbitrarily, it follows $f^\circ(x, u) \le \ell = \limsup_{y \to x, \, y \notin \Omega \cup \Omega_f} \nabla f(y)(u)$. Then for any realizing sequence (for which the lim sup is attained) $x_n \to x$, $x_n \notin \Omega \cup \Omega_f$, and for any $u \in \mathbb{R}^p \setminus \{0\}$, one has that

$$f^\circ(x, u) \le \left\langle \lim_{n \to \infty} \nabla f(x_n), u \right\rangle \le h_A(u) = h_{\mathrm{conv}\, A}(u),$$

where the last equality is easy to prove for any set A. Since $f^\circ(x, \cdot)$ is $h_{\partial_c f(x)}$, the conclusion follows from Proposition 5.1.2 (iii). $\qquad \square$

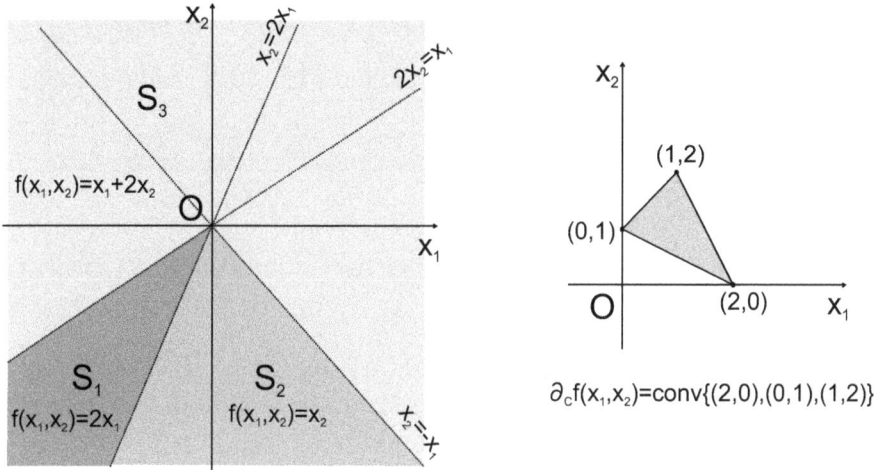

Figure 5.1: The domain of f and its Clarke subdifferential at $(0, 0)$.

Example 5.1.9. *Coming back to the problem (5.1.1) from the beginning of this chapter, observe that one can use the gradient formula in order to compute $\partial_c f(0, 0)$ for the function $f(x_1, x_2) = \max\{\min\{2x_1, x_2\}, x_1 + 2x_2\}$. The function f can equivalently be written as*

$$f(x_1, x_2) = \begin{cases} 2x_1, & \text{if } (x_1, x_2) \in S_1 := \{(x_1, x_2) \mid 2x_1 \le x_2 \text{ and } x_1 \ge 2x_2\} \\ x_2, & \text{if } (x_1, x_2) \in S_2 := \{(x_1, x_2) \mid 2x_1 \ge x_2 \text{ and } x_1 \le -x_2\} \\ x_1 + 2x_2, & \text{if } (x_1, x_2) \in S_3 := \{(x_1, x_2) \mid x_1 \le 2x_2 \text{ or } x_1 \ge -x_2\}. \end{cases}$$

$$(5.1.10)$$

Remark, moreover, that $S_1 \cup S_2 \cup S_3 = \mathbb{R}^2$, and the union of their boundaries, A, is a set of zero measure. Moreover, f is differentiable on $\mathbb{R}^2 \setminus A$, the gradient having three possible values: $(2, 0)$, $(0, 1)$ and $(1, 2)$. According to the gradient formula, one gets that $\partial_C f(0, 0)$ is the triangle obtained by the convex hull of these three points (see Figure 5.1).

Remark 5.1.10. *Observe that $\partial_C f(x)$ can fail to be a singleton even f is differentiable. An example is provided by the function $f : \mathbb{R} \to \mathbb{R}$,*

$$f(x) = \begin{cases} x^2 \sin \dfrac{1}{x}, & \text{if } x \neq 0 \\ 0, & \text{if } x = 0. \end{cases} \tag{5.1.11}$$

One can see that f is differentiable at 0 and $f'(0) = 0$. Moreover,

$$f'(x) = \begin{cases} 2x \sin \dfrac{1}{x} - \cos \dfrac{1}{x}, & \text{if } x \neq 0 \\ 0, & \text{if } x = 0, \end{cases}$$

which is not continuous. Observe also that, if $x_n \to 0$, then $2x_n \sin \dfrac{1}{x_n} \to 0$, hence for every such sequence, if it exists, $\lim\limits_{n\to\infty} f'(x_n) \in [-1, 1]$. Furthermore, for $y_n := \dfrac{1}{2n\pi} \to 0$ and $z_n := \dfrac{1}{(2n+1)\pi} \to 0$, one gets respectively $\lim\limits_{n\to\infty} f'(y_n) = -1$, and $\lim\limits_{n\to\infty} f'(z_n) = 1$. In conclusion, by the gradient formula, we obtain that $\partial_C f(0) = [-1, 1]$. This example shows that the generalized gradient extends the notion of continuous differentiability.

The next property will be useful in the development of the generalized gradient calculus.

Definition 5.1.11. *A function $f : \mathbb{R}^p \to \mathbb{R}$ is said to be regular at x if it is locally Lipschitz around x and admits directional derivatives $f'(x, u)$ (formally defined as in the convex case) satisfying $f'(x, u) = f^\circ(x, u)$ for any $u \in \mathbb{R}^p$.*

Theorem 5.1.7 shows that convex and C^1 functions are regular at any point. One can see that if f, g are regular functions at x and $\lambda \geq 0$, then λf and $f+g$ are regular at x. The first assertion easily follows, and the second one will be also proven (see the proof of Theorem 5.1.13). A first example of a function which is not regular is given by (5.1.11). This is because $f'(0, 1) = \nabla f(0)(1) = 0$, but $f^\circ(0, 1) = \max \{\xi \mid \xi \in \partial_C f(0) = [-1, 1]\} = 1$. Let us now provide another example of a function which is not regular.

Example 5.1.12. *Consider $f : \mathbb{R} \to \mathbb{R}$ given by $f(x) := -|x|$ for any $x \in \mathbb{R}$. This function is not regular at 0. Observe first that it is Lipschitz and $f'(0, u) = -|u|$ for any $u \in \mathbb{R}$. Moreover, since f is differentiable on $\mathbb{R} \setminus \{0\}$, with $\nabla f(x)$ being either -1 or 1 at any point different from 0, one gets that $\partial_C f(0) = [-1, 1]$ in view of the gradient formula. Formula (5.1.7) shows that $f^\circ(0, 1) = 1 \neq -1 = f'(0, 1)$.*

In fact, the example above is a particular instance of the following general case: every concave function that has a corner at x fails to be regular at x. This is because, for such a situation,

$$f'(x, u) \neq -f'(x, -u) = (-f)'(x, -u) = (-f)^\circ(x, -u) = f^\circ(x, u).$$

An important calculus rule is provided next.

Theorem 5.1.13 (sum rule). *Let $f, g : \mathbb{R}^p \to \mathbb{R}$ be two functions which are Lipschitz around x. Then*

$$\partial_C(f + g)(x) \subset \partial_C f(x) + \partial_C g(x). \tag{5.1.12}$$

Equality holds if f and g are regular at x.

Proof Observe that the definition of the upper limit allows us to write

$$
\begin{aligned}
(f + g)^\circ(x, u) &= \limsup_{y \to x, t \downarrow 0} \frac{(f + g)(y + tu) - (f + g)(y)}{t} \\
&\leq \limsup_{y \to x, t \downarrow 0} \frac{f(y + tu) - f(y)}{t} + \limsup_{y \to x, t \downarrow 0} \frac{g(y + tu) - g(y)}{t} \\
&= f^\circ(x, u) + g^\circ(x, u), \quad \forall u \in \mathbb{R}^p.
\end{aligned}
\tag{5.1.13}
$$

Now, take $\chi \in \partial_C(f + g)(x)$. This means that $\langle \chi, u \rangle \leq (f + g)^\circ(x, u) \leq f^\circ(x, u) + g^\circ(x, u)$, for any u. Using (5.1.7), there exists $\xi \in \partial_C f(x)$ such that $\langle \xi, u \rangle = f^\circ(x, u)$. Therefore, $\langle \chi - \xi, u \rangle \leq g^\circ(x, u)$ for any u, which proves that $\chi - \xi \in \partial_C g(x)$, i.e., $\chi \in \partial_C f(x) + \partial_C g(x)$.

Suppose now that f and g are regular. Then, using the regularity and the definitions of the Clarke generalized derivative and of the directional derivative, one gets

$$f^\circ(x, u) + g^\circ(x, u) = f'(x, u) + g'(x, u) = (f + g)'(x, u) \leq (f + g)^\circ(x, u), \quad \forall u \in \mathbb{R}^p.$$

Combined with (5.1.13), this proves that $f + g$ is regular at x. Now, for any $\xi \in \partial_C f(x)$, $\zeta \in \partial_C g(x)$, we have from definitions that $\langle \xi, u \rangle \leq f^\circ(x, u)$, $\langle \zeta, u \rangle \leq g^\circ(x, u)$ for any u, hence $\langle \xi + \zeta, u \rangle \leq f^\circ(x, u) + g^\circ(x, u) \leq (f + g)^\circ(x, u)$. It follows that $\xi + \zeta \in \partial_C(f + g)(x)$, which ends the proof. $\qquad\square$

Remark 5.1.14. *Since for any scalar $\lambda \in \mathbb{R}$, we have $\partial_C(\lambda f)(x) = \lambda \partial_C f(x)$, it follows that for any two locally Lipschitz functions around x, and for any scalars $\alpha, \beta \in \mathbb{R}$, one has*

$$\partial_C(\alpha f + \beta g)(x) \subset \alpha \partial_C f(x) + \beta \partial_C g(x). \tag{5.1.14}$$

Moreover, since the regularity is preserved by summation and multiplication with positive scalars, we deduce that for any two regular functions f, g, and any positive scalars $\alpha, \beta \geq 0$,

$$\partial_C(\alpha f + \beta g)(x) = \alpha \partial_C f(x) + \beta \partial_C g(x).$$

Remark 5.1.15. *An easy example of strict containment in (5.1.12) is given by the functions* $f, g : \mathbb{R} \to \mathbb{R}$, $f(x) := -|x|$, $g(x) := |x|$ *for any* x. *We saw before that* f *is not regular at 0 and* $\partial_c f(0) = [-1, 1]$. *One can easily prove (see Example 5.1.5) that* $\partial_c g(0) = [-1, 1]$. *Then*

$$\{0\} = \partial_c(f + g)(0) \subsetneq \partial_c f(0) + \partial_c g(0) = [-2, 2].$$

The next important result is a generalized Fermat rule.

Theorem 5.1.16 (Fermat rule for Clarke calculus). *Let* $f : \mathbb{R}^p \to \mathbb{R}$ *be a locally Lipschitz function. If* f *has a local minimum or maximum at* x, *then*

$$0 \in \partial_c f(x).$$

Proof One has, if x is a local minimum, that

$$f^\circ(x, u) = \limsup_{y \to x, \, t \downarrow 0} \frac{f(y + tu) - f(y)}{t} \geq \limsup_{t \downarrow 0} \frac{f(x + tu) - f(x)}{t} \geq 0.$$

On the other hand, if x is a local maximum,

$$f^\circ(x, u) = \limsup_{y \to x, \, t \downarrow 0} \frac{f(y + tu) - f(y)}{t} \geq \limsup_{t \downarrow 0} \frac{f(x) - f(x - tu)}{t} \geq 0.$$

Hence, in both cases, $f^\circ(x, u) \geq 0$ for any u, which gives the conclusion. $\quad\square$

The next result is a mean value theorem for locally Lipschitz functions due to Gérard Lebourg, see (Lebourg, 1975).

Theorem 5.1.17 (Lebourg). *Let* $x, y \in \mathbb{R}^p$, *and suppose that* $f : \mathbb{R}^p \to \mathbb{R}$ *is locally Lipschitz on a neighborhood of the line segment joining the points* x *and* y. *Then there exists* $z \in (x, y)$ *such that*

$$f(y) - f(x) \in \langle \partial_c f(z), y - x \rangle.$$

Proof Denote by $x_t := x + t(y - x)$ and consider the function $g : [0, 1] \to \mathbb{R}$ given by

$$g(t) := f(x_t).$$

Observe that for any $t_1, t_2 \in [0, 1]$,

$$|g(t_1) - g(t_2)| = |f(x + t_1(y - x)) - f(x + t_2(y - x))|$$
$$\leq \ell \, \|y - x\| \cdot |t_1 - t_2|,$$

where ℓ is the Lipschitz constant of f, hence g is Lipschitz on $[0, 1]$.

Now, for every $v \in \mathbb{R}$,

$$
\begin{aligned}
g^{\circ}(t, v) &= \limsup_{s \to t, \lambda \downarrow 0} \frac{g(s + \lambda v) - g(s)}{\lambda} \\
&= \limsup_{s \to t, \lambda \downarrow 0} \frac{f(x + (s + \lambda v)(y - x)) - f(x + s(y - x))}{\lambda} \\
&\leq \limsup_{z \to x_t, \lambda \downarrow 0} \frac{f(z + \lambda v(y - x)) - f(z)}{\lambda} = f^{\circ}(x_t, v(y - x)).
\end{aligned}
$$

But this means that for any $\xi \in \partial_C g(x)$ and any $v \in \mathbb{R}$,

$$
\xi v \leq g^{\circ}(t, v) \leq f^{\circ}(x_t, v(y - x)) = \max v \cdot \langle \partial_C f(x_t), y - x \rangle .
$$

Observe that $\langle \partial_C f(x_t), y - x \rangle$ is a compact interval of reals. Take above $v = \pm 1$ to deduce that there is $\chi \in \partial_C f(x_t)$ such that $\xi = \langle \chi, y - x \rangle$. This gives $\partial_C g(x) \subset \langle \partial_C f(x_t), y - x \rangle$.

Consider now the function $h : [0, 1] \to \mathbb{R}$ given as

$$
h(t) := g(t) - t(f(y) - f(x)).
$$

One easily gets that h is locally Lipschitz and, by Theorem 5.1.13 and Remark 5.1.14, that

$$
\partial_C h(t) \subset \partial_C g(t) - \left(f(y) - f(x) \right) \subset \langle \partial_C f(x_t), y - x \rangle - f(y) + f(x). \tag{5.1.15}
$$

Since $h(0) = h(1) = f(x)$, one gets by the continuity of h that there is $t_0 \in (0, 1)$ such that t_0 is a local minimum or a local maximum for h. In view of the Fermat rule (Theorem 5.1.16), one has

$$
0 \in \partial_C h(t_0),
$$

which gives, by the use of (5.1.15), the conclusion of the theorem with $z := x_{t_0}$. $\qquad \square$

For a function $f : \mathbb{R}^p \to \mathbb{R}^k$ and a vector $\xi \in \mathbb{R}^k$, define the function $\langle \xi, f \rangle : \mathbb{R}^p \to \mathbb{R}$ as

$$
\langle \xi, f \rangle (x) := \langle \xi, f(x) \rangle . \tag{5.1.16}
$$

Theorem 5.1.18 (chain rule 1). *Let $f := (f_1, \ldots, f_k) : \mathbb{R}^p \to \mathbb{R}^k$ and $g : \mathbb{R}^k \to \mathbb{R}$ be locally Lipschitz around x and $f(x)$, respectively. Then the function $h := g \circ f : \mathbb{R}^p \to \mathbb{R}$ is locally Lipschitz around x, and*

$$
\partial_C h(x) \subset \mathrm{cl\,conv} \left\{ \partial_C \langle y, f \rangle (x) \mid y \in \partial_C g(f(x)) \right\} . \tag{5.1.17}
$$

Moreover, if $f_1, \ldots, f_k$ are regular at x, g is also regular at $f(x)$, and $\partial_C g(f(x)) \subset \mathbb{R}_+^k$, then h is regular at x and equality holds in (5.1.17).

Proof The fact that h is locally Lipschitz around x is straightforward.

Pick $\xi \in \partial_C h(x)$, and take arbitrary $u \in \mathbb{R}^p$. Take some realizing sequences $y_n \to x$, $t_n \downarrow 0$ for $h^\circ(x, u)$, that is

$$h^\circ(x, u) = \lim_{n \to \infty} \frac{g(f(y_n + t_n u)) - g(f(y_n))}{t_n}.$$

By Theorem 5.1.17, for every n sufficiently large, there is $z_n \in \left[f(y_n), f(y_n + t_n u) \right]$ such that

$$g(f(y_n + t_n u)) - g(f(y_n)) \in \langle \partial_C g(z_n), f(y_n + t_n u) - f(y_n) \rangle.$$

It follows that there is $\chi_n \in \partial_C g(z_n)$ such that

$$\frac{h(y_n + t_n u) - h(y_n)}{t_n} = \frac{\langle \chi_n, f(y_n + t_n u) - f(y_n) \rangle}{t_n} = \left\langle \chi_n, \frac{f(y_n + t_n u) - f(y_n)}{t_n} \right\rangle.$$

Denote by $B(f(x), 2\varepsilon)$ a neighborhood of $f(x)$ where g is Lipschitz and by L_g its Lipschitz modulus. As $z_n \to f(x)$ for $n \to \infty$, it follows that for every n sufficiently large, $z_n \in B(f(x), \varepsilon)$, and g is Lipschitz with modulus L_g on $B(z_n, \varepsilon)$. Then, by relation (5.1.3) applied for z_n and the definition of the generalized gradient, one has for any n sufficiently large and any $v \in \mathbb{R}^k$ that

$$\langle \chi_n, v \rangle \leq g^\circ(z_n, v) \leq L_g \|v\|.$$

It follows that $\|\chi_n\| \leq L_g$ for any n sufficiently large, hence (χ_n) is bounded. Without loss of generality, we may suppose that $\chi_n \to \chi$. As $z_n \to f(x)$ and $\chi_n \in \partial_C g(z_n)$ for any n sufficiently large, we obtain that $\chi \in \partial_C g(f(x))$.

Also by Theorem 5.1.17, applied for $\langle \chi, f \rangle$, we get that there is $p_n \in [y_n, y_n + t_n u]$ such that

$$\langle \chi, f \rangle (y_n + t_n u) - \langle \chi, f \rangle (y_n) \in \langle \partial_C \langle \chi, f \rangle (p_n), t_n u \rangle,$$

hence there is $q_n \in \partial_C \langle \chi, f \rangle (p_n)$ such that

$$\left\langle \chi, \frac{f(y_n + t_n u) - f(y_n)}{t_n} \right\rangle = \frac{\langle \chi, f \rangle (y_n + t_n u) - \langle \chi, f \rangle (y_n)}{t_n} = \langle q_n, u \rangle.$$

Since $y_n \to x$ and $t_n \downarrow 0$, it follows that $p_n \to x$. Reasoning as above for the locally Lipschitz function $\langle \chi, f \rangle$, it follows that (q_n) is bounded, so we may suppose again, without loss of generality, that $q_n \to q$. Therefore, $q \in \partial_C \langle \chi, f \rangle (x)$.

Combining the previous relations, we get

$$\begin{aligned}
\frac{h(y_n + t_n u) - h(y_n)}{t_n} &= \left\langle \chi_n, \frac{f(y_n + t_n u) - f(y_n)}{t_n} \right\rangle \\
&= \left\langle \chi_n - \chi, \frac{f(y_n + t_n u) - f(y_n)}{t_n} \right\rangle + \left\langle \chi, \frac{f(y_n + t_n u) - f(y_n)}{t_n} \right\rangle \\
&= \left\langle \chi_n - \chi, \frac{f(y_n + t_n u) - f(y_n)}{t_n} \right\rangle + \langle q_n, u \rangle.
\end{aligned}$$

Observe that the Lipschitz property of f implies that $(f(y_n + t_n u) - f(y_n))/t_n$ is bounded, hence passing to the limit we get

$$h^\circ(x, u) = \langle q, u \rangle,$$

where $q \in \partial_C \langle \chi, f \rangle (x)$ and $\chi \in \partial_{C8}(f(x))$.

This means that for any $u \in \mathbb{R}^p$, $h^\circ(x, u) \leq h_A(u)$, where by A we have denoted the set cl conv $\{\partial_C \langle y, f \rangle (x) \mid y \in \partial_{C8}(f(x))\}$, and by h_A its support function. The conclusion follows by taking into account the equivalence (5.1.5) from Proposition 5.1.2.

Now, suppose that $f_1, \ldots, f_k$ are regular at x, g is regular at $f(x)$, and $\partial_{C8}(f(x)) \subset \mathbb{R}^k_+$. Take arbitrary $u \in \mathbb{R}^p$ and observe that

$$h'(x, u) = \lim_{t \downarrow 0} \frac{g(f(x + tu)) - g(f(x))}{t}$$

$$= -\lim_{t \downarrow 0} \frac{g(f(x) + tf'(x, u)) - g(f(x + tu))}{t} + g'(f(x), f'(x, u)),$$

where by $f'(x, u)$ we have denoted the vector whose components are $f_i'(x, u)$, $i = \overline{1, k}$.

Since

$$\left| \frac{g(f(x) + tf'(x, u)) - g(f(x + tu))}{t} \right| \leq \frac{1}{t} L_g \left\| tf'(x, u) - (f(x + tu) - f(x)) \right\|$$

$$= L_g \left\| f'(x, u) - \frac{f(x + tu) - f(x)}{t} \right\| \to 0$$

when $t \downarrow 0$, it means that

$$h'(x, u) = g'(f(x), f'(x, u)). \tag{5.1.18}$$

Now, take $y \in \partial_{C8}(f(x)) \subset \mathbb{R}^k_+$ and remark that the function $\langle y, f \rangle$ is regular at x, since

$$\langle y, f \rangle (x) = y_1 f_1(x) + \ldots + y_k f_k(x),$$

i.e., $\langle y, f \rangle$ is a positive linear combination of regular functions. Then

$$\max \{ \langle y, f \rangle^\circ (x, u) \mid y \in \partial_{C8}(f(x)) \} = \max \{ \langle y, f \rangle' (x, u) \mid y \in \partial_{C8}(f(x)) \}$$

$$= \max \{ \langle y, f'(x, u) \rangle \mid y \in \partial_{C8}(f(x)) \}$$

$$= g^\circ(f(x), f'(x, u))$$

$$= g'(f(x), f'(x, u)) \text{ (since } g \text{ is regular)}$$

$$= h'(x, u) \text{ (from (5.1.18))}$$

$$\leq h^\circ(x, u).$$

It means also from (5.1.5) that for any $y \in \partial_{C8}(f(x))$,

$$\partial_C \langle y, f \rangle (x) \subset \partial_C h(x).$$

But this shows that

$$\{\partial_C \langle y, f \rangle (x) \mid y \in \partial_C g(f(x))\} \subset \partial_C h(x),$$

and, as the set from the right-hand side is convex and closed, one has equality in (5.1.17). □

Theorem 5.1.19 (chain rule 2). *Let* $f : \mathbb{R}^p \to \mathbb{R}^k$ *be locally Lipschitz around* x *and* $g : \mathbb{R}^k \to \mathbb{R}$ *be continuously differentiable at* $f(x)$. *Then the function* $h := g \circ f : \mathbb{R}^p \to \mathbb{R}$ *is locally Lipschitz around* x, *and*

$$\partial_C h(x) = \partial_C \langle \nabla g(f(x)), f \rangle (x). \tag{5.1.19}$$

Proof Take y and $t > 0$ near x and 0, respectively, and observe that the Lagrange Theorem applied for g asserts that there is $z \in [f(y), f(y + tu)]$ such that

$$g(f(y + tu)) = g(f(y)) + \langle \nabla g(z), f(y + tu) - f(y) \rangle.$$

This gives us, for arbitrary $u \in \mathbb{R}^p$, that

$$h^\circ(x, u) = \limsup_{y \to x,\, t \downarrow 0} \frac{g(f(y + tu)) - g(f(y))}{t}$$

$$= \limsup_{y \to x,\, t \downarrow 0} \left\langle \nabla g(z), \frac{f(y + tu) - f(y)}{t} \right\rangle.$$

Observe that the Lipschitz property of f implies that $(f(y + tu) - f(y))/t$ is bounded. Moreover, $y \to x$, $t \downarrow 0$, combined with the same property of f, gives us $z \to f(x)$. Using also the C^1 property of g, one gets from above that

$$h^\circ(x, u) = \limsup_{y \to x,\, t \downarrow 0} \left\langle \nabla g(f(x)), \frac{f(y + tu) - f(y)}{t} \right\rangle$$

$$= \limsup_{y \to x,\, t \downarrow 0} \frac{\langle \nabla g(f(x)), f(y + tu) \rangle - \langle \nabla g(f(x)), f(y) \rangle}{t}$$

$$= \langle \nabla g(f(x)), f \rangle^\circ (x, u).$$

The relation (5.1.19) follows now from Proposition 5.1.2 (iii). □

Remark 5.1.20. *The inclusion "*$\subset$*" from (5.1.19) immediately follows from Theorem 5.1.18.*

Recall that for a differentiable function, its differential $\nabla f(x)$ can be identified with the Jacobian matrix.

Theorem 5.1.21 (chain rule 3). *Let* $f : \mathbb{R}^p \to \mathbb{R}^k$ *be continuously differentiable near* x *and* $g : \mathbb{R}^k \to \mathbb{R}$ *be locally Lipschitz around* $f(x)$. *Then the function* $h := g \circ f : \mathbb{R}^p \to \mathbb{R}$ *is locally Lipschitz around* x, *and*

$$\partial_C h(x) \subset \langle \partial_C g(f(x)), \nabla f(x) \rangle := \{\langle \xi, \nabla f(x) \rangle \mid \xi \in \partial_C g(f(x))\}. \tag{5.1.20}$$

Moreover, if $\nabla f(x) : \mathbb{R}^p \to \mathbb{R}^k$ is surjective, then equality holds.

Proof The proof of (5.1.20) follows taking into account in Theorem 5.1.18 that for any $y \in \partial_C g(f(x))$,

$$\partial_C \langle y, f \rangle (x) = \langle y, \nabla f(x) \rangle ,$$

hence, since $\partial_C g(f(x))$ is compact and convex and $\nabla f(x)$ is linear and bounded,

$$\mathrm{cl\,conv} \left\{ \partial_C \langle y, f \rangle (x) \mid y \in \partial_C g(f(x)) \right\} = \langle \partial_C g(f(x)), \nabla f(x) \rangle .$$

Suppose now that $\nabla f(x)$ is surjective. By the Lyusternik-Graves Theorem (i.e., Theorem 2.4.4), it follows that f is open, i.e., for any neighborhood U of x, $f(U)$ is a neighborhood of $f(x)$. This implies

$$\limsup_{y \to f(x),\, t\downarrow 0} \frac{g(y + t\nabla f(x)(u)) - g(y)}{t} = \limsup_{z \to x,\, t\downarrow 0} \frac{g(f(z) + t\nabla f(x)(u)) - g(f(z))}{t} .$$

Also, since f is C^1 near x,

$$\lim_{z \to x,\, t\downarrow 0} \frac{f(z + tu) - f(z) - t\nabla f(x)(u)}{t} = \lim_{z \to x,\, t\downarrow 0} \frac{f(z + tu) - f(z) - t\nabla f(z)(u)}{t}$$
$$+ \lim_{z \to x} \left(\nabla f(z) - \nabla f(x) \right) (u) = 0.$$

Therefore, it follows that

$$g^\circ(f(x), \nabla f(x)(u)) = \limsup_{y \to f(x),\, t\downarrow 0} \frac{g(y + t\nabla f(x)(u)) - g(y)}{t}$$
$$= \limsup_{z \to x,\, t\downarrow 0} \frac{g(f(z) + t\nabla f(x)(u)) - g(f(z))}{t}$$
$$= \limsup_{z \to x,\, t\downarrow 0} \frac{g(f(z + tu)) - g(f(z))}{t} = h^\circ(x, u), \ \forall u \in \mathbb{R}^p,$$

where for the third equality the local Lipschitz of g around $f(x)$ is used. Take now arbitrary $\chi \in \partial_C g(f(x))$ and $u \in \mathbb{R}^p$. Then

$$\langle \chi, \nabla f(x) \rangle (u) = \langle \chi, \nabla f(x)(u) \rangle \le g^\circ(f(x), \nabla f(x)(u)) = h^\circ(x, u).$$

It means that $\langle \chi, \nabla f(x) \rangle \in \partial_C h(x)$. Therefore, the conclusion follows. $\quad\square$

Finally, we present a result which will be useful in deriving optimality conditions.

Proposition 5.1.22. *Consider $\varphi : \mathbb{R}^p \to \mathbb{R}^k$ locally Lipschitz around x, $K \subset \mathbb{R}^k$ a compact set, and define*

$$F(x) := \max_{y \in K} \langle y, \varphi(x) \rangle .$$

Then F is locally Lipschitz around x. For any u around x, denote by

$$K(u) := \left\{ y \in K \mid F(u) = \langle y, \varphi(u) \rangle \right\}$$

the set of active indices at u. If $K(x)$ is a singleton set (i.e., $K(x) = \{y_0\}$), then

$$\partial_C F(x) \subset \partial_C \langle y_0, \varphi \rangle (x).$$

Proof Denote by L the Lipschitz constant of φ. The fact that F is locally Lipschitz around x is straightforward.

Pick $\xi \in \partial_C F(x)$ and take arbitrary $u \in \mathbb{R}^p$. Take some realizing sequences $y_n \to x$, $t_n \downarrow 0$ for $F^\circ(x, u)$. Moreover, take $y_n \in K(y_n + t_n u)$. Since K is compact, we may suppose, without loss of generality, that y_n converges to some $\bar{y}$. As $F(y_n + t_n u) = \langle y_n, \varphi(y_n + t_n u) \rangle$ for every n and $y_n + t_n u \to x$, we obtain by passing to the limit that $F(x) = \langle \bar{y}, \varphi(x) \rangle$, i.e., $\bar{y} \in K(x)$, which means that $\bar{y} = y_0$. It follows that

$$
\begin{aligned}
F^\circ(x, u) &= \lim_{n \to \infty} \frac{F(y_n + t_n u) - F(y_n)}{t_n} \leq \limsup_{n \to \infty} \frac{\langle y_n, \varphi(y_n + t_n u) \rangle - \langle y_n, \varphi(y_n) \rangle}{t_n} \\
&= \limsup_{n \to \infty} \frac{\langle y_0, \varphi \rangle (y_n + t_n u) - \langle y_0, \varphi \rangle (y_n)}{t_n} \\
&\quad + \limsup_{n \to \infty} \frac{\langle y_n - y_0, \varphi(y_n + t_n u) - \varphi(y_n) \rangle}{t_n} \\
&\leq \langle y_0, \varphi \rangle^\circ (x, u) + \limsup_{n \to \infty} \| y_n - y_0 \| \frac{\| \varphi(y_n + t_n u) - \varphi(y_n) \|}{t_n} \\
&\leq \langle y_0, \varphi \rangle^\circ (x, u) + \limsup_{n \to \infty} \| y_n - y_0 \| L \| u \| = \langle y_0, \varphi \rangle^\circ (x, u).
\end{aligned}
$$

Hence, we obtained $F^\circ(x, u) \leq \langle y_0, \varphi \rangle^\circ (x, u)$, which shows the inclusion $\partial_C F(x) \subset \partial_C \langle y_0, \varphi \rangle (x)$ by Proposition 5.1.2. $\qquad\square$

5.1.2 Clarke Tangent and Normal Cones

Throughout this section, we will discuss the notions of tangent and normal cones, as defined by Clarke. In the sequel, A denotes a nonempty closed subset of $\mathbb{R}^p$. Recall that the distance function associated to the set A is given by

$$d_A(x) = \inf_{y \in A} \| y - x \|,$$

and has the important property (in our context) that is 1–Lipschitz. The announced tangency notion will be constructed through the generalized gradient of this function, and the normal cone will be defined by polarity. One of the main reasons for which these notions were introduced by Clarke in the early 1980's is given by the next penalization principle. Notice that a local version of the first item of the next result was proved and used in the previous chapter (see Theorem 4.3.2).

Theorem 5.1.23 (penalization principle). *Let $U \subset \mathbb{R}^p$ be an open set which contains A, and $f : U \to \mathbb{R}$ be a Lipschitz function with modulus $L > 0$. Consider the minimization* *problem*

$$(P) \min f(x), \quad \text{subject to } x \in A.$$

(i) If $\bar{x} \in A$ is a global solution of (P), then for every $K \geq L$, the function $x \mapsto f(x) + Kd_A(x)$ attains an (unconstrained) minimum on U at $\bar{x}$.

(ii) Conversely, suppose that, for some $K > L$, the function $x \mapsto f(x) + Kd_A(x)$ attains its minimum on U at $\bar{x}$. Then $\bar{x} \in A$ and solves (P).

Proof Suppose first that $\bar{x} \in A$ is a solution of (P). Take $x \in U$ and $\varepsilon > 0$ arbitrarily. Pick $y_\varepsilon \in A$ such that $\|y_\varepsilon - x\| \leq d_A(x) + \varepsilon$. Using the Lipschitz property of f, one gets

$$f(\bar{x}) \leq f(y_\varepsilon) \leq f(x) + L\|y_\varepsilon - x\| \leq f(x) + Ld_A(x) + L\varepsilon.$$

Letting ε to tend at 0, one gets the conclusion of the first part for $K = L$. If $K > L$, the conclusion is now obvious.

For the converse, suppose by contradiction that $\bar{x} \notin A$, hence $x \in U \setminus A$. Since A is closed, it follows that $d_A(\bar{x}) > 0$. Then one can find $y \in A$ such that

$$\|y - \bar{x}\| < d_A(\bar{x}) + \left(\frac{K}{L} - 1\right) d_A(\bar{x}) = \frac{K}{L} d_A(\bar{x}).$$

Using the minimality assumption and the Lipschitz property of f, one gets

$$f(\bar{x}) + Kd_A(\bar{x}) \leq f(y) \leq f(\bar{x}) + L\|y - \bar{x}\| < f(\bar{x}) + Kd_A(\bar{x}),$$

a contradiction which shows that $\bar{x} \in A$. Then for every $x \in A$,

$$f(\bar{x}) = f(\bar{x}) + Kd_A(\bar{x}) \leq f(x) + Kd_A(x) = f(x),$$

hence $\bar{x}$ solves (P). $\qquad\square$

In view of the generalized Fermat rule (Theorem 5.1.16) and of sum rule (5.1.14), solutions of (P) satisfy

$$0 \in \partial_C(f + Kd_A)(x) \subset \partial_C f(x) + K\partial_C d_A(x). \tag{5.1.21}$$

It is interesting to study the geometric interpretation of the term $\partial_C d_A(x)$.

Definition 5.1.24. *Let $\bar{x} \in A$. The Clarke tangent and normal cones to A at $\bar{x}$ are respectively the sets*

$$T_C(A, \bar{x}) := \left\{u \in \mathbb{R}^p \mid d_A^\circ(\bar{x}, u) = 0\right\} = \left\{u \in \mathbb{R}^p \mid d_A^\circ(\bar{x}, u) \leq 0\right\} \tag{5.1.22}$$

$$N_C(A, \bar{x}) := T_C(A, \bar{x})^- = \left\{\xi \in \mathbb{R}^p \mid \langle \xi, x \rangle \leq 0, \, \forall x \in T_C(A, \bar{x})\right\}. \tag{5.1.23}$$

Remark that the second equality from (5.1.22) follows from the fact that one always has $d_A^\circ(\bar{x}, u) \geq 0$, since d_A attains a minimum at $\bar{x}$.

One may easily observe that $T_C(A, \bar{x}) = \mathbb{R}^p$ and $N_C(A, \bar{x}) = \{0\}$ if $\bar{x} \in \text{int} A$. The following theorem collects some other important properties of the Clarke tangent and normal cones.

Theorem 5.1.25. *Let $\bar{x} \in A$. Then:*

(i) $T_C(A, \bar{x})$ is a closed convex cone;

(ii) $u \in T_C(A, \bar{x})$ if and only if, for every $(x_n) \subset A$, $x_n \to \bar{x}$ and every $(t_n) \subset (0, \infty)$, $(t_n) \to 0$, there exists $(u_n) \to u$ such that $x_n + t_n u_n \in A$ for any n; as consequence, $T_C(A, \bar{x}) \subset T_B(A, \bar{x})$;

(iii) $N_C(A, \bar{x}) = \mathrm{cl}\,\mathrm{cone}\,\partial_C d_A(\bar{x})$;

(iv) $T_C(A, \bar{x}) = N_C(A, \bar{x})^-$;

(v) $N_C(A, \bar{x})$ is a closed convex cone containing $N_B(A, \bar{x})$.

Proof (i) We know by Proposition 5.1.1 that the function $u \mapsto d_A^\circ(\bar{x}, u)$ is a positively homogeneous convex function, which shows, in light of (5.1.22), that $T_C(A, \bar{x})$ is a convex cone. The continuity of the same function proves the closedness of $T_C(A, \bar{x})$.

(ii) Take $u \in T_C(A, \bar{x})$ and take arbitrary sequences $(x_n) \subset A$, $(x_n) \to \bar{x}$ and $(t_n) \subset (0, \infty)$, $(t_n) \to 0$. Since $d_A^\circ(\bar{x}, u) = 0$, one gets that

$$\inf_{r>0} \sup_{x \in B(\bar{x}, r),\, t \in (0, r)} \frac{d_A(y + tu) - d_A(y)}{t} = 0.$$

Take $\varepsilon > 0$. One can find $r > 0$ such that $\sup_{x \in B(\bar{x}, r),\, t \in (0, r)} \frac{d_A(y + tu) - d_A(y)}{t} < \varepsilon$. For this r, there is $n_0 \in \mathbb{N}$ such that, for any $n \geq n_0$, $x_n \in B(\bar{x}, r)$ and $t_n \in (0, r)$. This means that

$$\varepsilon > \sup_{x \in B(\bar{x}, r),\, t \in (0, r)} \frac{d_A(y + tu) - d_A(y)}{t} \geq \frac{d_A(x_n + t_n u) - d_A(x_n)}{t_n}$$
$$= \frac{d_A(x_n + t_n u)}{t_n} \geq 0,$$

for any $n \geq n_0$, which shows that

$$\lim_{n \to \infty} \frac{d_A(x_n + t_n u)}{t_n} = 0.$$

Take $y_n \in A$ such that $\|x_n + t_n u - y_n\| \leq d_A(x_n + t_n u) + \frac{t_n}{n}$, and define $u_n := \frac{y_n - x_n}{t_n}$. Using the relation above, we obtain that $u_n \to u$ and $x_n + t_n u_n = y_n \in A$, as claimed.

For the converse inclusion, take $u \in \mathbb{R}^p$ such that for every $(x_n) \subset A$, $(x_n) \to \bar{x}$ and every $(t_n) \subset (0, \infty)$, $(t_n) \to 0$, there exists $(u_n) \to u$ such that $x_n + t_n u_n \in A$ for any n. Fix some sequences $(y_n) \to \bar{x}$ and $(t_n) \downarrow 0$ such that

$$\lim_{n \to \infty} \frac{d_A(y_n + t_n u) - d_A(y_n)}{t_n} = d_A^\circ(\bar{x}, u).$$

Let $x_n \in A$ such that $\|x_n - y_n\| \leq d_A(y_n) + \frac{t_n}{n}$. It follows that $(x_n) \to \bar{x}$, hence from the property of u there is $(u_n) \to u$ such that $x_n + t_n u_n \in A$ for any n. But then, since d_A is 1–Lipschitz, one gets

$$d_A(y_n + t_n u) \leq d_A(x_n + t_n u_n) + \|x_n - y_n\| + t_n \|u_n - u\| \leq d_A(y_n) + t_n \left(\frac{1}{n} + \|u_n - u\| \right),$$

which implies that $d_A^\circ(\overline{x}, u) \le 0$, i.e., the conclusion.

(iii) Take nonzero $\xi_1, \xi_2 \in \partial_C d_A(\overline{x})$, $\lambda_1, \lambda_2 > 0$, and choose $t \in [0, 1]$. Define

$$\xi := \frac{(1-t)\lambda_1}{(1-t)\lambda_1 + t\lambda_2}\xi_1 + \frac{t\lambda_2}{(1-t)\lambda_1 + t\lambda_2}\xi_2 \in \partial_C d_A(\overline{x}),$$

and observe that $(1-t)\lambda_1\xi_1 + t\lambda_2\xi_2 = \left[(1-t)\lambda_1 + t\lambda_2\right]\xi \in \text{cone}\,\partial_C d_A(\overline{x})$, which means that $\text{cone}\,\partial_C d_A(\overline{x})$ is a convex cone. It follows that cl cone $\partial_C d_A(\overline{x})$ is a closed convex cone.

Take $u \in \partial_C d_A(\overline{x})^-$. If one would have $d_A^\circ(\overline{x}, u) = \sup\left\{\langle \xi, u \rangle \mid \xi \in \partial_C d_A(\overline{x})\right\} > 0$, then there exists $\xi \in \partial_C d_A(\overline{x})$ such that $\langle \xi, u \rangle > 0$, which is a contradiction that shows that $d_A^\circ(\overline{x}, u) \le 0$, i.e., $u \in T_C(A, \overline{x})$. The reverse inclusion is easy to prove in the same manner, hence

$$\partial_C d_A(\overline{x})^- = \left[\text{cl cone}\,\partial_C d_A(\overline{x})\right]^- = T_C(A, \overline{x}). \tag{5.1.24}$$

By applying the bipolar theorem (Theorem 2.1.6) in the last equality, one gets that $N_C(A, \overline{x}) = \text{cl cone}\,\partial_C d_A(\overline{x})$.

(iv) Follows from (5.1.24) and (iii).

(v) This is straightforward, taking into account that if $A \subset B$, then $A^- \supset B^-$. $\qquad\square$

The following proposition continues the idea from (5.1.21).

Proposition 5.1.26. *Let $f : \mathbb{R}^p \to \mathbb{R}$ be Lipschitz with modulus L around x and suppose x is a solution of (P). Then*

$$0 \in \partial_C(f + Ld_A)(x) \subset \partial_C f(x) + L\partial_C d_A(x) \subset \partial_C f(x) + N_C(A, x). \tag{5.1.25}$$

Proof It follows from (5.1.21) and (iv) from the previous theorem. $\qquad\square$

It is interesting to further study the structure of $N_C(A, x)$ in some special situations. To this end, a useful notion is introduced next.

Definition 5.1.27. *A set A is said to be regular at $x \in A$ if $T_C(A, x) = T_B(A, x)$.*

It admits the following characterization.

Proposition 5.1.28. *A is regular if and only if $N_C(A, x) = N_B(A, \overline{x})$.*

Proof For the necessity, observe that if A is regular, then

$$N_C(A, x) = T_C(A, x)^- = T_B(A, x)^- = N_B(A, x).$$

For the sufficiency, suppose that $N_C(A, x) = N_B(A, \overline{x})$. Then

$$T_C(A, x) = N_C(A, x)^- = N_B(A, x)^- = \left(T_B(A, x)^-\right)^- \supset T_B(A, x) \supset T_C(A, x),$$

which shows the desired conclusion. □

In the case of closed convex sets, we have the following result.

Theorem 5.1.29. *Suppose A is a closed convex set, and $x \in A$. Then A is regular and*

$$T_B(A, x) = T_C(A, x) = \operatorname{cl} \operatorname{cone}(A - x),$$
$$N_B(A, x) = N_C(A, x) = \{u \in \mathbb{R}^p \mid \langle u, y - x \rangle \le 0, \forall y \in A\}.$$

Proof In view of the Propositions 2.1.15 and 5.1.28, we only have to prove that A is regular.

Take $u \in A - x$, and choose arbitrary $(x_n) \subset A$, $(x_n) \to x$ and $(t_n) \downarrow 0$. Define $u_n := (x - x_n) + u$ for every n and remark that $(u_n) \to u$. Moreover, observe that

$$x_n + t_n u_n = (1 - t_n)x_n + t_n(u + x) \in A,$$

since A is convex. It follows from Theorem 5.1.25 that $u \in T_C(A, x)$. As $T_C(A, x)$ is a convex cone, we have $\operatorname{cl} \operatorname{cone}(A - x) \subset T_C(A, x)$. But we already know that

$$T_C(A, x) \subset T_B(A, x) = \operatorname{cl} \operatorname{cone}(A - x),$$

which ends the proof. □

Recall that if A is a nonempty closed set in $\mathbb{R}^p$, we denote by $\operatorname{pr}_A x$ the nonempty projection set $\{y \in A \mid d_A(x) = \|x - y\|\}$.

Theorem 5.1.30. *Let A be a nonempty closed subset of $\mathbb{R}^p$, and $\bar{x} \in \operatorname{bd} A$. Then*

$$N_C(A, \bar{x}) = \operatorname{cl} \operatorname{conv} \left\{ \lambda \lim_{n \to \infty} \frac{x_n - y_n}{\|x_n - y_n\|} \mid \lambda \ge 0, \ x_n \notin A, \ x_n \to \bar{x}, \ y_n \in \operatorname{pr}_A x_n \right\}. \tag{5.1.26}$$

Proof Observe first that if $x \in A$ is such that $\nabla d_A(x)$ exists, it follows that

$$\nabla d_A(x)(u) = \lim_{t \to 0} \frac{d_A(x + tu) - d_A(x)}{t} = \lim_{t \to 0} \frac{d_A(x + tu)}{t} \ge 0, \quad \forall u \in \mathbb{R}^p,$$

hence $\nabla d_A(x) = 0$.

Suppose $x \notin A$ and $\nabla d_A(x)$ exists. Take $y \in \operatorname{pr}_A x$ and $u \in \mathbb{R}^p$, and denote $g(x) := \|x - y\|$. Then

$$g^\circ(x, u) \ge \limsup_{t \downarrow 0} \frac{g(x + tu) - g(x)}{t} = \limsup_{t \downarrow 0} \frac{\|x + tu - y\| - \|x - y\|}{t}$$

$$\ge \limsup_{t \downarrow 0} \frac{d_A(x + tu) - d_A(x)}{t} = \nabla d_A(x)(u),$$

since $\|x - y\| = d_A(x)$ and $\nabla d_A(x)$ exists. But this means that $\nabla d_A(x) \in \partial_C g(x) = \left\{ \frac{x - y}{\|x - y\|} \right\}$, i.e., $\nabla d_A(x) = \frac{x - y}{\|x - y\|}$. Suppose now there are two points $y_1, y_2 \in \operatorname{pr}_A x$. It means that $\nabla d_A(x) = \frac{x - y_1}{d_A(x)} = \frac{x - y_2}{d_A(x)}$, i.e., $\operatorname{pr}_A x$ consists in only one point.

Take now $\overline{x} \in \operatorname{bd} A$. Using the gradient formula (5.1.8) for d_A, we have that

$$\partial_C d_A(\overline{x}) = \operatorname{conv} \left\{ \lim_{n \to \infty} \nabla d_A(x_n) \mid x_n \to \overline{x}, \ x_n \notin \Omega_{d_A} \right\},$$

where by Ω_{d_A} one denotes the set of points at which d_A fails to be differentiable. Taking into account the discussion above, one gets

$$\partial_C d_A(\overline{x}) \subset \operatorname{conv} \left\{ 0, \ \lim_{n \to \infty} \frac{x_n - y_n}{\|x_n - y_n\|} \mid x_n \notin A, \ x_n \to \overline{x}, \ y_n \in \operatorname{pr}_A x_n \right\}. \tag{5.1.27}$$

Now, observe that for any $x \in A$, the function d_A attains a minimum at x, hence by applying the generalized Fermat rule (Theorem 5.1.16), one gets that $0 \in \partial_C d_A(x)$.

Take now $x \notin A$ and $y \in \operatorname{pr}_A x$. Consider the function $f(z) := \|x - z\|$. This function, which is 1–Lipschitz, attains its minimum on A at y, hence by applying Proposition 5.1.26 it follows that

$$0 \in \partial_C f(y) + \partial_C d_A(y).$$

As $x \neq y$, f is C^1 around y, hence $\partial_C f(y) = \{\nabla f(y)\} = \left\{ -\frac{x - y}{\|x - y\|} \right\}$. It follows that $\frac{x - y}{\|x - y\|} \in \partial_C d_A(y)$. Hence,

$$\left\{ 0, \ \lim_{n \to \infty} \frac{x_n - y_n}{\|x_n - y_n\|} \mid x_n \notin A, \ x_n \to \overline{x}, \ y_n \in \operatorname{pr}_A x_n \right\} \subset \partial_C d_A(\overline{x})$$

and, as $\partial_C d_A(\overline{x})$ is a convex set, one has equality in (5.1.27). The result now follows from the fact that $N_C(A, \overline{x}) = \operatorname{cl} \operatorname{cone} \partial_C d_A(\overline{x})$. □

The next example emphasizes the utility of the formula (5.1.26), and at the same time exhibits a case where the Bouligand and the Clarke tangent cones are different.

Example 5.1.31. *Consider the set* $A := \left\{ (x_1, x_2) \in \mathbb{R}^2 \mid x_2 = |x_1| \right\}$. *Then it is easy to see, using Definition 2.1.9, that* $T_B(A, (0, 0)) = A$. *Moreover, using the polarity, one gets*

$$N_B(A, (0, 0)) = A^- = \left\{ (x_1, x_2) \in \mathbb{R}^2 \mid x_2 \leq -|x_1| \right\}.$$

Now, for the Clarke normal cone, one gets, by using formula (5.1.26), that

$$N_C(A, (0, 0)) = \operatorname{cl} \operatorname{conv} \left\{ (x_1, x_2) \in \mathbb{R}^2 \mid x_2 = |x_1| \ \text{or} \ x_2 \leq -|x_1| \right\} = \mathbb{R}^2,$$

hence $T_C(A, (0, 0)) = \{(0, 0)\}$.

The next theorem shows that the Clarke tangent and normal cones to the graphs of Lipschitz mappings are always linear subspaces. Among other things, this result emphasizes the importance of studying also other types of normal cones, such as the ones in the next chapter. In what follows, $\operatorname{gr} f$ denotes the graph of a function f.

Theorem 5.1.32 (Rockafellar). *Let $U \subset \mathbb{R}^p$ be an open set, and $f : U \to \mathbb{R}^k$ be a Lipschitz function. Consider $\overline{x} \in U$ and $\overline{y} := f(\overline{x})$. Then $T_C(\operatorname{gr} f, (\overline{x}, \overline{y}))$ and $N_C(\operatorname{gr} f, (\overline{x}, \overline{y}))$ are linear subspaces of $\mathbb{R}^p \times \mathbb{R}^k$.*

Proof According to Theorem 5.1.25 (ii), $(\xi, \eta) \in T_C(\operatorname{gr} f, (\overline{x}, \overline{y}))$ if and only if for every $(x_n, y_n) \subset \operatorname{gr} f$, $(x_n, y_n) \to (\overline{x}, \overline{y})$ and every $(t_n) \downarrow 0$, there exists $(\xi_n, \eta_n) \to (\xi, \eta)$ such that $(x_n, y_n) + t_n(\xi_n, \eta_n) \in \operatorname{gr} f$ for any n. Since f is continuous, this condition reduces to the following: for every $(x_n) \to \overline{x}$ and every $(t_n) \downarrow 0$, there is $\xi_n \to \xi$ such that

$$\frac{f(x_n + t_n \xi_n) - f(x_n)}{t_n} \to \eta. \tag{5.1.28}$$

Since f is Lipschitz, it follows that

$$\left\| f(x_n + t_n \xi_n) - f(x_n + t_n \xi) \right\| \le L t_n \left\| \xi_n - \xi \right\|,$$

where by L we have denoted the Lipschitz modulus of f. The limit from (5.1.28) is the same when one replaces ξ_n by ξ. It follows that $(\xi, \eta) \in T_C(\operatorname{gr} f, (\overline{x}, \overline{y}))$ if and only if for every $(x_n) \to \overline{x}$ and every $(t_n) \downarrow 0$, there is $\xi_n \to \xi$ such that

$$\frac{f(x_n + t_n \xi) - f(x_n)}{t_n} \to \eta,$$

that is,

$$\lim_{\substack{t \downarrow 0 \\ x \to \overline{x}}} \frac{f(x + t\xi) - f(x)}{t} = \eta,$$

i.e., the generalized directional derivative of f at $\overline{x}$ in the direction ξ is in fact a usual lim and equals η. This is also equivalent to

$$f^\circ(\overline{x}, \xi) = \eta = \lim_{\substack{t \downarrow 0 \\ x' \to \overline{x}}} \frac{f(x') - f(x' - t\xi)}{t},$$

where $x' := x + t\xi$, hence if $(\xi, \eta) \in T_C(\operatorname{gr} f, (\overline{x}, \overline{y}))$, one necessarily has also that $(-\xi, -\eta) \in T_C(\operatorname{gr} f, (\overline{x}, \overline{y}))$. Since we also know from Theorem 5.1.25 (i) that $T_C(\operatorname{gr} f, (\overline{x}, \overline{y}))$ is a closed convex cone, it follows that $T_C(\operatorname{gr} f, (\overline{x}, \overline{y}))$ is actually a (closed) linear subspace of $\mathbb{R}^p \times \mathbb{R}^k$. Since the dual set of a linear subspace is also a linear subspace, it follows that $N_C(\operatorname{gr} f, (\overline{x}, \overline{y}))$ is also a (closed) linear subspace of $\mathbb{R}^p \times \mathbb{R}^k$. $\square$

We have seen that the distance function serves as a link between the analytical theory of generalized gradients and the geometric constructions given by the tangent and normal cones. The next result shows other interesting connections, this time for a general function; the final equivalence in the section about the Fréchet and Mordukhovich normal cones arises from very definitions, proving that in fact all the normal constructions we use can be seen in a unifying way.

Theorem 5.1.33. *Let $f : \mathbb{R}^p \to \mathbb{R}$ be Lipschitz around $\overline{x}$. Then:*
  (i) $T_C(\mathrm{epi}\, f, (\overline{x}, f(\overline{x}))) = \mathrm{epi}\, f^\circ(\overline{x}, \cdot)$;
  (ii) $\xi \in \partial_C f(\overline{x}) \Leftrightarrow (\xi, -1) \in N_C(\mathrm{epi}\, f, (\overline{x}, f(\overline{x}))).$

Proof (i) Take $(u, r) \in T_C(\mathrm{epi}\, f, (\overline{x}, f(\overline{x})))$. Moreover, take some realizing sequences for $f^\circ(\overline{x}, u)$: we have $(x_n) \to \overline{x}$ and $(t_n) \downarrow 0$ such that

$$\lim_{n \to \infty} \frac{f(x_n + t_n u) - f(x_n)}{t_n} = f^\circ(\overline{x}, u).$$

Since $(x_n, f(x_n))$ belongs to epi f and converges to $(\overline{x}, f(\overline{x}))$, we have, due to Theorem 5.1.25 (ii), that there is $(u_n, r_n) \to (u, r)$ such that $(x_n, f(x_n)) + t_n(u_n, r_n) \in \mathrm{epi}\, f$ for every n, i.e.,

$$f(x_n + t_n u_n) \leq f(x_n) + t_n r_n,$$

and using the Lipschitz property of f (with constant ℓ), it follows that

$$\frac{f(x_n + t_n u) - f(x_n)}{t_n} \leq r_n + \ell \|u_n - u\|.$$

Passing to the limit for $n \to \infty$, we get $f^\circ(\overline{x}, u) \leq r$, i.e., $(u, r) \in \mathrm{epi}\, f^\circ(\overline{x}, \cdot)$.

For the converse inclusion, we prove that for every $\delta \geq 0$, we have $(u, f^\circ(\overline{x}, u) + \delta) \in T_C(\mathrm{epi}\, f, (\overline{x}, f(\overline{x})))$. Take arbitrary $(x_n, r_n) \subset \mathrm{epi}\, f$, $(x_n, r_n) \to (\overline{x}, f(\overline{x}))$, and $(t_n) \downarrow 0$. Define $u_n := u$ for every n, and take

$$s_n := \max \left\{ f^\circ(\overline{x}, u) + \delta, \frac{f(x_n + t_n u) - f(x_n)}{t_n} \right\}.$$

Since $\limsup_{n \to \infty} \frac{f(x_n + t_n u) - f(x_n)}{t_n} \leq f^\circ(\overline{x}, u)$, it follows that $s_n \to f^\circ(\overline{x}, u) + \delta$, hence $(u_n, s_n) \to (u, f^\circ(\overline{x}, u) + \delta)$. In order to show by Theorem 5.1.25 (ii) the desired assertion, we must prove that $(x_n, r_n) + t_n(u_n, s_n) \in \mathrm{epi}\, f$ for every n, i.e.,

$$f(x_n + t_n u) \leq r_n + t_n s_n, \quad \forall n.$$

The definition of s_n and the fact that $(x_n, r_n) \in \mathrm{epi}\, f$ for every n, imply that

$$r_n + t_n s_n \geq r_n + f(x_n + t_n u) - f(x_n) \geq f(x_n + t_n u),$$

so the proof of (i) is now finished.

  (ii) Remark that $\xi \in \partial_C f(\overline{x})$ if and only if $f^\circ(\overline{x}, u) \geq \langle \xi, u \rangle$ for every u, that is, for every $u \in \mathbb{R}^p$ and every $r \geq f^\circ(\overline{x}, u)$, one has $\langle (\xi, -1), (u, r) \rangle \leq 0$. Using (i), the last assertion is equivalent to: for every $(u, r) \in \mathrm{epi}\, f^\circ(\overline{x}, \cdot) = T_C(\mathrm{epi}\, f, (\overline{x}, f(\overline{x})))$, one has $\langle (\xi, -1), (u, r) \rangle \leq 0$, i.e., $(\xi, -1) \in T_C(\mathrm{epi}\, f, (\overline{x}, f(\overline{x})))^- = N_C(\mathrm{epi}\, f, (\overline{x}, f(\overline{x})))$. $\qquad \square$

5.1.3 Optimality Conditions in Lipschitz Optimization

In this section, we will derive necessary optimality conditions in Fritz John form for the minimization problem

$$(MP) \quad \min f(x) \text{ subject to } g(x) \le 0, \ h(x) = 0, \ x \in A,$$

where $f, g = (g_1, \ldots, g_n)$ and $h = (h_1, \ldots, h_m)$ are locally Lipschitz functions which map from $\mathbb{R}^p$ into $\mathbb{R}$, $\mathbb{R}^n$ and $\mathbb{R}^m$, respectively, and the set A is closed. Notice that the (MP) problem combines both functional and geometrical constraints, which were separately analyzed in the smooth case, since the presence of the set A along the smooth functions introduces a nonsmooth behaviour, which cannot be investigated with the tools of classical differentiation.

Theorem 5.1.34 (Fritz John conditions for Clarke calculus). *Let $\overline{x}$ be a solution of (MP), where f, g and h are Lipschitz around $\overline{x}$. Then there exist $\lambda_0 \ge 0$, $\lambda = (\lambda_1, \ldots, \lambda_n) \in \mathbb{R}^n$ and $\mu = (\mu_1, \ldots, \mu_m) \in \mathbb{R}^m$, with $\lambda_0 + \|\lambda\| + \|\mu\| \ne 0$, such that*

$$0 \in \lambda_0 \partial_C f(\overline{x}) + \sum_{i=1}^{n} \lambda_i \partial_C g_i(\overline{x}) + \sum_{j=1}^{m} \mu_j \partial_C h_j(\overline{x}) + N_C(A, \overline{x}) \tag{5.1.29}$$

and

$$\lambda_i \ge 0, \ \lambda_i g_i(\overline{x}) = 0 \text{ for any } i \in \overline{1, n}. \tag{5.1.30}$$

Proof Suppose, without loss of generality, that the functions f, g and h are Lipschitz on a neighborhood of the set A (otherwise, replace A by $A \cap B(\overline{x}, \delta)$ for some $\delta > 0$, and observe that neither the assumptions, nor the conclusion change).

Fix $\varepsilon \in (0, 1)$, denote the compact set $S(0, 1) \cap [\mathbb{R}_+ \times \mathbb{R}_+^n \times \mathbb{R}^m]$ by S, and define the function

$$F_\varepsilon(x) := \max_{(\lambda_0, \lambda, \mu) \in S} \left\{ \lambda_0(f(x) - f(\overline{x}) + \varepsilon) + \sum_{i=1}^{n} \lambda_i g_i(x) + \sum_{j=1}^{m} \mu_j h_j(x) \right\}.$$

Suppose there is an x such that $F_\varepsilon(x) \le 0$. It follows that

$$\lambda_0(f(x) - f(\overline{x}) + \varepsilon) + \sum_{i=1}^{n} \lambda_i g_i(\overline{x}) + \sum_{j=1}^{m} \mu_j h_j(\overline{x}) \le 0,$$

for any $(\lambda_0, \lambda, \mu) \in S$. By taking successively one of the scalars $\lambda_0, \lambda_1, \ldots, \lambda_n, \mu_1, \ldots, \mu_m$ equal to 1, and the rest equal to 0, one gets

$$f(x) \le f(\overline{x}) - \varepsilon, \ g(\overline{x}) = 0, \ h(\overline{x}) \le 0,$$

a contradiction which shows that the assumption made was false. Consequently, $F_\varepsilon(x) > 0$ for any $x \in A$. Moreover, being the max of Lipschitz functions, F_ε is Lipschitz on a neighborhood of A, with a constant we denote by L which may be chosen such that it does not depend on ε. Consider the closed set A and apply the Ekeland Variational Principle (see Remark 3.1.13) to get the existence of $x_\varepsilon \in A$ such that

$$F_\varepsilon(x_\varepsilon) \leq F_\varepsilon(\overline{x}) - \sqrt{\varepsilon}\, \|x_\varepsilon - \overline{x}\|$$

and

$$F_\varepsilon(x_\varepsilon) \leq F_\varepsilon(x) + \sqrt{\varepsilon}\, \|x_\varepsilon - x\| \text{ for any } x \in A.$$

Observe that $F_\varepsilon(\overline{x}) = \varepsilon$. One gets from the first relation that $\|x_\varepsilon - \overline{x}\| \leq \sqrt{\varepsilon}$. On the other hand, from the second relation, one sees that the Lipschitz function $x \mapsto F_\varepsilon(x) + \sqrt{\varepsilon}\, \|x_\varepsilon - x\|$ attains its minimum on A at x_ε, hence by applying Proposition 5.1.26, one gets the existence of $L > 0$ such that

$$0 \in \partial_C F_\varepsilon(x_\varepsilon) + \sqrt{\varepsilon}D(0, 1) + L\partial_C d_A(x_\varepsilon). \tag{5.1.31}$$

Denote by

$$S(x_\varepsilon) := \left\{ y \in S \mid F_\varepsilon(x_\varepsilon) = \langle y, \varphi(x_\varepsilon) \rangle \right\},$$

where $y := (\lambda_0, \lambda, \mu)$ and $\varphi(x) := \left(f(x) - f(\overline{x}) + \varepsilon, g(x), h(x) \right)$. In order to apply Proposition 5.1.22, we want to prove that $S(x_\varepsilon)$ is a singleton. Suppose the contrary. Then there exist $y^l := (\lambda_0^l, \lambda^l, \mu^l)$ $(l = 1, 2)$, two different points in $S(x_\varepsilon)$. Remember that S is a subset of the unit sphere in $\mathbb{R}^{n+m+1}$, and choose $t > 1$ such that $\overline{y} := \frac{t}{2}(y^1 + y^2) \in S$. Then, since $F_\varepsilon(x_\varepsilon) > 0$,

$$F_\varepsilon(x_\varepsilon) \geq \langle \overline{y}, \varphi(x_\varepsilon) \rangle = \frac{t}{2}\left(\langle y^1, \varphi(x_\varepsilon) \rangle + \langle y^2, \varphi(x_\varepsilon) \rangle \right) = tF_\varepsilon(x_\varepsilon) > F_\varepsilon(x_\varepsilon),$$

a contradiction. Hence, $S(x_\varepsilon) = \left\{ (\lambda_0^\varepsilon, \lambda^\varepsilon, \mu^\varepsilon) \right\}$.

We can apply now Proposition 5.1.22, to get that

$$\partial_C F_\varepsilon(x_\varepsilon) \subset \partial_C \left(\lambda_0^\varepsilon f + \sum_{i=1}^{n} \lambda_i^\varepsilon g_i + \sum_{j=1}^{m} \mu_j^\varepsilon h_j \right)(x_\varepsilon).$$

By the use of the sum rule (Theorem 5.1.13), we get from (5.1.31) that

$$0 \in \lambda_0^\varepsilon \partial_C f(x_\varepsilon) + \sum_{i=1}^{n} \lambda_i^\varepsilon \partial_C g_i(x_\varepsilon) + \sum_{j=1}^{m} \mu_j^\varepsilon \partial_C h_j(x_\varepsilon) + \sqrt{\varepsilon}D(0, 1) + L\partial_C d_A(x_\varepsilon).$$

Set now ε from the above equal to $\frac{1}{k}$ for any $k \geq 2$. One gets hence x_k such that $\|x_k - \overline{x}\| \leq \frac{1}{\sqrt{k}}$, and also $(\lambda_0^k, \lambda^k, \mu^k) \in S$, as well as $\xi^k \in D\left(0, \frac{1}{\sqrt{k}} \right)$ such that

$$\xi^k \in \lambda_0^k \partial_C f(x_k) + \sum_{i=1}^{n} \lambda_i^k \partial_C g_i(x_k) + \sum_{j=1}^{m} \mu_j^k \partial_C h_j(x_k) + N_C(A, x_n).$$

Observe that we may suppose again, since S is compact, that $(\lambda_0^k, \lambda^k, \mu^k)$ converges to some $(\lambda_0, \lambda, \mu) \in S$. As $x_k \to \overline{x}$ and $\xi^k \to 0$, by passing to the limit in the relation above (see Proposition 5.1.6), one gets (5.1.29).

Suppose now that there is $i_0 \in \overline{1, n}$ such that $g_{i_0}(\overline{x}) < 0$. If $\lambda_{i_0} > 0$, we will have

$$F(x_k) = \lambda_0^k \left(f(x_k) - f(\overline{x}) + \frac{1}{k} \right) + \sum_{i=1}^{n} \lambda_i^k g_i(x_k) + \sum_{j=1}^{m} \mu_j^k h_j(x_k) \to \sum_{i=1}^{n} \lambda_i g_i(\overline{x}) < 0,$$

but this contradicts the fact that $F(x_k) > 0$ for every k. $\qquad\square$

Notice again that specific constraint qualification conditions could be imposed in order to avoid the situation $\lambda_0 = 0$. We mention that, for this case of nonsmooth calculus, the qualification conditions are beyond the aim of this book, but the interested reader is invited to consult (Clarke, 1983).

Example 5.1.35. *Let us examine again the problem (5.1.1). We are now close to our initial aim, which was to provide optimality conditions and, furthermore, to find the minimum.*

Observe first that, by Weierstrass Theorem, this problem admits at least one solution, since the function f is continuous, and the set of restrictions is compact. If (x_1, x_2) is a minimum point, we know by Theorem 5.1.34 that there exists $(\lambda_0, \lambda_1) \neq (0, 0)$ such that

$$\begin{aligned}
(0, 0) &\in \lambda_0 \partial_C f(x_1, x_2) + \lambda_1 \partial_C g(x_1, x_2) \\
&= \lambda_0 \partial_C f(x_1, x_2) + \lambda_1 \nabla g(x_1, x_2) \\
&= \lambda_0 \partial_C f(x_1, x_2) + 2\lambda_1 \cdot (x_1, x_2),
\end{aligned}$$

as $g(x_1, x_2) = x_1^2 + x_2^2 - 1$ is continuously differentiable and $\nabla g(x_1, x_2) = 2(x_1, x_2)$.

Suppose $\lambda_0 = 0$. Then, again by Theorem 5.1.34, $\lambda_1 > 0$ and $g(x_1, x_2) = 0$, i.e., $x_1^2 + x_2^2 = 1$. On the other hand, one must have $(x_1, x_2) = (0, 0)$, which gives us a contradiction which means that we can take $\lambda_0 = 1$.

Suppose $\lambda_1 = 0$. Then $(0, 0) \in \partial_C f(x_1, x_2)$. We know from Example 5.1.9 that $\partial_C f(0, 0) = \text{conv} \{(2, 0), (0, 1), (1, 2)\}$, hence (x_1, x_2) cannot be $(0, 0)$. Recall that the function f can equivalently be written as in (5.1.10). Suppose (x_1, x_2) lies in the interior of one of the sets S_1, S_2 and S_3. But in this case $\partial_C f(x_1, x_2) = \{\nabla f(x_1, x_2)\}$, and the gradient $\nabla f(x_1, x_2)$ equals, respectively, $(2, 0), (0, 1)$ and $(1, 2)$, so $(0, 0) \in \partial_C f(x_1, x_2)$ cannot hold. Suppose now the nonzero vector (x_1, x_2) is on the boundary between two of the sets S_1, S_2 and S_3. For instance, take $(x_1, x_2) \in S_1 \cap S_2 \setminus \{(0, 0)\}$. Then $\partial_C f(x_1, x_2) = \text{conv} \{(2, 0), (0, 1)\}$, so the inclusion $(0, 0) \in \partial_C f(x_1, x_2)$ cannot hold. The same happens in the other two similar cases. In conclusion, our assumption, that $\lambda_1 = 0$, must be false. Therefore, we can write:

$$(x_1, x_2) \in -\frac{1}{2\lambda_1} \partial_C f(x_1, x_2), \quad \lambda_1 > 0.$$

We know from Theorem 5.1.34, that $x_1^2 + x_2^2 = 1$. Suppose now (x_1, x_2) lies in the interior of one of the sets S_1, S_2 and S_3. If $(x_1, x_2) \in \text{int } S_1 =$

$\{(x_1, x_2) \mid 2x_1 < x_2 \text{ and } x_1 > 2x_2\}$, *one must have, since* $\partial_C f(x_1, x_2) = \{(2, 0)\}$, *that*

$$(x_1, x_2) = \left(-\frac{1}{\lambda_1}, 0 \right),$$

which is obviously a contradiction. If $(x_1, x_2) \in \text{int}\,S_2 = \{(x_1, x_2) \mid 2x_1 > x_2 \text{ and } x_1 < -x_2\}$, *one would obtain* $\left(0, -\frac{1}{2\lambda_1}\right) \in \text{int}\,S_2$, *again a contradiction. Finally, if* $(x_1, x_2) \in \text{int}\,S_3 = \{(x_1, x_2) \mid x_1 < 2x_2 \text{ or } x_1 > -x_2\}$, *it should happen that* $-\frac{1}{2\lambda_1}(1, 2) \in \text{int}\,S_3$ *for some positive* λ_1, *which is false. Only three points remain as candidates for the minimum points: the points from the unit sphere which lie on the boundary between two of the sets* S_1, S_2 *and* S_3. *Suppose* $(x_1, x_2) \in S_2 \cap S_3 \setminus \{(0, 0)\}$. *Then* (x_1, x_2) *must be of the form* $(\varepsilon, -\varepsilon)$, *with* $\varepsilon > 0$, *and should be an element of the set* $-\frac{1}{2\lambda_1}\,\text{conv}\,\{(0, 1), (1, 2)\}$ *for some positive* λ_1, *which is false. If* $(x_1, x_2) \in S_1 \cap S_3 \setminus \{(0, 0)\}$, *then* $(x_1, x_2) = (-2\varepsilon, -\varepsilon)$ *for some* $\varepsilon > 0$. *It should satisfy* $(-2\varepsilon, -\varepsilon) \in -\frac{1}{2\lambda_1}\,\text{conv}\,\{(2, 0), (1, 2)\}$, *and taking into account that* $x_1^2 + x_2^2 = 1$, *one gets the point* $\left(-\frac{2}{\sqrt{5}}, -\frac{1}{\sqrt{5}}\right)$ *as a candidate for the minimum. Finally, if* $(x_1, x_2) \in S_1 \cap S_2 \setminus \{(0, 0)\}$, *then* (x_1, x_2) *must be of the form* $(-\varepsilon, -2\varepsilon)$, *for some* $\varepsilon > 0$, *and should be an element of* $-\frac{1}{2\lambda_1}\,\text{conv}\,\{(2, 0), (0, 1)\}$ *for some positive* λ_1, *which set equals in fact* $-\mathbb{R}_+^2$. *The equality* $x_1^2 + x_2^2 = 1$ *gives the point* $\left(-\frac{1}{\sqrt{5}}, -\frac{2}{\sqrt{5}}\right)$ *as another candidate for providing us the minimum. Observe, though, that for sufficiently small* δ, $\left(-\frac{1}{\sqrt{5}} + 3\delta, -\frac{2}{\sqrt{5}} - \delta\right) \in S_2$ *and is feasible. Moreover, for* $\delta \to 0$, *it tends to* $\left(-\frac{1}{\sqrt{5}}, -\frac{2}{\sqrt{5}}\right)$, *and* $f\left(-\frac{1}{\sqrt{5}} + 3\delta, -\frac{2}{\sqrt{5}} - \delta\right) = -\frac{2}{\sqrt{5}} - \delta < -\frac{2}{\sqrt{5}} = f\left(-\frac{1}{\sqrt{5}}, -\frac{2}{\sqrt{5}}\right)$, *hence* $\left(-\frac{1}{\sqrt{5}}, -\frac{2}{\sqrt{5}}\right)$ *is not a local minimum. Analogously, one can prove that* $\left(-\frac{1}{\sqrt{5}}, -\frac{2}{\sqrt{5}}\right)$ *is not a local maximum, too. It means that the minimum point we are looking for is* $\left(-\frac{2}{\sqrt{5}}, -\frac{1}{\sqrt{5}}\right)$.

5.2 Mordukhovich Generalized Calculus

We saw in the previous section an approach which provided us with useful generalized differentiation tools, in order to solve some nonconvex and nonsmooth optimization problems. We saw that the Clarke tangent and normal cones, as well as the associated subdifferential, have the property that they are convex. In order to pass to a fully non-convex case, we present in this section some other (even more general) differentiation objects, which have the advantage, besides allowing us to work in the fully nonconvex case, to exhibit robust (exact) calculus rules.

5.2.1 Fréchet and Mordukhovich Normal Cones

Let us begin our presentation by introducing two important constructions.

Definition 5.2.1. *Let A be a nonempty subset of $\mathbb{R}^p$.*
    (i) Given $x \in A$, the Fréchet normal cone to A at x is the set

$$N_F(A, x) := \left\{ \xi \in \mathbb{R}^p \mid \limsup_{u \xrightarrow{A} x} \frac{\langle \xi, u - x \rangle}{\|u - x\|} \le 0 \right\}, \tag{5.2.1}$$

where by $u \xrightarrow{A} x$ we understand $u \to x$ with $u \in A$. An element $\xi \in N_F(A, x)$ is called Fréchet normal to A at x. If $x \notin A$, we define $N_F(A, x) := \emptyset$.
    (ii) Let $\overline{x} \in A$. The Mordukhovich (or limiting, or basic) normal cone to A at $\overline{x}$ is

$$N_M(A, \overline{x}) := \left\{ \xi \in \mathbb{R}^p \mid \exists x_n \xrightarrow{A} \overline{x}, \ \xi_n \to \xi, \ \xi_n \in N_F(A, x_n), \forall n \in \mathbb{N} \right\}. \tag{5.2.2}$$

If $\overline{x} \notin A$, we put $N_M(A, x) := \emptyset$.

Remark 5.2.2. *Observe that $\xi \in N_F(A, \overline{x})$ if and only if for every $\varepsilon > 0$, there exists a neighborhood V of $\overline{x}$ such that, for every $x \in A \cap V$, one has*

$$\langle \xi, x - \overline{x} \rangle \le \varepsilon \|x - \overline{x}\|. \tag{5.2.3}$$

In the next proposition we collect some basic properties of the Fréchet and Mordukhovich normal cones.

Proposition 5.2.3. *Let A be a nonempty subset of $\mathbb{R}^p$ and $\overline{x} \in A$. Then:*
    (i) One has

$$N_F(A, \overline{x}) = N_F(\operatorname{cl} A, \overline{x}) \ and \ N_M(A, \overline{x}) \subset N_M(\operatorname{cl} A, \overline{x}). \tag{5.2.4}$$

(ii) If $\overline{x} \in \operatorname{int} A$, one has $N_F(A, \overline{x}) = N_M(A, \overline{x}) = \{0\}$.
(iii) $N_F(A, \overline{x})$ is a closed convex cone, and $N_M(A, \overline{x})$ is a closed cone.
(iv) If A is a convex set, then

$$N_F(A, \overline{x}) = N_M(A, \overline{x}) = \left\{ \xi \in \mathbb{R}^p \mid \langle \xi, x - \overline{x} \rangle \le 0 \ for \ every \ x \in A \right\}, \tag{5.2.5}$$

i.e., it coincides with the normal cone $N(A, \overline{x})$.
    (v) If B is a nonempty subset of $\mathbb{R}^k$ and $\overline{y} \in B$, then

$$N_F(A \times B, (\overline{x}, \overline{y})) = N_F(A, \overline{x}) \times N_F(B, \overline{y}),$$
$$N_M(A \times B, (\overline{x}, \overline{y})) = N_M(A, \overline{x}) \times N_M(B, \overline{y}).$$

Proof (i) Observe first that for every two sets $A \subset B$ with $x \in X$, one has $N_F(B, x) \subset N_F(A, x)$. This follows immediately from the fact that $A \subset B$ implies

$\left[u \xrightarrow{A} x \Rightarrow u \xrightarrow{B} x \right]$, which gives us that $\limsup\limits_{u \xrightarrow{A} x} \dfrac{\langle \xi, u - x \rangle}{\|u - x\|} \leq \limsup\limits_{u \xrightarrow{B} x} \dfrac{\langle \xi, u - x \rangle}{\|u - x\|}$. Con-

sequently, one has that $N_F(\mathrm{cl}\, A, \overline{x}) \subset N_F(A, \overline{x})$.

For the reverse inclusion, take $\xi \in N_F(A, \overline{x})$. Then the equivalence from Remark 5.2.2 is valid, where the neighborhood V can be taken as being open, without loss of generality. Fix $y \in \mathrm{cl}\, A \cap V$. If $y \in A$, then it satisfies (5.2.3). If $y \in \mathrm{cl}\, A \backslash A$, then there exists $(y_n) \subset A$, $y_n \to y$. Since V is open, one has that $y_n \in A \cap V$ for every n sufficiently large. Then y_n satisfies (5.2.3), which implies when passing to the limit that $\langle \xi, y - \overline{x} \rangle \leq \varepsilon \|y - \overline{x}\|$. Consequently, $\limsup\limits_{y \xrightarrow{\mathrm{cl}\, A} \overline{x}} \dfrac{\langle \xi, y - \overline{x} \rangle}{\|y - \overline{x}\|} \leq \varepsilon$ for every $\varepsilon > 0$, which

shows that $\xi \in N_F(\mathrm{cl}\, A, \overline{x})$.

For the second relation from (5.2.4), observe that

$$\left[x_n \xrightarrow{A} \overline{x} \Rightarrow x_n \xrightarrow{\mathrm{cl}\, A} \overline{x} \right] \text{ and } \left[N_F(A, x_n) = N_F(\mathrm{cl}\, A, x_n) \right]$$

give us the desired inclusion.

(ii) If $\overline{x} \in \mathrm{int}\, A$ and $\xi \in N_F(A, \overline{x})$, then for every $\varepsilon > 0$, there exists $\delta > 0$ such that $D(\overline{x}, \delta) \subset A$ and $\langle \xi, x - \overline{x} \rangle < \varepsilon \|x - \overline{x}\|$ for every $x \in D(\overline{x}, \delta)$. It follows that $\langle \xi, u \rangle < \varepsilon \|u\|$, for every $u \in D(0, 1)$, hence $\|\xi\| < \varepsilon$. Since ε was arbitrarily chosen, we have that $\xi = 0$. The formula (5.2.2) and the fact that $x_n \xrightarrow{A} \overline{x}$ imply that $x_n \in \mathrm{int}\, A$ for every n sufficiently large gives us that $N_M(A, \overline{x}) = \{0\}$.

(iii) The fact that $N_F(A, \overline{x})$ and $N_M(A, \overline{x})$ are cones easily follows from their definitions. Taking into account, for instance, the equivalence from Remark 5.2.2, one easily gets the convexity and closedness of $N_F(A, \overline{x})$.

Consider now $(\xi_n) \subset N_M(A, \overline{x})$, $\xi_n \to \xi$. Then, for every $\varepsilon > 0$, there is $n_0 \in \mathbb{N}$ such that $\|\xi_n - \xi\| < 2^{-1}\varepsilon$ for every $n \geq n_0$. Fix $n \geq n_0$. Since $\xi_n \in N_M(A, \overline{x})$, there exist $(x_k^n)_k \to \overline{x}$ and $(\xi_k^n)_k \to \xi$ such that $\xi_k^n \in N_F(A, x_k^n)$, for every k. Consequently, one can find $k_n^1, k_n^2 \geq n \geq n_0$ such that

$$\left\| x_k^n - \overline{x} \right\| < \varepsilon, \ \forall k \geq k_n^1 \text{ and}$$
$$\left\| \xi_k^n - \xi_n \right\| < 2^{-1}\varepsilon, \ \forall k \geq k_n^2.$$

Define $k_n := \max\left\{ k_n^1, k_n^2 \right\}$, $y_n := x_{k_n}^n$ and $\eta_n := \xi_{k_n}^n$. One has

$$\|y_n - \overline{x}\| < \varepsilon \text{ and}$$
$$\|\eta_n - \xi\| \leq \|\eta_n - \xi_n\| + \|\xi_n - \xi\| < \varepsilon,$$

hence, as n was taken arbitrary such that $n \geq n_0$, it follows that $y_n \to \overline{x}$ and $\eta_n \to \xi$. Moreover, since $\eta_n \in N_F(A, y_n)$ for every $n \geq n_0$, we obtain that $\xi \in N_M(A, \overline{x})$. Hence, $N_M(A, \overline{x})$ is a closed set.

(iv) We prove first that

$$N_F(A, \overline{x}) = \left\{ \xi \in \mathbb{R}^p \mid \langle \xi, x - \overline{x} \rangle \leq 0 \text{ for every } x \in A \right\}. \tag{5.2.6}$$

The "⊃" inclusion from (5.2.6) immediately follows for an arbitrary set A. Suppose now A is a convex set, and take $\xi \in N_F(A, \bar{x})$ and $x \in A$. Then $x_\lambda := (1 - \lambda)\bar{x} + \lambda x \in A$ for every $\lambda \in [0, 1]$, and $x_\lambda \to \bar{x}$ for $\lambda \downarrow 0$. It follows that for every $\varepsilon > 0$, there exists $\lambda > 0$ sufficiently small such that

$$\langle \xi, x_\lambda - \bar{x} \rangle < \varepsilon \|x_\lambda - \bar{x}\|,$$

which means, successively, that

$$\langle \xi, \lambda(x - \bar{x}) \rangle < \varepsilon \|\lambda(x - \bar{x})\|,$$
$$\frac{\langle \xi, x - \bar{x} \rangle}{\|x - \bar{x}\|} < \varepsilon, \ \forall \varepsilon > 0, \forall x \in A \backslash \{\bar{x}\},$$
$$\langle \xi, x - \bar{x} \rangle \leq 0, \ \forall x \in A.$$

Now, since $N_F(A, \bar{x}) \subset N_M(A, \bar{x})$, take any $\xi \in N_M(A, \bar{x})$. There exist $(x_n) \overset{A}{\to} \bar{x}$ and $(\xi_n) \to \xi$ such that $\xi_n \in N_F(A, x_n)$ for every n. It means, in view of (5.2.6), that

$$\langle \xi_n, x - x_n \rangle \leq 0, \ \forall n \in \mathbb{N}, \forall x \in A.$$

Passing to the limit for $n \to \infty$, one gets the conclusion.
(v) Both equalities follow straightforward. $\qquad\square$

The following representation of the Mordukhovich normal cone can be useful in computations.

Theorem 5.2.4. *Suppose $A \subset \mathbb{R}^p$ is a closed set and let $\bar{x} \in A$. Then*

$$N_M(A, \bar{x}) = \left\{ \xi \in \mathbb{R}^p \ \middle| \ \begin{array}{c} \exists (\lambda_n) \subset [0, \infty), \ (x_n) \to \bar{x}, \ (y_n) \subset \mathbb{R}^p \ such\ that \\ y_n \in \mathrm{pr}_A\, x_n \ for\ every\ n\ and\ \lambda_n(x_n - y_n) \to \xi \end{array} \right\} \qquad (5.2.7)$$

Proof We will prove first that for every $x \in A$, the next inclusion holds:

$$N_F(A, x) \subset \left\{ \xi \in \mathbb{R}^p \ \middle| \ \begin{array}{c} \exists (\lambda_n) \subset [0, \infty), \ (x_n) \to x, \ (y_n) \subset \mathbb{R}^p \ such\ that \\ y_n \in \mathrm{pr}_A\, x_n \ for\ every\ n\ and\ \lambda_n(x_n - y_n) \to \xi \end{array} \right\}. \qquad (5.2.8)$$

Take $x \in A$ and $\xi \in N_F(A, x)$. Set $x_n := x + \frac{1}{n}\xi$ and pick $y_n \in \mathrm{pr}_A\, x_n$ for any n. Observe that $y_n \in \mathrm{pr}_A\, x_n$ if and only if

$$\|x_n - y_n\| \leq \|x_n - v\|, \ \forall v \in A,$$

that is

$$0 \leq \|x_n - v\|^2 - \|x_n - y_n\|^2 = \langle x_n - v, x_n - y_n \rangle$$
$$+ \langle x_n - v, y_n - v \rangle - \langle x_n - y_n, v - y_n \rangle - \langle x_n - y_n, x_n - v \rangle$$
$$= \langle 2x_n - v - y_n, y_n - v \rangle = -2 \langle x_n - y_n, v - y_n \rangle + \|v - y_n\|^2.$$

By replacing v with x and using the form of x_n, one gets

$$2 \left\langle x + \frac{1}{n}\xi - y_n, x - y_n \right\rangle \leq \|x - y_n\|^2 ,$$

$$n \|x - y_n\| \leq \frac{2 \langle \xi, y_n - x \rangle}{\|y_n - x\|}.$$

Moreover, because $x \in A$ and $y_n \in \mathrm{pr}_A x_n$, we know that

$$\|y_n - x\| \leq \|y_n - x_n\| + \|x - x_n\|$$

$$\leq \|x - x_n\| + \|x - x_n\| = \frac{2}{n} \|\xi\| \to 0,$$

hence, since $y_n \xrightarrow{A} x$ and $\xi \in N_F(A, x)$, one has that $\dfrac{2 \langle \xi, y_n - x \rangle}{\|y_n - x\|} \to 0$. Consequently, $n\|x - y_n\| \to 0$ and

$$n(x_n - y_n) = n(x - y_n) + \xi \to \xi,$$

which proves (5.2.8).

To prove now the "$\subset$" in (5.2.7), take $\xi \in N_M(A, \overline{x})$. Then there exists $x_n \xrightarrow{A} \overline{x}$, $\xi_n \to \xi$ such that $\xi_n \in N_F(A, x_n)$ for every n. For every $\varepsilon > 0$, there exists then $n_0 \in \mathbb{N}$ such that, for every $n \geq n_0$,

$$\|x_n - \overline{x}\| < 2^{-1}\varepsilon,$$

$$\|\xi_n - \xi\| < 2^{-1}\varepsilon.$$

Fix $n \geq n_0$. Since $x_n \in A$, one has using (5.2.8) that there exist $(\lambda_k^n)_k \subset [0, \infty)$, $(x_k^n)_k \to x_n$, $(y_k^n)_k$ such that $y_k^n \in \mathrm{pr}_A x_k^n$ for every k, and $\lambda_k^n(x_k^n - y_k^n) \to \xi_n$ for $k \to \infty$. For the ε chosen before, there are $k_n^1, k_n^2 \geq n \geq n_0$ such that

$$\|x_k^n - x_n\| < 2^{-1}\varepsilon, \ \forall k \geq k_n^1,$$

$$\|\lambda_k^n(x_k^n - y_k^n) - \xi_n\| < 2^{-1}\varepsilon, \ \forall k \geq k_n^2.$$

Take $k_n := \max\{k_n^1, k_n^2\}$ and set $u_n := x_{k_n}^n$, $v_n := y_{k_n}^n$, $\alpha_n := \lambda_{k_n}^n$. One has then

$$\|u_n - \overline{x}\| \leq \|u_n - x_n\| + \|x_n - \overline{x}\| < \varepsilon,$$

$$\|\alpha_n(u_n - v_n) - \xi\| \leq \|\alpha_n(u_n - v_n) - \xi_n\| + \|\xi_n - \xi\| < \varepsilon,$$

and since $v_n \in \mathrm{pr}_A u_n$, we have proved the desired inclusion in (5.2.8).

Let us prove the "$\supset$" inclusion in (5.2.8). Now consider $\xi \in \mathbb{R}^p$ such that there exist $(\lambda_n) \subset [0, \infty)$ and $(x_n) \to \overline{x}$ such that $\lambda_n(x_n - y_n) \to \xi$, where $y_n \in \mathrm{pr}_A x_n$ for every n.

Take arbitrary $v \in A$ and use the above characterization of the projection $y_n \in \mathrm{pr}_A x_n$ to deduce successively that

$$\langle x_n - y_n, v - y_n \rangle \leq \frac{1}{2} \|v - y_n\|^2 ,$$

$$\frac{\langle \lambda_n(x_n - y_n), v - y_n \rangle}{\|v - y_n\|} \le \frac{\lambda_n}{2} \|v - y_n\|, \ \forall v \in A.$$

Taking the lim sup for fixed n in the previous inequality and setting $\xi_n := \lambda_n(x_n - y_n)$, we get that

$$\limsup_{v \xrightarrow{A} y_n} \frac{\langle \xi_n, v - y_n \rangle}{\|v - y_n\|} \le 0,$$

hence $\xi_n \in N_F(A, y_n)$ for every n. Moreover, since $x_n \to \bar{x}$ and $y_n \in \mathrm{pr}_A x_n$, one can prove as above that $y_n \to \bar{x}$. It follows that $\xi \in N_M(A, \bar{x})$ and the reverse inclusion is now proved. $\qquad\square$

Let us provide some examples which contain the calculus of the Fréchet and Mordukhovich normal cones to some sets.

Example 5.2.5. *Consider the set* $A := \{(x_1, x_2) \in \mathbb{R}^2 \mid x_2 = -|x_1|\}$. *Let us compute first* $N_F(A, (0, 0))$. *Take arbitrary* $\varepsilon > 0$. *One should find a neighborhood V of* $(0, 0)$ *such that, for every* $(u, v) \in A \cap V$,

$$\langle (\xi, \eta), (u, v) \rangle < \varepsilon \|(u, v)\|.$$

The form of the set A allows us to deduce that

$$\xi u - \eta |u| < \sqrt{2}\varepsilon |u|,$$

for every u close to 0. One gets from the above inequality that $\eta \ge |\xi|$. *As for every such element* (ξ, η) *and* $(u, v) \in A$ *one has* $\langle (\xi, \eta), (u, v) \rangle \le 0$, *we deduce that*

$$N_F(A, (0, 0)) = \left\{ (\xi, \eta) \in \mathbb{R}^2 \mid \eta \ge |\xi| \right\}.$$

Let us use now formula (5.2.7) to calculate $N_M(A, (0, 0))$. *For this, let us observe first how the projection set of different elements from* $\mathbb{R}^2$ *on A looks. If* $(x_1, x_2) \in \mathbb{R}^2$ *is such that* $x_2 \ge |x_1|$, *then* $\mathrm{pr}_A(x_1, x_2) = \{(0, 0)\}$. *If* $(x_1, x_2) \in \{(u, v) \mid u \ge 0, v \le u\}$, *then* $\mathrm{pr}_A(x_1, x_2)$ *is the unique point from* $\{(\xi, -\xi) \mid \xi \ge 0\}$ *such that the line passing through* (x_1, x_2) *is orthogonal on it. Similarly, if* $(x_1, x_2) \in \{(u, v) \mid u \le 0, v \le -u\}$, *then* $\mathrm{pr}_A(x_1, x_2)$ *is the unique point from* $\{(\xi, \xi) \mid \xi \le 0\}$ *such that the line passing through* (x_1, x_2) *is orthogonal to this line. It follows that when* $(x_n, y_n) \to (0, 0)$ *and* $\lambda_n \ge 0$, *the all possible limits of sequences of the type* $\lambda_n((x_n, y_n) - \mathrm{pr}_A(x_n, y_n))$ *lie in the set*

$$\left\{ (\xi, \eta) \in \mathbb{R}^2 \mid \eta \ge |\xi| \text{ or } \eta = -|\xi| \right\},$$

which is equal to $N_M(A, (0, 0))$.

It can be observed from the previous example that the Mordukhovich normal cone is nonconvex in general, which shows that it cannot be the polar to any tangent cone, since polarity always implies convexity. It is not the case with the Fréchet normal cone, as the next result shows.

Theorem 5.2.6. *Let $A \subset \mathbb{R}^p$ be a nonempty set and $\overline{x} \in A$. Then*

$$N_F(A, \overline{x}) = T_B(A, \overline{x})^-.$$

Proof Take $\xi \in N_F(A, \overline{x})$ and choose arbitrary $v \in T_B(A, \overline{x})$. There exist sequences $(t_n) \downarrow 0$ and $(v_n) \to v$ such that $\overline{x} + t_n v_n \in A$ for every n. Using the definition of the Fréchet normal cone, it follows that for every $\varepsilon > 0$, exists $n_0 \in \mathbb{N}$ such that

$$t_n \langle \xi, v_n \rangle < \varepsilon t_n \|v_n\|, \ \forall n \geq n_0.$$

This shows, after dividing by t_n and passing to the limit above, that $\langle \xi, v \rangle \leq \varepsilon \|v\|$. Letting $\varepsilon \to 0$, one gets that $\langle \xi, v \rangle \leq 0$ and, furthermore, taking into account the arbitrariness of v, that $\xi \in T_B(A, \overline{x})^-$.

For the reverse inclusion, suppose $\langle \xi, v \rangle \leq 0$ for every $v \in T_B(A, \overline{x})$. Now, choose an sequence $x_n \overset{A}{\to} \overline{x}$ for which the lim sup from the definition of the Fréchet normal cone is attained:

$$\limsup_{x \overset{A}{\to} \overline{x}} \frac{\langle \xi, x - \overline{x} \rangle}{\|x - \overline{x}\|} = \lim_{n \to \infty} \frac{\langle \xi, x_n - \overline{x} \rangle}{\|x_n - \overline{x}\|}.$$

Since $\overline{x} + \|x_n - \overline{x}\| \dfrac{x_n - \overline{x}}{\|x_n - \overline{x}\|} = x_n \in A$ for every n, $\|x_n - \overline{x}\| \downarrow 0$, and without loss of generality we may suppose that $\dfrac{x_n - \overline{x}}{\|x_n - \overline{x}\|}$ converges to some u. It follows that $u \in T_B(A, \overline{x})$, and hence

$$\lim_{n \to \infty} \left\langle \xi, \frac{x_n - \overline{x}}{\|x_n - \overline{x}\|} \right\rangle = \langle \xi, u \rangle \leq 0,$$

which shows that $\xi \in N_F(A, \overline{x})$ and ends the proof. $\square$

Theorem 5.1.32 shows that, in case of a Lipschitz function f, $N_C(\mathrm{gr}\, f, \overline{x})$ is always a linear subspace, which drives us to the idea that, in many situations, especially when dealing with graphical sets, considering the Mordukhovich normal cone could be more appropriate.

Let us now formulate a result which gives us a smooth variational description of Fréchet normals, useful in many instances, including calculus rules for the Fréchet subdifferential of the difference of two functions.

Theorem 5.2.7. *Let A be a nonempty subset of $\mathbb{R}^p$ and $\overline{x} \in A$. For $\xi \in \mathbb{R}^p$, suppose that there exists a function $s : U \to \mathbb{R}$ defined on a neighborhood of $\overline{x}$, which is Fréchet differentiable at $\overline{x}$, and such that $\nabla s(\overline{x}) = \xi$ and s achieves a local maximum relative to A at $\overline{x}$. Then $\xi \in N_F(A, \overline{x})$. Conversely, for every $\xi \in N_F(A, \overline{x})$, there is a function $s : \mathbb{R}^p \to \mathbb{R}$ such that $s(x) \leq s(\overline{x}) = 0$ for every $x \in A$ and s is Fréchet differentiable at $\overline{x}$ with $\nabla s(\overline{x}) = \xi$.*

Proof For the direct implication, we know that there is a continuous function α with $\lim_{x \to \overline{x}} \frac{\alpha(\|x - \overline{x}\|)}{\|x - \overline{x}\|} = 0$ such that

$$s(x) = s(\overline{x}) + \langle \xi, x - \overline{x} \rangle + \alpha(\|x - \overline{x}\|) \leq s(\overline{x}),$$

for every $x \in A$ near $\overline{x}$. It means that $\langle \xi, x - \overline{x} \rangle + \alpha(\|x - \overline{x}\|) \leq 0$ for every $x \in A$ near $\overline{x}$, which implies for such x that

$$\frac{\langle \xi, x - \overline{x} \rangle}{\|x - \overline{x}\|} \leq -\frac{\alpha(\|x - \overline{x}\|)}{\|x - \overline{x}\|}.$$

Passing to lim sup for $x \xrightarrow{A} \overline{x}$, we deduce that $\xi \in N_F(A, \overline{x})$.

For the converse, consider the function $s : \mathbb{R}^p \to \mathbb{R}$ given by

$$s(x) := \begin{cases} \min\{0, \langle \xi, x - \overline{x} \rangle\}, & \text{if } x \in A \\ \langle \xi, x - \overline{x} \rangle, & \text{otherwise.} \end{cases}$$

Remark that $s(x) \leq 0 = s(\overline{x})$ for every $x \in A$ and $s(x) \leq \langle \xi, x - \overline{x} \rangle$ for every $x \in \mathbb{R}^p$. Then

$$\left| \frac{s(x) - s(\overline{x}) - \langle \xi, x - \overline{x} \rangle}{\|x - \overline{x}\|} \right| = \frac{\langle \xi, x - \overline{x} \rangle - s(x)}{\|x - \overline{x}\|}$$

$$= \begin{cases} 0, & \text{if } x \notin A \text{ or } x \in A \text{ and } \langle \xi, x - \overline{x} \rangle \leq 0 \\ \dfrac{\langle \xi, x - \overline{x} \rangle}{\|x - \overline{x}\|}, & \text{if } x \in A \text{ and } \langle \xi, x - \overline{x} \rangle > 0. \end{cases}$$

Since $\xi \in N_F(A, \overline{x})$, for every $\varepsilon > 0$ there exists a neighborhood V of $\overline{x}$ such that $\dfrac{\langle \xi, x - \overline{x} \rangle}{\|x - \overline{x}\|} < \varepsilon$ for every $x \in A \cap V$, which means that

$$\left| \frac{s(x) - s(\overline{x}) - \langle \xi, x - \overline{x} \rangle}{\|x - \overline{x}\|} \right| < \varepsilon, \ \forall x \in V.$$

But this means that s is Fréchet differentiable at $\overline{x}$ with $\nabla s(\overline{x}) = \xi$. □

We close this section with a result useful for the next.

Proposition 5.2.8. *Suppose $A \subset \mathbb{R}^p$ is a closed set and $\overline{x} \in \text{bd}\, A$. Then*

$$N_M(A, \overline{x}) \subset N_M(\text{bd}\, A, \overline{x}).$$

Proof Pick $\xi \in N_M(A, \overline{x}) \backslash \{0\}$. According to the definition of the Mordukhovich normal cone, there exist $(x_n) \xrightarrow{A} \overline{x}$ and $(\xi_n) \to \xi$ such that $\xi_n \in N_F(A, x_n)$ for every n. Due to the continuity of the norm, it follows that $\|\xi_n\| \to \|\xi\| > 0$, hence $x_n \notin \text{int}\, A$, i.e., $x_n \in \text{bd}\, A$ for every n sufficiently large, because otherwise one would have $\xi_n = 0$. From this,

$$\limsup_{x \xrightarrow{\text{bd}\, A} x_n} \frac{\langle \xi_n, x - x_n \rangle}{\|x - x_n\|} \leq \limsup_{x \xrightarrow{A} x_n} \frac{\langle \xi_n, x - x_n \rangle}{\|x - x_n\|} \leq 0,$$

hence $\xi_n \in N_F(\text{bd}\, A, x_n)$ for every n, so $\xi \in N_M(\text{bd}\, A, \overline{x})$. □

5.2.2 Fréchet and Mordukhovich Subdifferentials

Let us introduce now the subdifferentials associated to the normal cones defined before. We begin with the Fréchet subdifferential.

Definition 5.2.9. *Let $f : \mathbb{R}^p \to \mathbb{R}$. The set*

$$\partial_F f(\overline{x}) := \left\{ \xi \in \mathbb{R}^p \mid (\xi, -1) \in N_F(\text{epi}\, f, (\overline{x}, f(\overline{x}))) \right\}$$

is called the Fréchet subdifferential of f at $\overline{x}$, and its elements are called Fréchet subgradients of f at $\overline{x}$.

Observe that, since the Fréchet normal cone is a closed convex cone, the Fréchet subdifferential is a closed convex set in $\mathbb{R}^p$. The next example shows that the subdifferential can be empty, even for simple Lipschitz functions.

Example 5.2.10. *Consider the function $f : \mathbb{R} \to \mathbb{R}$, $f(x) = -|x|$. Then $\text{epi}\, f = \left\{ (x_1, x_2) \mid x_2 \geq -|x_1| \right\}$. One can easily see that $T_B(\text{epi}\, f, (0, 0)) = \text{epi}\, f$, hence*

$$N_F(\text{epi}\, f, (0, 0)) = (\text{epi}\, f)^- = \left\{ (0, 0) \right\}.$$

It follows that $\partial_F f(0, 0) = \emptyset$.

Let us provide an analytical characterization of the Fréchet subgradients.

Theorem 5.2.11. *Let $f : \mathbb{R}^p \to \mathbb{R}$. Then*

$$\partial_F f(\overline{x}) = \left\{ \xi \in \mathbb{R}^p \mid \liminf_{x \to \overline{x}} \frac{f(x) - f(\overline{x}) - \langle \xi, x - \overline{x} \rangle}{\|x - \overline{x}\|} \geq 0 \right\}. \tag{5.2.9}$$

Proof Take ξ from the right-hand side set. We want to prove that $\xi \in \partial_F f(\overline{x})$. Observe that, from the choice of ξ, for every $\varepsilon > 0$, there exists a neighborhood V of $\overline{x}$ such that, for every $x \in V$, one has

$$f(x) - f(\overline{x}) - \langle \xi, x - \overline{x} \rangle \geq -\varepsilon \|x - \overline{x}\|.$$

Pick such x and fix $\alpha \geq f(x)$. Then

$$
\begin{aligned}
\langle (\xi, -1), (x, \alpha) - (\overline{x}, f(\overline{x})) \rangle &= \langle \xi, x - \overline{x} \rangle + f(\overline{x}) - \alpha \\
&\leq \langle \xi, x - \overline{x} \rangle + f(\overline{x}) - f(x) \\
&\leq \varepsilon \|x - \overline{x}\| \\
&\leq \varepsilon \|(x, \alpha) - (\overline{x}, f(\overline{x}))\|,
\end{aligned}
$$

which proves that $(\xi, -1) \in N_F(\text{epi}\, f, (\overline{x}, f(\overline{x})))$.

For the converse inclusion, take $\xi \in \partial_F f(\overline{x})$ and suppose that there is $\varepsilon > 0$ such that

$$\sup_{V \in \mathcal{V}(\overline{x})} \inf_{x \in V} \frac{f(x) - f(\overline{x}) - \langle \xi, x - \overline{x} \rangle}{\|x - \overline{x}\|} < -\varepsilon.$$

It means that one can construct a sequence $x_n \to \overline{x}$ such that

$$\frac{f(x_n) - f(\overline{x}) - \langle \xi, x_n - \overline{x} \rangle}{\|x_n - \overline{x}\|} < -\varepsilon,$$

which implies that $(x_n, \alpha_n) \in \text{epi} f$, where $\alpha_n := f(\overline{x}) + \langle \xi, x_n - \overline{x} \rangle - \varepsilon \|x_n - \overline{x}\|$. We have

$$\left\| (x_n, \alpha_n) - (\overline{x}, f(\overline{x})) \right\| \leq \|x_n - \overline{x}\| + |\langle \xi, x_n - \overline{x} \rangle - \varepsilon \|x_n - \overline{x}\||$$

$$\leq (1 + \|\xi\| + \varepsilon) \|x_n - \overline{x}\|,$$

whence

$$\frac{\langle (\xi, -1), (x_n, \alpha_n) - ((\overline{x}, f(\overline{x}))) \rangle}{\left\| (x_n, \alpha_n) - ((\overline{x}, f(\overline{x}))) \right\|} = \frac{\varepsilon \|x_n - \overline{x}\|}{\left\| (x_n, \alpha_n) - ((\overline{x}, f(\overline{x}))) \right\|}$$

$$\geq \frac{\varepsilon}{1 + \|\xi\| + \varepsilon}.$$

It follows that $(\xi, -1) \notin N_F(\text{epi} f, (\overline{x}, f(\overline{x})))$, a contradiction. The proof is now complete. □

The next proposition partially justifies the name of Fréchet subgradients for the elements of $\partial_F f(\overline{x})$.

Proposition 5.2.12. *Let $f : \mathbb{R}^p \to \mathbb{R}$ be Fréchet differentiable at $\overline{x}$. Then $\partial_F f(\overline{x}) = \{\nabla f(\overline{x})\}$.*

Proof It is obvious from the definition of Fréchet differentiability and the analytical characterization of the Fréchet subgradients that $\nabla f(\overline{x}) \in \partial_F f(\overline{x})$.

Suppose $\xi \in \partial_F f(\overline{x})$. Using (5.2.9), one gets that

$$\liminf_{x \to \overline{x}} \frac{\langle \nabla f(\overline{x}) - \xi, x - \overline{x} \rangle}{\|x - \overline{x}\|} \geq 0,$$

which means that for every $\varepsilon > 0$, there exists $\delta > 0$ such that, for every $x \in D(\overline{x}, \delta)$,

$$\langle \xi - \nabla f(\overline{x}), x - \overline{x} \rangle \leq \varepsilon \|x - \overline{x}\|.$$

It follows that

$$\left\| \xi - \nabla f(\overline{x}) \right\| \leq \varepsilon$$

for every positive ε, i.e., $\xi = \nabla f(\overline{x})$, and hence the conclusion. □

The next smooth variational description of Fréchet subgradients follows from Theorem 5.2.7.

Theorem 5.2.13. *Let $f : \mathbb{R}^p \to \mathbb{R}$ be a function. Then $\xi \in \partial_F f(\overline{x})$ if and only if there exists a function $s : \mathbb{R}^p \to \mathbb{R}$, Fréchet differentiable at $\overline{x}$, which satisfies:*

$$s(\overline{x}) = f(\overline{x}), \ s(x) \leq f(x), \ \forall x \in \mathbb{R}^p, \ and \ \nabla s(\overline{x}) = \xi. \tag{5.2.10}$$

Proof Suppose $\xi \in \partial_F f(\overline{x})$. Consider the function $s : \mathbb{R}^p \to \mathbb{R}$ given as

$$s(x) := \min \left\{ f(x), f(\overline{x}) + \langle \xi, x - \overline{x} \rangle \right\}, \ \forall x \in \mathbb{R}^p.$$

Then, obviously, $s(\overline{x}) = f(\overline{x})$ and $s(x) \leq f(x)$ for every $x \in \mathbb{R}^p$ from the definition. Also from the definition of s, it immediately follows that

$$\limsup_{x \to \overline{x}} \frac{s(x) - s(\overline{x}) - \langle \xi, x - \overline{x} \rangle}{\|x - \overline{x}\|} \leq 0. \tag{5.2.11}$$

Moreover, one has

$$\frac{s(x) - s(\overline{x}) - \langle \xi, x - \overline{x} \rangle}{\|x - \overline{x}\|} = \begin{cases} 0, & \text{if } f(x) \geq f(\overline{x}) + \langle \xi, x - \overline{x} \rangle \\ \frac{f(x) - f(\overline{x}) - \langle \xi, x - \overline{x} \rangle}{\|x - \overline{x}\|}, & \text{otherwise,} \end{cases}$$

which implies, using $\xi \in \partial_F f(\overline{x})$, that

$$\liminf_{x \to \overline{x}} \frac{s(x) - s(\overline{x}) - \langle \xi, x - \overline{x} \rangle}{\|x - \overline{x}\|} \geq 0. \tag{5.2.12}$$

Relations (5.2.11) and (5.2.12) mean that s is Fréchet differentiable at $\overline{x}$ and $\nabla s(\overline{x}) = \xi$.

Conversely, suppose $\xi \in \mathbb{R}^p$ and there exists an $s : \mathbb{R}^p \to \mathbb{R}$, Fréchet differentiable at $\overline{x}$ and satisfying (5.2.10). Then

$$\liminf_{x \to \overline{x}} \frac{f(x) - f(\overline{x}) - \langle \xi, x - \overline{x} \rangle}{\|x - \overline{x}\|} \geq \liminf_{x \to \overline{x}} \frac{s(x) - s(\overline{x}) - \langle \xi, x - \overline{x} \rangle}{\|x - \overline{x}\|} = 0,$$

hence $\xi \in \partial_F f(\overline{x})$. □

If $f : \mathbb{R}^p \to \mathbb{R}$ is convex, then the Fréchet subdifferential reduces to the usual convex subdifferential given in the previous chapter.

Proposition 5.2.14. *Suppose $f : \mathbb{R}^p \to \mathbb{R}$ is convex. Then*

$$\partial_F f(\overline{x}) = \partial_M f(\overline{x}) = \partial f(\overline{x}).$$

Proof This immediately follows from the fact that $\operatorname{epi} f$ is a convex set and from the formula (5.2.5). □

Let us proceed now in introducing the Mordukhovich subdifferential. For this, we need first the next result.

Proposition 5.2.15. *Let $f : \mathbb{R}^p \to \mathbb{R}$ be a function and $(\overline{x}, \overline{\alpha}) \in \operatorname{epi} f$. Then $\lambda \geq 0$ for every $(\xi, -\lambda) \in N_M(\operatorname{epi} f, (\overline{x}, \overline{\alpha}))$, hence there exist the subsets D and D^∞ of $\mathbb{R}^p$ such that*

$$N_M(\operatorname{epi} f, (\overline{x}, \overline{\alpha})) = \left\{ \lambda(\xi, -1) \mid \xi \in D, \lambda > 0 \right\} \cup \left\{ (\xi, 0) \mid \xi \in D^\infty \right\}.$$

Proof Choose arbitrary $(\xi, -\lambda) \in N_M(\text{epi}\, f, (\bar{x}, \bar{a}))$. Then by formula (5.2.2), it follows that $\exists (x_n, \alpha_n) \stackrel{\text{epi}\, f}{\to} (\bar{x}, \bar{a})$, $\xi_n \to \xi$, and $\lambda_n \to \lambda$, such that

$$\limsup_{(x,a) \stackrel{\text{epi}\, f}{\to} (x_n, a_n)} \frac{\langle \xi_n, x - x_n \rangle - \lambda_n(\alpha - \alpha_n)}{\|(x, \alpha) - (x_n, \alpha_n)\|} \leq 0, \ \forall n.$$

Then, for every $\varepsilon > 0$, there exists $\delta_{\varepsilon,n} > 0$ such that, for every $(x, \alpha) \in D((x_n, \alpha_n), \delta)$, one has

$$\frac{\langle \xi_n, x - x_n \rangle - \lambda_n(\alpha - \alpha_n)}{\|(x, \alpha) - (x_n, \alpha_n)\|} < \varepsilon.$$

By taking $x := x_n$ and $\alpha := \alpha_n + \delta$, it follows that $(x, \alpha) \in D((x_n, \alpha_n), \delta)$, hence

$$\frac{-\lambda_n(\alpha - \alpha_n)}{|\alpha - \alpha_n|} = -\lambda_n < \varepsilon.$$

Passing to the limit for $n \to \infty$, one gets that $-\lambda \leq \varepsilon$ for every positive ε, hence the conclusion. $\qquad\square$

Definition 5.2.16. *Let $f : \mathbb{R}^p \to \mathbb{R}$ be a function and $\bar{x} \in \mathbb{R}^p$.*
(i) The set

$$\partial_M f(\bar{x}) := \left\{ \xi \in \mathbb{R}^p \mid (\xi, -1) \in N_M(\text{epi}\, f, (\bar{x}, f(\bar{x}))) \right\}$$

is called the Mordukhovich (or limiting, or basic) subdifferential of f at $\bar{x}$, and its elements are called the basic subgradients of f at $\bar{x}$.
(ii) The set

$$\partial^\infty f(\bar{x}) := \left\{ \xi \in \mathbb{R}^p \mid (\xi, 0) \in N_M(\text{epi}\, f, (\bar{x}, f(\bar{x}))) \right\}$$

is called the singular subdifferential of f at $\bar{x}$, and its elements are called the singular subgradients of f at $\bar{x}$.

It follows from the definitions that for every function $f : \mathbb{R}^p \to \mathbb{R}$, and every $\bar{x} \in \mathbb{R}^p$,

$$\partial_F f(\bar{x}) \subset \partial_M f(\bar{x}). \tag{5.2.13}$$

Example 5.2.17. *Consider again the function $f : \mathbb{R} \to \mathbb{R}$, $f(x) := -|x|$. Since $\text{epi}\, f = \{(x_1, x_2) \mid x_2 \geq -|x_1|\}$, one easily deduces, using the formula (5.2.7), that*

$$N_M(\text{epi}\, f, (0, 0)) = \{(x, x) \mid x \leq 0\} \cup \{(x, -x) \mid x > 0\}.$$

Hence, $\partial_M f(0) = \{-1, 1\}$ and $\partial^\infty f(0) = \{0\}$. Recall that $\partial_F f(0) = \emptyset$, hence the inclusion (5.2.13) can be strict. Note that the Mordukhovich subdifferential is a nonconvex set, and this should not be surprising, since its definition is based on the Mordukhovich normal cone, which is nonconvex in general.

The next result emphasizes estimates for the subdifferential of locally Lipschitz functions.

Theorem 5.2.18. *Let $f : \mathbb{R}^p \to \mathbb{R}$. If f is locally Lipschitz around $\bar{x}$ with modulus $\ell \geq 0$, then*

$$\|\xi\| \leq \ell, \ \forall \xi \in \partial_M f(\bar{x}) \tag{5.2.14}$$

and

$$\partial^\infty f(\bar{x}) = \{0\}. \tag{5.2.15}$$

Proof Take $(\xi, \lambda) \in N_M(\text{epi}\, f, (\bar{x}, f(\bar{x})))$. Since f is continuous, epi f is closed, and since $\text{gr}\, f = \text{bd}\, \text{epi}\, f$, it follows from Proposition 5.2.8 that $(\xi, \lambda) \in N_M(\text{gr}\, f, (\bar{x}, f(\bar{x})))$. Then there exist $(x_n) \to \bar{x}$, $(\xi_n) \to \xi$ and $(\lambda_n) \to \lambda$ such that $(\xi_n, \lambda_n) \in N_F(\text{gr}\, f, (x_n, f(x_n)))$ for every n.

Now, from the Lipschitz property of f, we know that there exists $\delta > 0$ such that, for every $x, u \in D(\bar{x}, 2\delta)$, one has

$$|f(x) - f(u)| \leq \ell \|x - u\|. \tag{5.2.16}$$

Moreover, for every $\varepsilon > 0$, and for every n, there is $y \leq \min\{\delta, \ell\delta\}$ such that, for every $u \in D(x_n, y)$,

$$\langle \xi_n, u - x_n \rangle + \lambda_n(f(u) - f(x_n)) \leq \varepsilon \left(\|u - x_n\| + |f(u) - f(x_n)|\right). \tag{5.2.17}$$

Take n sufficiently large such that $x_n \in D(\bar{x}, \delta)$ and $u \in D(x_n, y\ell^{-1})$. It follows that

$$\|u - \bar{x}\| \leq \|u - x_n\| + \|x_n - \bar{x}\| \leq y\ell^{-1} + \delta \leq 2\delta.$$

One can employ now (5.2.17) and (5.2.16) for u and x_n in order to get that

$$\begin{aligned}
\langle \xi_n, u - x_n \rangle &\leq -\lambda_n(f(u) - f(x_n)) + \varepsilon \left(\|u - x_n\| + |f(u) - f(x_n)|\right) \\
&\leq (|\lambda_n|\ell + \varepsilon + \varepsilon\ell)\|u - x_n\| \\
&\leq |\lambda_n|y + \varepsilon y\ell^{-1} + \varepsilon y.
\end{aligned}$$

Since the previous relation holds for every $u \in D(x_n, y\ell^{-1})$, it follows that

$$y\ell^{-1} \|\xi_n\| \leq y \left(|\lambda_n| + \varepsilon\ell^{-1} + \varepsilon\right).$$

Dividing by y, passing to the limit for $n \to \infty$, and then for $\varepsilon \to 0$, one gets that

$$\ell^{-1} \|\xi\| \leq |\lambda|. \tag{5.2.18}$$

Now, if $\xi \in \partial_M f(\bar{x})$, it means that λ from above equals -1, hence from (5.2.18) one gets (5.2.14). If $\xi \in \partial^\infty f(\bar{x})$, it means that λ from above is 0 and (5.2.15) follows. $\qquad \square$

Remark 5.2.19. *In view of relation (5.2.13), the previous theorem also shows that for every locally Lipschitz function f around a point $\bar{x}$, having the Lipschitz modulus $\ell \geq 0$, one has*

$$\|\xi\| \leq \ell, \ \forall \xi \in \partial_F f(\bar{x}). \tag{5.2.19}$$

We now emphasize a case when the Mordukhovich subdifferential reduces to the usual Fréchet differential.

Proposition 5.2.20. *Suppose $f : \mathbb{R}^p \to \mathbb{R}$ is C^1 around $\bar{x}$. Then*

$$\partial_M f(\bar{x}) = \{\nabla f(\bar{x})\} .$$

Proof Pick $\xi \in \partial_M f(\bar{x})$. Then $(\xi, -1) \in N_M(\text{epi} f, (\bar{x}, f(\bar{x}))) \subset N_M(\text{gr} f, (\bar{x}, f(\bar{x})))$, which means that there exist $(x_n) \to \bar{x}$, $(\xi_n) \to \xi$ and $(\lambda_n) \to 1$ such that for every $\varepsilon > 0$, there exists $\delta > 0$ such that for every $x \in D(x_n, \delta)$,

$$\langle \xi_n, x - x_n \rangle - \lambda_n(f(x) - f(x_n)) \leq \varepsilon \left(\|x - x_n\| + |f(x) - f(x_n)| \right) .$$

Using the Taylor expansion for the function f at every such point x_n, one gets that there is y_n on the segment $[x, x_n]$ such that

$$f(x) - f(x_n) = \langle \nabla f(y_n), x - x_n \rangle .$$

It means that

$$\langle \xi_n - \lambda_n \nabla f(y_n), x - x_n \rangle \leq \varepsilon \left(\|x - x_n\| + |f(x) - f(x_n)| \right)$$
$$\leq (\varepsilon + \varepsilon \ell) \|x - x_n\|$$
$$\leq (\varepsilon + \varepsilon \ell)\delta, \ \forall x \in D(x_n, \delta),$$

where $\ell > 0$ is a Lipschitz modulus for the C^1 function f. Hence,

$$\delta \left\| \xi_n - \lambda_n \nabla f(y_n) \right\| \leq (\varepsilon + \varepsilon \ell)\delta.$$

Passing to the limit for $\varepsilon \to 0$ and $n \to \infty$, and taking into account that $y_n \to \bar{x}$, hence $\nabla f(y_n) \to \nabla f(\bar{x})$ since f is C^1, one gets that $\xi = \nabla f(\bar{x})$. $\square$

The representation of singular subgradients will be useful in the following sections. The notation $x_n \xrightarrow{f} \bar{x}$ means $x_n \to \bar{x}$ and $f(x_n) \to f(\bar{x})$.

Theorem 5.2.21 (representation of singular subgradients). *Let $f : \mathbb{R}^p \to \mathbb{R}$ be lower semicontinuous around $\bar{x}$. Then*

$$\partial_M f(\bar{x}) = \left\{ \xi \in \mathbb{R}^p \mid \exists x_n \xrightarrow{f} \bar{x}, \ \xi_n \to \xi \text{ such that } \xi_n \in \partial_F f(x_n), \ \forall n \in \mathbb{N} \right\}. \tag{5.2.20}$$

Proof The "⊃" inclusion follows immediately. Suppose $\xi \in \partial_M f(\overline{x})$, i.e., $(\xi, -1) \in N_M(\text{epi} f, (\overline{x}, f(\overline{x})))$. Since f is lower semicontinuous around $\overline{x}$, epi f is locally closed around $(\overline{x}, f(\overline{x}))$, hence by employing Proposition 5.2.8, $(\xi, -1) \in N_M(\text{gr} f, (\overline{x}, f(\overline{x})))$. There exist then $(x_n, f(x_n)) \to (\overline{x}, f(\overline{x}))$ and $(\xi_n, \lambda_n) \to (\xi, 1)$ such that $(\xi_n, -\lambda_n) \in N_F(\text{gr} f, (x_n, f(x_n)))$ for every n. It means that for every $\varepsilon > 0$, there exists $\delta > 0$ such that

$$\langle \xi_n, x - x_n \rangle - \lambda_n(f(x) - f(x_n)) \leq \varepsilon \left(\|x - x_n\| + |f(x) - f(x_n)| \right)$$

for every $x \in D(x_n, \delta)$ and $f(x) \in D(f(x_n), \delta)$. Since $\lambda_n \to 1$, for every $\varepsilon > 0$, and every n such that $|\lambda_n - 1| < \varepsilon$, one has from above that

$$\langle \xi_n, x - x_n \rangle - (f(x) - f(x_n)) \leq \varepsilon \left(\|x - x_n\| + |f(x) - f(x_n)| \right) + |\lambda_n - 1| \cdot |f(x) - f(x_n)|$$
$$\leq 2\varepsilon \left(\|x - x_n\| + |f(x) - f(x_n)| \right),$$

for every $x \in D(x_n, \delta)$ and $f(x) \in D(f(x_n), \delta)$. It means that $(\xi_n, -1) \in N_F(\text{gr} f, (x_n, f(x_n)))$ for every n sufficiently large. Suppose, by contradiction, that $(\xi_n, -1) \notin N_F(\text{epi} f, (x_n, f(x_n)))$. Pick $y \in (0, 1)$ and sequences $(u_j, \alpha_j) \overset{\text{epi} f}{\to} (x_n, f(x_n))$ such that

$$\langle \xi_n, u_j - x_n \rangle - (\alpha_j - f(x_n)) > y \left\| (u_j, \alpha_j) - (x_n, f(x_n)) \right\|. \tag{5.2.21}$$

Since f is lower semicontinuous at x_n, we have

$$f(x_n) = \lim_{j \to \infty} \alpha_j \geq \limsup_{j \to \infty} f(u_j) \geq \liminf_{j \to \infty} f(u_j) \geq f(x_n),$$

hence $f(u_j) \to f(x_n)$. Moreover,

$$\left\| (u_j, f(u_j)) - (x_n, f(x_n)) \right\| \leq \left\| (u_j, \alpha_j) - (x_n, f(x_n)) \right\| + \alpha_j - f(u_j).$$

It follows from (5.2.21) that

$$\langle \xi_n, u_j - x_n \rangle - (\alpha_j - f(x_n)) > y \left\| (u_j, f(u_j)) - (x_n, f(x_n)) \right\| - y \left(\alpha_j - f(u_j) \right)$$
$$\geq y \left\| (u_j, f(u_j)) - (x_n, f(x_n)) \right\| - \left(\alpha_j - f(u_j) \right),$$

i.e.,

$$\langle \xi_n, u_j - x_n \rangle - (f(u_j) - f(x_n)) > y \left\| (u_j, f(u_j)) - (x_n, f(x_n)) \right\|,$$

with $u_j \to x_n$ and $f(u_j) \to f(x_n)$, which contradicts $(\xi_n, -1) \in N_F(\text{gr} f, (x_n, f(x_n)))$. Hence, $(\xi_n, -1) \in N_F(\text{epi} f, (x_n, f(x_n)))$, i.e., $\xi_n \in \partial_F f(x_n)$, and the desired inclusion is now proved. □

Example 5.2.22. *Let us consider the function $f : \mathbb{R} \to \mathbb{R}$ given as*

$$f(x) = \begin{cases} x^2 \sin \frac{1}{x}, & \text{if } x \neq 0 \\ 0, & \text{if } x = 0. \end{cases}$$

Using formula (5.2.20), one gets that $\partial_M f(0) = [-1, 1]$, while $\nabla f(0) = 0$. This shows that the Mordukhovich subdifferential does not reduce in general to the gradient, in case of a differentiable function.

The following result is yet another nonsmooth version of the Fermat rule.

Theorem 5.2.23 (Fermat rule for Mordukhovich calculus). *Let $f : \mathbb{R}^p \to \mathbb{R}$. If f has a local minimum at $\overline{x}$, then*

$$0 \in \partial_F f(\overline{x}) \subset \partial_M f(\overline{x}).$$

Proof Since $f(x) \geq f(\overline{x})$ for every x close to $\overline{x}$, it follows that

$$\liminf_{x \to \overline{x}} \frac{f(x) - f(\overline{x}) - \langle 0, x - \overline{x} \rangle}{\|x - \overline{x}\|} \geq 0,$$

hence $0 \in \partial_F f(\overline{x})$ due to the analytical characterization of Fréchet subgradients. The inclusion between the two subdifferentials is true in general. □

We continue by presenting some considerations about the links between the Fréchet and Mordukhovich normal cones and the corresponding subdifferentials of the distance function.

Proposition 5.2.24. *For any set $A \subset \mathbb{R}^p$ and $\overline{x} \in A$, one has*

$$\partial_F d_A(\overline{x}) = N_F(A, \overline{x}) \cap D(0, 1), \qquad N_F(A, \overline{x}) = \bigcup_{\lambda > 0} \lambda \partial_F d_A(\overline{x}). \tag{5.2.22}$$

Proof Take $\xi \in \partial_F d_A(\overline{x})$. Using the analytical description of the Fréchet subgradients given by (5.2.9), we get that for every $\varepsilon > 0$, there is $\delta > 0$ such that, for any $x \in D(\overline{x}, \delta)$, one has

$$d_A(x) - d_A(\overline{x}) - \langle \xi, x - \overline{x} \rangle \geq -\varepsilon \|x - \overline{x}\|. \tag{5.2.23}$$

This means that for any $x \in A \cap D(\overline{x}, \delta)$, one has $\langle \xi, x - \overline{x} \rangle \leq \varepsilon \|x - \overline{x}\|$, which shows that $\xi \in N_F(A, \overline{x})$. Moreover, since d_A is 1–Lipschitz, we get from (5.2.23) that

$$\langle \xi, x - \overline{x} \rangle \leq (\varepsilon + 1) \|x - \overline{x}\| \leq (\varepsilon + 1)\delta, \ \forall x \in D(\overline{x}, \delta),$$

hence $\delta \|\xi\| \leq (\varepsilon + 1)\delta$, i.e., $\|\xi\| \leq 1 + \varepsilon$ for every $\varepsilon > 0$. The "$\subset$" inclusion from the first relation in (5.2.22) holds.

Take now $\xi \in N_F(A, \overline{x})$ such that $\|\xi\| \leq 1$. For $x \notin A$, we find $u \in A$ such that

$$\|x - u\| \leq d_A(x) + \|x - \overline{x}\|^2.$$

Then

$$\begin{aligned}
\|u - \overline{x}\| &\leq \|u - x\| + \|x - \overline{x}\| \\
&\leq d_A(x) + \|x - \overline{x}\|^2 + \|x - \overline{x}\| \\
&\leq 3 \|x - \overline{x}\|,
\end{aligned}$$

for any x close to $\overline{x}$. Hence

$$\liminf_{\substack{x\to\overline{x}\\x\notin A}} \frac{d_A(x)-d_A(\overline{x})-\langle\xi,x-\overline{x}\rangle}{\|x-\overline{x}\|} \geq \liminf_{\substack{x\to\overline{x}\\x\notin A}} \frac{\|x-u\|-\|x-\overline{x}\|^2-\langle\xi,x-\overline{x}\rangle}{\|x-\overline{x}\|}$$

$$\geq \liminf_{\substack{x\to\overline{x}\\x\notin A}} \frac{\|x-u\|-\langle\xi,x-u\rangle}{\|x-\overline{x}\|} - \limsup_{\substack{x\to\overline{x}\\x\notin A}} \frac{\langle\xi,u-\overline{x}\rangle}{\|x-\overline{x}\|}$$

$$\geq \liminf_{\substack{x\to\overline{x}\\x\notin A}} \frac{(1-\|\xi\|)\|x-u\|}{\|x-\overline{x}\|} - \limsup_{\substack{x\to\overline{x}\\x\notin A}} \frac{\langle\xi,u-\overline{x}\rangle}{\|x-\overline{x}\|} \geq -\limsup_{\substack{x\to\overline{x}\\x\notin A}} \frac{\langle\xi,u-\overline{x}\rangle}{\|x-\overline{x}\|}.$$

If $\langle\xi,u-\overline{x}\rangle > 0$ (hence $u\neq\overline{x}$), then

$$\frac{\langle\xi,u-\overline{x}\rangle}{\|x-\overline{x}\|} \leq \frac{3\langle\xi,u-\overline{x}\rangle}{\|u-\overline{x}\|}.$$

This implies that $-\limsup_{\substack{x\to\overline{x}\\x\notin A}} \frac{\langle\xi,u-\overline{x}\rangle}{\|x-\overline{x}\|} \geq 0$, since $\xi\in N_F(A,\overline{x})$. Because in case that $\langle\xi,u-\overline{x}\rangle \leq 0$ one trivially has the same inequality, we have hence

$$\liminf_{\substack{x\to\overline{x}\\x\notin A}} \frac{d_A(x)-d_A(\overline{x})-\langle\xi,x-\overline{x}\rangle}{\|x-\overline{x}\|} \geq 0.$$

Moreover, from $\xi\in N_F(A,\overline{x})$, we get that

$$\liminf_{\substack{x\to\overline{x}\\x\in A}} \frac{d_A(x)-d_A(\overline{x})-\langle\xi,x-\overline{x}\rangle}{\|x-\overline{x}\|} = -\limsup_{\substack{x\to\overline{x}\\x\notin A}} \frac{\langle\xi,x-\overline{x}\rangle}{\|x-\overline{x}\|} \geq 0.$$

This implies that $\xi\in\partial_F d_A(\overline{x})$, and the first equality from (5.2.22) is completely proved.

To finish the proof, observe that the second equality from (5.2.22) easily follows from the first one. $\qquad\square$

Theorem 5.2.25. *Let $A\subset\mathbb{R}^p$ be a closed set and $\overline{x}\in A$. Then*

$$N_M(A,\overline{x}) = \bigcup_{\lambda>0}\lambda\partial_M d_A(\overline{x}). \tag{5.2.24}$$

Proof For the direct inclusion, take $\xi\in N_M(A,\overline{x})\setminus\{0\}$, and find by definition sequences $(x_n)\xrightarrow{A}\overline{x}$ and $(\xi_n)\to\xi$ such that $\xi_n\in N_F(A,x_n)$ for any n. Since (ξ_n) is bounded, there is a bounded sequence of $\lambda_n>0$ such that $\frac{\|\xi_n\|}{\lambda_n}\leq 1$ for any n. It means that $\frac{\xi_n}{\lambda_n}\in\partial_F d_A(x_n)$ for any n, by the first relation from (5.2.22). Without loss of generality, because $\lambda_n\geq\|\xi_n\|$, we may suppose that λ_n converges to some $\lambda>0$. Moreover, using the representation of the Mordukhovich subgradients (5.2.20), we get that $\xi\in\lambda\partial_M d_A(\overline{x})$.

Suppose now A is closed and prove the opposite inclusion in (5.2.24). Take $\xi\in\partial_M d_A(\overline{x})$, and find by (5.2.20) some sequences $(x_n)\to\overline{x}$ and $(\xi_n)\to\xi$ such that $\xi_n\in$

$\partial_F d_A(x_n)$ for any n. If there is (x_k), a subsequence of (x_n), which belongs to A, then the desired inclusion follows by passing to the limit for $k \to \infty$ and taking into account that $\xi_k \in \partial_F d_A(x_k) = N_F(A, x_k) \cap D(0, 1)$ for any k. Suppose next that $x_n \notin A$ for all $n \in \mathbb{N}$. Since

$$\liminf_{x \to x_n} \frac{d_A(x) - d_A(x_n) - \langle \xi_n, x - x_n \rangle}{\|x - x_n\|} \geq 0 > -\frac{1}{n}, \tag{5.2.25}$$

it follows that there is $\eta_n \downarrow 0$ such that

$$\langle \xi_n, x - x_n \rangle \leq \frac{1}{n} \|x - x_n\|, \quad \text{for any } x \in D(x_n, \eta_n) \cap A, \ n \in \mathbb{N}.$$

Choose $\rho_n \downarrow 0$ such that $\rho_n < \min\left\{\eta_n^2, \frac{1}{n} d_A(x_n)\right\}$, and take $v_n \downarrow 1$ such that $(v_n - 1)d_A(x_n) < \rho_n^2$. We can find then $\tilde{x}_n \in A$ such that $\|\tilde{x}_n - x_n\| \leq v_n d_A(x_n)$, and because of (5.2.25) we get that

$$\langle \xi_n, u \rangle \leq d_A(x_n + u) - d_A(x_n) + \frac{1}{n}\|u\|$$

$$\leq d_A(x_n + u) - v_n^{-1}\|\tilde{x}_n - x_n\| + \frac{1}{n}\|u\|$$

$$\leq d_A(\tilde{x}_n + u) + (1 - v_n^{-1})\|\tilde{x}_n - x_n\| + \frac{1}{n}\|u\|$$

whenever $\|u\| \leq \eta_n$. It follows that

$$\left\langle \xi_n, x - \tilde{x}_n \right\rangle \leq (1 - v_n^{-1})\|\tilde{x}_n - x_n\| + \frac{1}{n}\|x - \tilde{x}_n\|$$

whenever $x \in D(\tilde{x}_n, \eta_n) \cap A$, hence

$$0 \leq \varphi_n(x) := -\left\langle \xi_n, x - \tilde{x}_n \right\rangle + \frac{1}{n}\|x - \tilde{x}_n\| + y_n^2, \quad x \in D(\tilde{x}_n, \eta_n) \cap A,$$

where $y_n^2 := (1 - v_n^{-1})\|\tilde{x}_n - x_n\|$. This implies that

$$y_n^2 = \varphi_n(\tilde{x}_n) \leq \inf_{x \in D(\tilde{x}_n, \eta_n) \cap A} \varphi_n(x) + y_n^2$$

for any n. We can apply hence the Ekeland Variational Principle to the continuous function φ_n on the closed set $D(\tilde{x}_n, \eta_n) \cap A$ (see Remark 3.1.13), in order to get $\hat{x}_n \in D(\tilde{x}_n, \eta_n) \cap A$ such that $\|\hat{x}_n - \tilde{x}_n\| \leq y_n$ and

$$-\left\langle \xi_n, \hat{x}_n - \tilde{x}_n \right\rangle + \frac{1}{n}\|\hat{x}_n - \tilde{x}_n\| \leq -\left\langle \xi_n, x - \tilde{x}_n \right\rangle + \frac{1}{n}\|x - \tilde{x}_n\| + y_n\|x - \hat{x}_n\| \tag{5.2.26}$$

for all $x \in D(\tilde{x}_n, \eta_n) \cap A$. By the fact that $y_n^2 \leq v_n(1 - v_n^{-1})d_A(x_n) < \rho_n^2$, defining $r_n := \rho_n - y_n > 0$, we get that

$$\|x - \hat{x}_n\| \leq r_n \Rightarrow \|x - \tilde{x}_n\| \leq \|x - \hat{x}_n\| + y_n \leq \rho_n \leq \eta_n.$$

It follows from (5.2.26) that

$$\langle \xi_n, x - \hat{x}_n \rangle \leq \frac{1}{n}\left(\|x - \tilde{x}_n\| - \|\hat{x}_n - \tilde{x}_n\|\right) + y_n\|x - \hat{x}_n\| \leq \left(\frac{1}{n} + y_n\right)\|x - \hat{x}_n\|$$

whenever $x \in D(\widehat{x}_n, r_n) \cap A$.

Since A is a closed set, for each n we form $\widehat{x}_n + \alpha \xi_n$ with some parameter $\alpha > 0$, and select $w_n \in \mathrm{pr}_A(\widehat{x}_n + \alpha \xi_n)$. It means that

$$\left\| \widehat{x}_n + \alpha \xi_n - w_n \right\|^2 \le \alpha^2 \left\| \xi_n \right\|^2 ,$$

because $\widehat{x}_n \in A$. We have that

$$\left\| \widehat{x}_n + \alpha \xi_n - w_n \right\|^2 = \left\| \widehat{x}_n - w_n \right\|^2 + 2\alpha \left\langle \xi_n, \widehat{x}_n - w_n \right\rangle + \alpha^2 \left\| \xi_n \right\|^2 ,$$

hence we obtain that

$$\left\| \widehat{x}_n - w_n \right\|^2 \le 2\alpha \left\langle \xi_n, w_n - \widehat{x}_n \right\rangle \text{ for any } \alpha > 0. \tag{5.2.27}$$

Using the convergence $w_n \to \widehat{x}_n$ if $\alpha \downarrow 0$, we find from above a sequence $(\alpha_n) \downarrow 0$ such that

$$\left\langle \xi_n, w_n - \widehat{x}_n \right\rangle \le \left(\frac{1}{n} + y_n \right) \left\| w_n - \widehat{x}_n \right\| .$$

But this implies that $\left\| w_n - \widehat{x}_n \right\| \le 2\alpha_n \left(\frac{1}{n} + y_n \right)$ due to (5.2.27). Hence, $w_n \to \overline{x}$ for $n \to \infty$. Moreover, by defining

$$\chi_n := \xi_n + \frac{1}{\alpha_n} (\widehat{x}_n - w_n),$$

we get that $\| \chi_n - \xi_n \| \le 2 \left(\frac{1}{n} + y_n \right)$, hence $\chi_n \to \xi$ for $n \to \infty$.

To end the proof of the theorem, it remains to show that $\chi_n \in N_F(A, w_n)$ for any n. Indeed, for every fixed $x \in A$ we have that

$$
\begin{aligned}
0 &\le \left\| \widehat{x}_n + \alpha_n \xi_n - x \right\|^2 - \left\| \widehat{x}_n + \alpha_n \xi_n - w_n \right\|^2 \\
&= \left\langle \alpha_n \xi_n + \widehat{x}_n - x, \alpha_n \xi_n + \widehat{x}_n - w_n \right\rangle + \left\langle \alpha_n \xi_n + \widehat{x}_n - x, w_n - x \right\rangle \\
&\quad - \left\langle \alpha_n \xi_n + \widehat{x}_n - w_n, x - w_n \right\rangle - \left\langle \alpha_n \xi_n + \widehat{x}_n - w_n, \alpha_n \xi_n + \widehat{x}_n - x \right\rangle \\
&= -2\alpha_n \left\langle \chi_n, x - w_n \right\rangle + \left\| x - w_n \right\|^2 .
\end{aligned}
$$

It means that

$$\left\langle \chi_n, x - w_n \right\rangle \le \frac{1}{2\alpha_n} \left\| x - w_n \right\|^2 \text{ for all } x \in A,$$

which shows that $\chi_n \in N_F(A, w_n)$ for any n, and this ends the proof of the theorem. $\square$

5.2.3 The Extremal Principle

We dedicate this section to an important concept concerning the local extremality of sets, which will provide us, by different variants of the associated Extremal Principle, powerful tools of understanding and proving further results.

Definition 5.2.26 (local extremality of sets). *Let $A_1, \ldots, A_k$ be nonempty subsets of $\mathbb{R}^p$ for $k \geq 2$, and let $\bar{x} \in A_1 \cap \ldots \cap A_k$. We say that $\bar{x}$ is a local extremal point of the set system $\{A_1, \ldots, A_k\}$ if there are sequences $(x_{in}) \subset \mathbb{R}^p, i = 1, \ldots, k$, and a neighborhood U of $\bar{x}$ such that $x_{in} \to 0$ for $n \to \infty$ and*

$$\bigcap_{i=1}^{k}(A_i - x_{in}) \cap U = \emptyset \text{ for sufficiently large } n \in \mathbb{N}.$$

In this case $\{A_1, \ldots, A_k, \bar{x}\}$ is called an extremal system in $\mathbb{R}^p$.

The local extremality of sets at a common point means, intuitively, that these sets can be locally "pushed apart" by a small translation of one of them. If $n = 2$, the local extremality of $\{A_1, A_2, \bar{x}\}$ can be equivalently written as follows: there exists a neighborhood U of $\bar{x}$ such that for any $\varepsilon > 0$, there is $x \in D(0, \varepsilon)$ such that $(A_1 + x) \cap A_2 \cap U = \emptyset$. For instance, if one takes the sets $A_1 := \{(x, x) \mid x \in \mathbb{R}\}$ and $A_2 := \{(x, -x) \mid x \in \mathbb{R}\}$, then $\{A_1, A_2, (0, 0)\}$ is not an extremal system in $\mathbb{R}^2$. In this example, the condition $A_1 \cap A_2 = \{\bar{x}\}$ does not mean in general that $\{A_1, A_2, \bar{x}\}$ is an extremal system. It is the case, though, if $A_2 = \{\bar{x}\}$ and $\bar{x} \in \text{bd } A_1$.

Another important example of extremal system follows. Consider the constrained minimization problem

$$\min f(x) \text{ for } x \in S \subset \mathbb{R}^p,$$

where $f : \mathbb{R}^p \to \mathbb{R}$, and suppose $\bar{x}$ is a local solution of this problem. If one takes $A_1 := \text{epi} f$ and $A_2 := S \times \{f(\bar{x})\}$, then $\{A_1, A_2, (\bar{x}, f(\bar{x}))\}$ is an extremal system in $\mathbb{R}^{p+1}$. This can be seen from the fact that V, a neighborhood of $\bar{x}$, exists and $f(x) \geq f(\bar{x})$, for every $x \in V \cap S$, hence if $(\alpha_n) \subset (-\infty, 0)$ converges to 0, $\alpha - \alpha_n \geq f(x) - \alpha_n > f(\bar{x})$ for every $(x, \alpha) \in \text{epi} f \cap [V \cap S \times \mathbb{R}]$. Consequently, taking $x_{1n} =: (0, \alpha_n)$, $x_{2n} := (0, 0)$ and $U := V \times \mathbb{R}$, one has

$$(A_1 - x_{1n}) \cap (A_2 - x_{2n}) \cap U = \left\{ (x, \alpha - \alpha_n) \middle| \begin{array}{c} (x, \alpha) \in \text{epi} f \cap [V \cap S \times \mathbb{R}] \\ \alpha - \alpha_n = f(\bar{x}) \end{array} \right\}$$
$$= \emptyset.$$

We proceed by introducing two versions of the extremal principle, which can be seen, as we will justify in the next, as local extensions of the separation theorems for nonconvex sets around the extremal points (see Theorems 4.1.3, 4.1.4). We say that a set $A \subset \mathbb{R}^p$ is closed around $\bar{x} \in A$ if there is $\varepsilon > 0$ such that $A \cap D(\bar{x}, \varepsilon)$ is closed. Notice that every closed set, as well as every open set, is locally closed.

Definition 5.2.27. *Let $\{A_1, \ldots, A_k, \bar{x}\}$ be an extremal system in $\mathbb{R}^p$.*

(i) $\{A_1, \ldots, A_k, \bar{x}\}$ satisfies the approximate extremal principle if for every $\varepsilon > 0$, there are $x_i \in A_i \cap D(\bar{x}, \varepsilon)$ and $\xi_i \in N_F(A_i, x_i) + D(0, \varepsilon), i = 1, \ldots, k$ such that

$$\xi_1 + \ldots + \xi_k = 0, \quad \|\xi_1\| + \ldots + \|\xi_k\| = 1. \tag{5.2.28}$$

(ii) $\{A_1, \ldots, A_k, \overline{x}\}$ *satisfies the exact extremal principle if there are* $\xi_i \in N_M(A_i, \overline{x})$, $i = 1, \ldots, k$ *such that (5.2.28) holds.*

We say that the corresponding version of the extremal principle holds in $\mathbb{R}^p$ *if it holds for every extremal system* $\{A_1, \ldots, A_k, \overline{x}\}$ *, where all the sets* A_i *are closed around* $\overline{x}$.

If $\{A_1, \ldots, A_k, \overline{x}\}$ satisfies the exact extremal principle, then it also satisfies the approximate extremal principle. This follows immediately: from the definition of the Mordukhovich normal cones, since $\xi_i \in N_M(A_i, \overline{x})$, there are the sequences $(x_{in}) \xrightarrow{A_i} \overline{x}$ and $(\xi_{in}) \to \xi_i$ such that $\xi_{in} \in N_F(A_i, x_i)$, $\forall i = 1, \ldots, k$. Then for every $\varepsilon > 0$, one can choose n sufficiently large such that $x_{in} \in A_i \cap D(\overline{x}, \varepsilon)$, and also $\xi_i = \xi_{in} + (\xi_i - \xi_{in}) \in N_F(A_i, x_i) + D(0, \varepsilon)$, for every $i = 1, \ldots, k$.

Observe also that in case $n = 2$, and A_1 and A_2 are convex sets, due to the form of the normal cones in the convex case, if the exact extremal principle holds, then it reduces to the existence of

$$\xi \in N_M(A_1, \overline{x}) \cap \left(-N_M(A_2, \overline{x}) \right) \setminus \{0\},$$

which yields

$$\langle \xi, x_1 \rangle \leq \langle \xi, x_2 \rangle \text{ for every } x_1 \in A_1, \text{ and } x_2 \in A_2,$$

i.e., exactly the separation property of A_1 and A_2.

The main result of this section, proven by Mordukhovich by the method of metric approximations, is the following:

Theorem 5.2.28 (exact extremal principle). *The exact extremal principle holds in* $\mathbb{R}^p$.

Proof Consider the extremal system $\{A_1, \ldots, A_k, \overline{x}\}$, where we suppose without loss of generality that all the sets A_i are closed. Also, without loss of generality, suppose that the neighborhood U of $\overline{x}$ from Definition 5.2.27 is $\mathbb{R}^p$, and consider the sequences $(x_{in}) \subset \mathbb{R}^p$, $i = 1, \ldots, k$ such that $x_{in} \to 0$ and

$$\bigcap_{i=1}^{k} (A_i - x_{in}) = \emptyset \text{ for all } n \in \mathbb{N}. \tag{5.2.29}$$

Moreover, for every n, consider the minimization problem

$$\min d_n(x) := \left[\sum_{i=1}^{k} d_{A_i}^2 (x + x_{in}) \right]^{1/2} + \|x - \overline{x}\|^2, \quad x \in \mathbb{R}^p.$$

By using Theorem 3.1.7, the above problem admits a solution x_n. Denote

$$\alpha_n := \left[\sum_{i=1}^{k} d_{A_i}^2 (x_n + x_{in}) \right]^{1/2}.$$

If α_n would be 0, this would mean, due to the closedness of the sets A_i, that $x_n \in \bigcap_{i=1}^{k}(A_i - x_{in})$, which obviously contradicts (5.2.29). Hence,

$$0 < d_n(x_n) = \alpha_n + \|x_n - \overline{x}\|^2 \le \left[\sum_{i=1}^{k} d_{A_i}^2(\overline{x} + x_{in})\right]^{1/2}$$

$$\le \left[\sum_{i=1}^{k} x_{in}^2\right]^{1/2} \downarrow 0,$$

which shows that $x_n \to \overline{x}$ and $\alpha_n \downarrow 0$.

Now, since the sets A_i are closed, consider the elements u_{in} from the nonempty projection sets $\mathrm{pr}_{A_i}(x_n + x_{in}) : u_{in} \in A_i$ and $d_{A_i}(x_n + x_{in}) = \|x_n + x_{in} - u_{in}\|$. Also, for all n, consider the minimization problem

$$\min \rho_n(x) := \left[\sum_{i=1}^{k} \|x + x_{in} - u_{in}\|^2\right]^{1/2} + \|x - \overline{x}\|^2, \quad x \in \mathbb{R}^p. \tag{5.2.30}$$

Because

$$\rho_n(x) \ge d_n(x) \ge d_n(x_n) = \rho_n(x_n),$$

it follows that the problem (5.2.30) has the optimal solution x_n. Moreover, since $\alpha_n > 0$, the functions ρ_n are C^1 around x_n. Due to the classical Fermat rule, this means that

$$\nabla \rho_n(x_n) = \sum_{i=1}^{k} \xi_{in} + 2(x_n - \overline{x}) = 0, \tag{5.2.31}$$

where $\xi_{in} := \dfrac{1}{\alpha_n}(x_n + x_{in} - u_{in})$, $i = 1, \dots, k$. Remark that

$$\|\xi_{1n}\|^2 + \dots + \|\xi_{kn}\|^2 = 1. \tag{5.2.32}$$

It follows that all ξ_{in} belong to the unit ball of $\mathbb{R}^p$, which is a compact set. We may suppose then, without loss of generality, that there are elements ξ_i in $D(0, 1)$ such that $\xi_{in} \to \xi_i$ for every $i = 1, \dots, k$. Passing to the limit in (5.2.31), we get that $\xi_1 + \dots + \xi_k = 0$. Moreover, due to the fact that $u_{in} \in \mathrm{pr}_{A_i}(x_n + x_{in})$, $\frac{1}{\alpha_n} \subset (0, \infty)$, $x_n + x_{in} \to \overline{x}$ and $\xi_{in} \to \xi_i$, it means, based on formula (5.2.7), that $\xi_i \in N_M(A_i, \overline{x})$ for all $i = 1, \dots, k$. Passing to the limit in (5.2.32), it means that not all ξ_i are 0, which gives us

$$\|\xi_1\| + \dots + \|\xi_k\| = 1,$$

eventually by replacing ξ_i with $(\|\xi_1\| + \dots + \|\xi_k\|)^{-1} \cdot \xi_i$. $\qquad\square$

As consequence, we get the next result. We say that a set is proper if it is nonempty and different from the whole space.

Corollary 5.2.29. *For every proper closed set $A \subset \mathbb{R}^p$ and $\overline{x} \in A$, one has*

$$N_M(A, \overline{x}) \neq \{0\} \iff \overline{x} \in \mathrm{bd}\,A.$$

Proof It follows immediately that if $\overline{x} \notin \mathrm{bd}\,A$, then $\overline{x} \in \mathrm{int}\,A$, hence $N_M(A, \overline{x}) = \{0\}$. If $\overline{x} \in \mathrm{bd}\,A$, then $\{A, \{\overline{x}\}, \overline{x}\}$ is an extremal system in $\mathbb{R}^p$, hence using Theorem 5.2.28, one has that there are $\xi_1 \in N_M(A, \overline{x})$ and $\xi_2 \in N_M(\{\overline{x}\}, \overline{x}) = \mathbb{R}^p$, not equal both to 0, such that $\xi_1 = -\xi_2$. It follows that $0 \neq \xi_1 \in N_M(A, \overline{x})$. $\qquad\square$

5.2.4 Calculus Rules

One of the main advantages of the generalized differentiation objects (i.e., normal cones, subdifferentials) presented in this chapter is their rich associated calculus. As we presented in the introduction of this chapter, since it would be outside the aim of this book, we do not present generalized differentiation objects related to set-valued mappings (derivatives and coderivatives), which are closely interrelated in the calculus rules mentioned before. In what follows, we will concentrate on rules involving subdifferentials of sums and differences of functions. Moreover, such sum rules have the advantage to fully employ some important results previously given, in such a way that one can draw an intuitive picture on how the whole machinery of calculus associated to Mordukhovich differentiation objects works. For many other results involving calculus rules for Mordukhovich generalized differentiation objects, the reader could consult (Mordukhovich, 2006).

We begin our exposition with a sum rule where one of the functions is differentiable.

Proposition 5.2.30. *Let $f : \mathbb{R}^p \to \mathbb{R}$.*

(i) For any $g : \mathbb{R}^p \to \mathbb{R}$ Fréchet differentiable at $\overline{x}$, one has

$$\partial_F(f + g)(\overline{x}) = \partial_F f(\overline{x}) + \nabla g(\overline{x}). \tag{5.2.33}$$

(ii) If f is lower semicontinuous around $\overline{x}$, for any $g : \mathbb{R}^p \to \mathbb{R}$ continuously differentiable around $\overline{x}$, one has

$$\partial_M(f + g)(\overline{x}) = \partial_M f(\overline{x}) + \nabla g(\overline{x}). \tag{5.2.34}$$

(iii) For any $g : \mathbb{R}^p \to \mathbb{R}$ Lipschitz continuous around $\overline{x}$, one has

$$\partial^\infty(f + g)(\overline{x}) = \partial^\infty f(\overline{x}). \tag{5.2.35}$$

Proof (i) Let us prove the "$\subset$" inclusion in (5.2.33). Suppose $\xi \in \partial_F(f + g)(\overline{x})$. Fix arbitrary $\varepsilon > 0$. Using the analytical characterization of the Fréchet subgradients (i.e., Theorem 5.2.11), there exists $\delta > 0$ such that, for every $x \in D(\overline{x}, \delta)$,

$$f(x) + g(x) - f(\overline{x}) - g(\overline{x}) - \langle \xi, x - \overline{x} \rangle \geq -\frac{\varepsilon}{2} \|x - \overline{x}\|.$$

Also, from the Fréchet differentiability of g at $\overline{x}$, there exists $y \in (0, \delta)$ such that, for every $x \in D(x, y)$,

$$-g(x) + g(\overline{x}) + \langle \nabla g(\overline{x}), x - \overline{x} \rangle \geq -\frac{\varepsilon}{2} \|x - \overline{x}\|.$$

Adding the previous relations, one gets that $\xi - \nabla g(\overline{x}) \in \partial_F f(\overline{x})$.

For the reverse inclusion, take $\xi \in \partial_F f(\overline{x})$. Then $\xi \in \partial_F ((f + g) - g)(\overline{x}) \subset \partial_F(f + g)(\overline{x}) - \nabla g(\overline{x})$ from the already proven inclusion. It means that $\xi + \nabla g(\overline{x}) \in \partial_F(f + g)(\overline{x})$.

(ii) Consider now $\xi \in \partial_M(f+g)(\overline{x})$, i.e., $(\xi, -1) \in N_M(\mathrm{epi}(f+g), (\overline{x}, (f+g)(\overline{x})))$. Since $f+g$ is a lower semicontinuous function, the set epi f is closed, hence using Proposition 5.2.8 we know that $(\xi, -1) \in N_M(\mathrm{gr}(f + g), (\overline{x}, (f + g)(\overline{x})))$. This means, based on the definition of basic subgradients, that there exist $(x_n) \to \overline{x}$, $(\xi_n) \to \xi$ and $(\lambda_n) \to 1$ such that $(\xi_n, -\lambda_n) \in N_F(\mathrm{gr}(f + g), (x_n, (f + g)(x_n)))$ for every n. This means that for every $\varepsilon > 0$, there exists $\delta > 0$ such that for every $(x, (f + g)(x)) \in D((x_n, (f + g)(x_n)), \delta)$, one has

$$\langle \xi_n, x - x_n \rangle - \lambda_n(f(x) + g(x) - f(x_n) - g(x_n)) \leq \varepsilon(\|x - x_n\| \tag{5.2.36}$$
$$+ |f(x) - f(x_n)| + |g(x) - g(x_n)|).$$

On the other hand, for every fixed such x and x_n, due to the fact that g is C^1, there exists y_n on the segment which joins x and x_n such that

$$g(x) - g(x_n) = \langle \nabla g(y_n), x - x_n \rangle.$$

Combining relation (5.2.36) with the Lipschitz property of g (with modulus $\ell \geq 0$), one gets that

$$\langle \xi_n - \lambda_n \nabla g(y_n), x - x_n \rangle - \lambda_n(f(x) - f(x_n)) \leq \varepsilon(1 + \ell)\left(\|x - x_n\| + |f(x) - f(x_n)|\right). \tag{5.2.37}$$

Since if x is close to x_n, and $(f + g)(x)$ is close to $(f + g)(x_n)$, one gets from the Lipschitz property of g that $f(x)$ must be close to $f(x_n)$. The previous relation shows that $(\xi_n - \lambda_n \nabla g(y_n), -\lambda_n) \in N_F(\mathrm{gr}\, f, (x_n, f(x_n)))$. Set $\eta_n := \xi_n - \lambda_n \nabla g(y_n)$ and suppose, by contradiction, that $(\eta_n, -\lambda_n) \notin N_F(\mathrm{epi}\, f, (x_n, f(x_n)))$. Then there exist $y \in (0, \lambda_n)$ and sequences $(u_j, \alpha_j) \overset{\mathrm{epi}\, f}{\to} (x_n, f(x_n))$ such that

$$\langle \eta_n, u_j - x_n \rangle - \lambda_n (\alpha_j - f(x_n)) > y \|(u_j, \alpha_j) - (x_n, f(x_n))\|.$$

On the other hand, since $\alpha_j \geq f(u_j)$, one has

$$\|(u_j, f(u_j)) - (x_n, f(x_n))\| \leq \|(u_j, \alpha_j) - (x_n, f(x_n))\| + (\alpha_j - f(u_j)),$$

hence

$$\langle \eta_n, u_j - x_n \rangle - \lambda_n (\alpha_j - f(x_n)) > y \|(u_j, f(u_j)) - (x_n, f(x_n))\| - y (\alpha_j - f(u_j))$$
$$> y \|(u_j, f(u_j)) - (x_n, f(x_n))\| - \lambda_n (\alpha_j - f(u_j)).$$

It means that

$$\langle \eta_n, u_j - x_n \rangle - \lambda_n \left(f(u_j) - f(x_n) \right) > y \left\| (u_j, f(u_j)) - (x_n, f(x_n)) \right\|,$$

which contradicts the fact that $(\eta_n, -\lambda_n) \in N_F(\mathrm{gr}\, f, (x_n, f(x_n)))$. It follows that $(\xi_n - \lambda_n \nabla g(y_n), -\lambda_n) \in N_F(\mathrm{epi}\, f, (x_n, f(x_n)))$ for every n. Since $\xi_n - \lambda_n \nabla g(y_n) \to \xi - \nabla g(\overline{x})$ due to the fact that g is C^1, and $(x_n, f(x_n)) \overset{\mathrm{epi}\, f}{\to} (\overline{x}, f(\overline{x}))$ due to the fact that $x_n \to \overline{x}$, $g(x_n) \to g(\overline{x})$ and $(f+g)(x_n) \to (f+g)(\overline{x})$, it follows that $\xi - \nabla g(\overline{x}) \in \partial_M f(\overline{x})$. We proved the "$\subset$" inclusion in (5.2.34). The other inclusion can be deduced as in (i).

(iii) Let us prove now the "$\subset$" inclusion in (5.2.35). Take $\xi \in \partial^\infty (f + g)(\overline{x})$. There exist the sequences $(x_n, \alpha_n) \overset{\mathrm{epi}(f+g)}{\to} (\overline{x}, (f+g)(\overline{x}))$, $(\xi_n) \to \xi$ and $(\lambda_n) \to 0$ such that, for every $\varepsilon > 0$, there exists $\delta > 0$ such that

$$\langle \xi_n, x - x_n \rangle - \lambda_n (\alpha - \alpha_n) \leq \varepsilon \left(\| x - x_n \| + |\alpha - \alpha_n| \right)$$

for every $(x, \alpha) \in \mathrm{epi}(f + g)$ such that $x \in D(x_n, \delta)$ and α such that $|\alpha - \alpha_n| < \delta$. Suppose ℓ is the Lipschitz modulus of g around $\overline{x}$, and take $\beta_n := \alpha_n - g(x_n)$. We have that $(x_n, \beta_n) \overset{\mathrm{epi}\, f}{\to} (\overline{x}, f(\overline{x}))$ and if $(x, \alpha) \in \mathrm{epi}\, f$, $x \in D\left(x_n, \frac{\delta}{2(\ell+1)}\right)$, and $|\alpha - \beta_n| \leq \frac{\delta}{2(\ell+1)}$, one has $(x, g(x) + \alpha) \in \mathrm{epi}(f + g)$, and

$$\left| (g(x) + \alpha) - \alpha_n \right| \leq \left| g(x) - g(x_n) \right| + \left| \alpha - (\alpha_n - g(x_n)) \right|$$
$$\leq \ell \| x - x_n \| + |\alpha - \beta_n| \leq \frac{\delta}{2} < \delta,$$

hence

$$\langle \xi_n, x - x_n \rangle - \lambda_n (\alpha - \beta_n) = \langle \xi_n, x - x_n \rangle - \lambda_n((\alpha + g(x)) - \alpha_n) - \lambda_n(g(x_n) - g(x))$$
$$\leq \varepsilon \left(\| x - x_n \| + \left| (\alpha + g(x)) - \alpha_n \right| \right) + \lambda_n \left| g(x_n) - g(x) \right|$$
$$\leq \varepsilon \left(\| x - x_n \| + \left| (\alpha + g(x_n)) - \alpha_n \right| \right) + (\varepsilon + \lambda_n)\ell \| x - x_n \|$$
$$\leq (\varepsilon + \varepsilon\ell + \lambda_n\ell) \left(\| x - x_n \| + |\alpha - \beta_n| \right),$$

for every $(x, \alpha) \in \mathrm{epi}\, f$, $x \in D\left(x_n, \frac{\delta}{2(\ell+1)}\right)$, and $|\alpha - \beta_n| \leq \frac{\delta}{2(\ell+1)}$. Since $\varepsilon' := \varepsilon + \varepsilon\ell + \lambda_n\ell$ can be taken positive and arbitrarily small due to the fact that $\lambda_n \to 0$ and $\varepsilon > 0$ is arbitrary, it follows from above that $(\xi_n, -\lambda_n) \in N_F(\mathrm{epi}\, f, (x_n, \beta_n))$ for every n, hence $(\xi, 0) \in N_M(\mathrm{epi}\, f, (\overline{x}, f(\overline{x})))$. The other inclusion can be proved similarly to the final part of (i). □

Lemma 5.2.31. *Consider $f_1, f_2 : \mathbb{R}^p \to \mathbb{R}$ such that f_1 is Lipschitz continuous around $\overline{x}$, and f_2 is lower semicontinuous around this point. If $f_1 + f_2$ attains a local minimum at $\overline{x}$, then for every $\varepsilon > 0$, there are $x_i \in D(\overline{x}, \varepsilon)$ with $\left| f_i(x_i) - f_i(\overline{x}) \right| \leq \varepsilon$, $i = 1, 2$, such that*

$$0 \in \partial_F f_1(x_1) + \partial_F f_2(x_2) + D(0, \varepsilon). \tag{5.2.38}$$

Proof Fix $\varepsilon > 0$. Without loss of generality, suppose $\overline{x} = 0$, $f_1(\overline{x}) = f_2(\overline{x}) = 0$, f_1 is Lipschitz on $D(0, \varepsilon)$ with modulus $\ell > 0$, and f_2 is lower semicontinuous on $D(0, \varepsilon)$. Then the sets

$$A_1 := \text{epi} f_1 \text{ and } A_2 := \{(x, \alpha) \mid f_2(x) \le -\alpha\}$$

are closed around $(0, 0)$. Moreover, since $f_1 + f_2$ attains a local minimum at 0, and $(f_1 + f_2)(0) = 0$, it follows that $(0, 0)$ is an extremal point for $\{A_1, A_2\}$. Using Theorem 5.2.28, it follows that the exact extremal principle, hence also the approximate extremal principle hold for the system $\{A_1, A_2, (0, 0)\}$.

Then for every $y > 0$, there are $(x_i, \alpha_i) \in A_i$, and $(\xi_i, \lambda_i) \in \mathbb{R}^{p+1}$, $i = 1, 2$, such that

$$(\xi_1, -\lambda_1) \in N_F(A_1, (x_1, \alpha_1)), \quad (-\xi_2, \lambda_2) \in N_F(A_2, (x_2, \alpha_2)), \tag{5.2.39}$$

$$\left\|(x_i, \alpha_i)\right\| < y, \quad \frac{1}{2} - y < \left\|(\xi_i, \lambda_i)\right\| < \frac{1}{2} + y, \ i = 1, 2 \tag{5.2.40}$$

$$\left\|(\xi_1, -\lambda_1) + (-\xi_2, \lambda_2)\right\| \le y. \tag{5.2.41}$$

Due to the form of the sets A_1, A_2 and (5.2.39), it follows that $\lambda_1, \lambda_2 \ge 0$. Choose $y \in \left(0, \min\left\{\frac{1}{4(2+\ell)}, \frac{\varepsilon}{4(1+\ell)^2}\right\}\right)$ such that relations (5.2.39)–(5.2.41) hold. Using the fact that f_1 is Lipschitz continuous on $D(0, \varepsilon)$, one gets (see the proof of Theorem 5.2.18) from $(\xi_1, -\lambda_1) \in N_F(A_1, (x_1, \alpha_1))$ with $\left\|(x_1, \alpha_1)\right\| < y < \varepsilon$ that $\|\xi_1\| \le \ell\lambda_1$. Hence,

$$\left\|(\xi_1, \lambda_1)\right\| \le (\ell + 1)\lambda_1.$$

One can now employ (5.2.40) and (5.2.41) to get

$$\lambda_1 \ge \frac{1}{\ell + 1} \left\|(\xi_1, \lambda_1)\right\| > \frac{1}{2(\ell + 1)} - \frac{y}{\ell + 1} > 0$$

and

$$\lambda_1 - \lambda_2 \le |\lambda_1 - \lambda_2| \le \left\|(\xi_1, -\lambda_1) + (-\xi_2, \lambda_2)\right\| \le y,$$

hence

$$\lambda_2 \ge \lambda_1 - y > \frac{1}{2(\ell + 1)} - \frac{y}{\ell + 1} - y > \frac{1}{4(\ell + 1)} > 0$$

due to the choice of y. Since $\alpha_1 \ge f_1(x_1)$, suppose $\alpha_1 > f_1(x_1)$. From $\left(\frac{\xi_1}{\lambda_1}, -1\right) \in N_F(A_1, (x_1, \alpha_1))$, one gets that for every $\delta \in (0, 1)$, there exists $\tau > 0$ such that for every $(u, \alpha) \in D((x_1, \alpha_1), \tau) \cap \text{epi} f_1$, one has

$$\left\langle \frac{\xi_1}{\lambda_1}, u - x_1 \right\rangle - (\alpha - \alpha_1) < \delta \left\|(u - x_1, \alpha - \alpha_1)\right\|. \tag{5.2.42}$$

Choosing $u := x_1$, one can find from $\alpha_1 > f_1(x_1)$ that there is $\alpha \in (f(x_1), \alpha_1)$ arbitrarily close to α_1. For such (u, α), it follows from (5.2.42) that $\alpha_1 - \alpha < \delta(\alpha_1 - \alpha)$, hence $\delta > 1$, a contradiction. Then, $\alpha_1 = f_1(x_1)$. Similarly, one can prove that $\alpha_2 = -f_2(x_2)$. Hence,

$$\eta_1 := \frac{\xi_1}{\lambda_1} \in \partial_F f_1(x_1) \text{ and } \eta_2 := -\frac{\xi_2}{\lambda_2} \in \partial_F f_2(x_2).$$

By (5.2.40), we have that

$$\|x_i\| \le y < \varepsilon \text{ and } \left|f_i(x_i) - f_i(\overline{x})\right| = |\alpha_i| \le y < \varepsilon, \quad i = 1, 2.$$

Moreover,

$$\|\eta_1 + \eta_2\| = \left\| \frac{\xi_1(\lambda_1 - \lambda_2)}{\lambda_1 \lambda_2} - \frac{\xi_1 - \xi_2}{\lambda_2} \right\| \le \frac{\|\xi_1\|}{\lambda_1} \left(\frac{|\lambda_1 - \lambda_2|}{\lambda_2} \right) + \frac{\|\xi_1 - \xi_2\|}{\lambda_2}$$

$$\le \ell \frac{y}{\lambda_2} + \frac{y}{\lambda_2} = \frac{y}{\lambda_2}(\ell + 1) \le 4y(\ell + 1)^2 < \varepsilon.$$

We have hence (5.2.38). $\qquad\qquad\square$

Theorem 5.2.32 (sum rule). *Consider $f_1, f_2 : \mathbb{R}^p \to \mathbb{R}$ such that f_1 is Lipschitz continuous around $\overline{x}$, and f_2 is lower semicontinuous around this point. Then:*
(i) for every $\varepsilon > 0$, one has

$$\partial_F(f_1 + f_2)(\overline{x}) \subset \bigcup \{ \partial_F f_1(x_1) + \partial_F f_2(x_2) \mid x_i \in D(\overline{x}, \varepsilon), \qquad (5.2.43)$$
$$\left|f_i(x_i) - f_i(\overline{x})\right| \le \varepsilon, \ i = 1, 2 \} + D(0, \varepsilon).$$

(ii) one has

$$\partial_M(f_1 + f_2)(\overline{x}) \subset \partial_M f_1(\overline{x}) + \partial_M f_2(\overline{x}). \qquad (5.2.44)$$

Proof (i) Fix $\varepsilon > 0$ and take $\xi \in \partial_F(f_1 + f_2)(\overline{x})$. Then one can find η such that

$$0 < \eta < \min \left\{ \frac{\varepsilon}{4}, \frac{\varepsilon}{\|\xi\| + \varepsilon + 2} \right\}.$$

By using the analytical characterization of the Fréchet subgradients given by (5.2.9), one deduces that the function

$$\left(f_1(x) - \langle \xi, x - \overline{x} \rangle + \varepsilon \|x - \overline{x}\|\right) + f_2(x)$$

attains a local minimum at $\overline{x}$. One may apply Lemma 5.2.31 for the chosen η to find $x_i \in D(\overline{x}, \eta)$ and $\xi_i \in \mathbb{R}^p$, $i = 1, 2$, such that

$$\left|f_1(x_1) - \langle \xi, x_1 - \overline{x} \rangle + \varepsilon \|x_1 - \overline{x}\| - f_1(\overline{x})\right| \le \eta, \quad \left|f_2(x_2) - f_2(\overline{x})\right| \le \eta$$
$$\xi_1 \in \partial_F \left(f_1 - \langle \xi, \cdot \rangle + \eta \|\cdot - \overline{x}\|\right)(x_1), \quad \xi_2 \in \partial_F f_2(x_2),$$
$$-\xi_1 - \xi_2 \in D(0, \eta).$$

By (5.2.33), it follows that $\xi_1' := \xi + \xi_1 \in \partial_F (f_1 + \eta \|\cdot - \overline{x}\|) (x_1)$, and hence $\xi - \xi_1' - \xi_2 \in D(0, \eta)$. Moreover,

$$\left| f_1(x_1) - f_1(\overline{x}) \right| \leq \eta + (\|\xi\| + \varepsilon) \|x_1 - \overline{x}\| < \eta (\|\xi\| + \varepsilon + 1).$$

Now, due to the analytical characterization of the Fréchet subgradients given by (5.2.9), we have from $\xi_1' \in \partial_F (f_1 + \eta \|\cdot - \overline{x}\|) (x_1)$ that the function $f_1 + \varphi$ with

$$\varphi(x) := \eta \|x - \overline{x}\| - \langle \xi_1', x - x_1 \rangle + \eta \|x - x_1\|$$

attains a local minimum at x_1. Remark that, because φ is the sum of three convex (hence, continuous) functions on $\mathbb{R}^p$, one has by Proposition 5.2.14 that $\partial_M \varphi(x) = \partial \varphi(x) \subset -\xi_1' + D(0, 2\eta)$ for any $x \in \mathbb{R}^p$. Using again Lemma 5.2.31 for the chosen η and for the sum $f_1 + \varphi$, we find $x_1' \in D(x_1, \eta)$ such that

$$\left| f_1(x_1') - f_1(x_1) \right| \leq \eta \text{ and } \xi_1' \in \partial_F f_1(x_1') + D(0, 3\eta).$$

Concluding, we found

$$\xi \in \partial_F f_1(x_1') + \partial_F f_2(x_2) + D(0, 4\eta),$$

with $\|x_1' - \overline{x}\| \leq 2\eta < \varepsilon$, $\|x_2 - \overline{x}\| \leq \eta < \varepsilon$, $\left| f_1(x_1') - f_1(\overline{x}) \right| \leq \eta (\|\xi\| + \varepsilon + 2) < \varepsilon$, $\left| f_2(x_2) - f_2(\overline{x}) \right| \leq \eta < \varepsilon$. Hence, (5.2.43) is completely proved.

(ii) Take arbitrary $\xi \in \partial_M(f_1 + f_2)(\overline{x})$. Since $f_1 + f_2$ is lower semicontinuous, it follows by (5.2.20) that there are sequences $(x_n) \to \overline{x}$ and $(\xi_n) \to \xi$ such that $f_1(x_n) + f_2(x_n) \to f_1(\overline{x}) + f_2(\overline{x})$ and $\xi_n \in \partial_F(f_1 + f_2)(x_n)$ for every n. Using now (5.2.43) for $\varepsilon := \frac{1}{n}$, we can find sequences $(x_{in}) \to \overline{x}$ and $(\xi_{in}) \in \mathbb{R}^p$ with $f_i(x_{in}) \to f_i(\overline{x})$, $\xi_{in} \in \partial_F f_i(x_{in})$, $i = 1, 2$, and

$$\|\xi_n - \xi_{1n} - \xi_{2n}\| \leq \frac{1}{n}, \quad \forall n \geq 1. \tag{5.2.45}$$

Because f_1 is Lipschitz continuous around $\overline{x}$, we know that the sequence (ξ_{1n}) is bounded by the Lipschitz modulus of f_1 due to (5.2.19). Also, (ξ_n) is bounded because it is convergent. It follows from (5.2.45) that (ξ_{2n}) is also bounded. Without loss of generality, we may suppose hence that there are $\xi_1, \xi_2 \in \mathbb{R}^p$ such that $\xi_{in} \to \xi_i$, $i = 1, 2$. From (5.2.45), we know that $\xi = \xi_1 + \xi_2$. Moreover, due to (5.2.20), we have that $\xi_i \in \partial_M f_i(x_i)$, which ends the proof of (5.2.44). $\square$

We continue our exposition dedicated to calculus rules by providing a difference rule for the Fréchet subgradients. As mentioned before, its proof uses the smooth variational description of the Fréchet subgradients, given in Theorem 5.2.13.

Theorem 5.2.33. *Let $f, g : \mathbb{R}^p \to \mathbb{R}$ be two functions, and $\overline{x} \in \mathbb{R}^p$. Then*

$$\partial_F(f - g)(\overline{x}) \subset \bigcap_{\xi \in \partial_F g(\overline{x})} \left[\partial_F f(\overline{x}) - \xi \right] \subset \partial_F f(\overline{x}) - \partial_F g(\overline{x}), \tag{5.2.46}$$

provided $\partial_F g(\overline{x}) \neq \emptyset$.

Proof Take $\chi \in \partial_F(f - g)(\overline{x})$ arbitrarily and $\xi \in \partial_F g(\overline{x})$. Using Theorem 5.2.13 for $\xi \in \partial_F g(\overline{x})$, one gets the existence of $s : \mathbb{R}^p \to \mathbb{R}$, Fréchet differentiable at $\overline{x}$, which satisfies:

$$s(\overline{x}) = g(\overline{x}), \ s(x) \leq g(x) \ \forall x \in \mathbb{R}^p, \text{ and } \nabla s(\overline{x}) = \xi. \tag{5.2.47}$$

For any $\varepsilon > 0$, by applying the definition of the Fréchet subgradients for $\chi \in \partial_F(f - g)(\overline{x})$, there is $\delta > 0$ such that

$$\langle \chi, x - \overline{x} \rangle \leq f(x) - g(x) - \big(f(\overline{x}) - g(\overline{x})\big) + \varepsilon \|x - \overline{x}\|$$
$$\leq f(x) - s(x) - \big(f(\overline{x}) - s(\overline{x})\big) + \varepsilon \|x - \overline{x}\|$$

whenever $\|x - \overline{x}\| < \delta$. Using now (5.2.47) and (5.2.33), one gets that

$$\chi \in \partial_F(f - s)(\overline{x}) = \partial_F f(\overline{x}) - \nabla s(\overline{x}) = \partial_F f(\overline{x}) - \xi,$$

which ends the proof. □

By Proposition 5.2.30, (5.2.46) becomes equality if one of the functions f, g is Fréchet differentiable at $\overline{x}$. There exist as well other cases when the equality holds: take $f, g : \mathbb{R} \to \mathbb{R}$ given as $f(x) = \sqrt{|x|}$ and $g(x) = |x|$. Then $\partial_F f(0) = \mathbb{R}$, $\partial_F g(0) = [-1, 1]$, and $\partial_F(f - g)(0) = \mathbb{R}$, hence equality holds in (5.2.46). Also, observe that the assumption that $\partial_F g(\overline{x}) \neq \emptyset$ is essential for the fulfillment of the result: take, for instance, $f, g : \mathbb{R} \to \mathbb{R}$, $f(x) = |x|$ and $g(x) = -|x|$. Then $\partial_F(f - g)(0) = [-2, 2]$, but $\partial_F g(\overline{x}) = \emptyset$.

As corollaries of the Theorem 5.2.33, we obtain some other interesting facts. The first one concerns the so-called DC-functions, i.e., differences of convex functions. We denote, as always, by $\partial f(\overline{x})$ the convex subdifferential.

Corollary 5.2.34. *Let $f, g : \mathbb{R}^p \to \mathbb{R}$ be convex functions, and $\overline{x} \in \mathbb{R}^p$. Then*

$$\partial_F(f - g)(\overline{x}) = \bigcap_{\xi \in \partial g(\overline{x})} \big[\partial f(\overline{x}) - \xi\big]$$

Proof It follows from Theorem 5.2.33 and Proposition 5.2.14. □

The second corollary gives us a calculus rule for the Mordukhovich subdifferential of the difference of two functions.

Corollary 5.2.35. *Let $f : \mathbb{R}^p \to \mathbb{R}$ be lower semicontinuous around $\overline{x} \in \mathbb{R}^p$, and $g : \mathbb{R}^p \to \mathbb{R}$ be locally Lipschitz around $\overline{x}$, and such that $\partial_M g(x)$ is nonempty for every x close to $\overline{x}$. Then*

$$\partial_M(f - g)(\overline{x}) \subset \partial_M f(\overline{x}) - \partial_M g(\overline{x}).$$

Proof Take $\xi \in \partial_M(f-g)(\overline{x})$ arbitrarily. By using Theorem 5.2.21, there exist sequences $(x_n) \overset{f-g}{\to} \overline{x}$ and $(\xi_n) \to \xi$ such that $\xi_n \in \partial_F(f-g)(x_n)$ for every n. Applying Theorem 5.2.33 for the previous relation, one gets that there are sequences (χ_n), (η_n) such that $\chi_n \in \partial_F f(x_n)$ and $\eta_n \in \partial_F g(x_n)$ for every n and $\xi_n = \chi_n - \eta_n$, for every n. Since g is locally Lipschitz around $\overline{x}$, we know by (5.2.19) that the sequence (η_n) is bounded, so we may suppose, without loss of generality, that there is $\eta \in \mathbb{R}^p$ such that $\eta_n \to \eta$. The Lipschitz property of g implies $g(x_n) \to g(\overline{x})$, and by $f(x_n)-g(x_n) \to f(\overline{x})-g(\overline{x})$ one gets that $f(x_n) \to f(\overline{x})$. Moreover, we have that $\eta \in \partial_M g(\overline{x})$. Furthermore, since $\xi_n = \chi_n - \eta_n$, for every n, $\xi_n \to \xi$, and $\eta_n \to \eta$, we get that $\chi_n \to \xi + \eta$, hence $\xi + \eta \in \partial_M f(\overline{x})$. But this means that $\xi \in \partial_M f(\overline{x}) - \partial_M g(\overline{x})$, and the proof is finished. $\qquad\square$

Observe that all the assumptions on g from the previous corollary are automatically satisfied if g is convex.

Recall that for a function $f : \mathbb{R}^p \to \mathbb{R}^k$ and a vector $\xi \in \mathbb{R}^k$, the function $\langle \xi, f \rangle : \mathbb{R}^p \to \mathbb{R}$ is given by

$$\langle \xi, f \rangle (x) = \langle \xi, f(x) \rangle .$$

We can present a chain rule for the Fréchet subdifferential.

Theorem 5.2.36 (chain rule). *Let $f : \mathbb{R}^p \to \mathbb{R}^k$ be Lipschitz around $\overline{x} \in \mathbb{R}^p$, $g : \mathbb{R}^k \to \mathbb{R}$ be a function, and denote $\overline{y} := f(\overline{x})$. Then*

$$\partial_F(g \circ f)(\overline{x}) \subset \bigcap_{-\xi \in \partial_F(-g)(\overline{y})} \partial_F \langle \xi, f \rangle (\overline{x}), \tag{5.2.48}$$

provided $\partial_F(-g)(\overline{y}) \neq \emptyset$. Moreover, (5.2.48) holds as equality if g is Fréchet differentiable at $\overline{y}$.

Proof Take $\chi \in \partial_F(g \circ f)(\overline{x})$ and $-\xi \in \partial_F(-g)(\overline{y})$. Using Theorem 5.2.13, one gets the existence of $s : \mathbb{R}^p \to \mathbb{R}$, Fréchet differentiable at $\overline{y}$, which satisfies:

$$s(\overline{y}) = g(\overline{y}), \; g(y) \leq s(y) \; \forall y \in \mathbb{R}^k, \text{ and } \nabla s(\overline{y}) = \xi. \tag{5.2.49}$$

Now, from the definition of the Fréchet subgradients, applied for $\chi \in \partial_F(g \circ f)(\overline{x})$, and then from (5.2.49), one gets that for any $\varepsilon > 0$, there is $\delta > 0$ such that

$$\begin{aligned}
\langle \chi, x - \overline{x} \rangle &\leq g(f(x)) - g(f(\overline{x})) + \varepsilon \|x - \overline{x}\| \\
&\leq s(f(x)) - s(f(\overline{x})) + \varepsilon \|x - \overline{x}\| \\
&= \langle \xi, f(x) - f(\overline{x}) \rangle + \alpha(\|x - \overline{x}\| + \|f(x) - f(\overline{x})\|) + \varepsilon \|x - \overline{x}\| ,
\end{aligned}$$

whenever $\|x - \overline{x}\| < \delta$, where α is a function such that $\lim_{x \to \overline{x}} \frac{\alpha(\|x-\overline{x}\| + \|f(x)-f(\overline{x})\|)}{\|x-\overline{x}\| + \|f(x)-f(\overline{x})\|} = 0$. Hence, for every $\varepsilon > 0$,

$$\langle \chi, x - \overline{x} \rangle - \langle \xi, f(x) - f(\overline{x}) \rangle \leq 2\varepsilon \left(\|x - \overline{x}\| + \|f(x) - f(\overline{x})\| \right) ,$$

whenever $\|x - \overline{x}\| < \delta$, which implies that

$$\langle \xi, f \rangle (x) - \langle \xi, f \rangle (\overline{x}) - \langle \chi, x - \overline{x} \rangle \geq -2\varepsilon(\ell + 1) \|x - \overline{x}\|$$

whenever $\|x - \overline{x}\| < \delta$, since f is Lipschitz around $\overline{x}$ (with modulus $\ell \geq 0$). This implies that

$$\liminf_{x \to \overline{x}} \frac{\langle \xi, f \rangle (x) - \langle \xi, f \rangle (\overline{x}) - \langle \chi, x - \overline{x} \rangle}{\|x - \overline{x}\|} \geq -2\varepsilon(\ell + 1).$$

Since $\varepsilon > 0$ was arbitrarily chosen, we have that $\chi \in \partial_F \langle \xi, f \rangle (\overline{x})$, and (5.2.48) is proved.

Suppose now that g is Fréchet differentiable. Then $\partial_F(-g)(\overline{y}) = -\nabla g(\overline{y}) =: -\xi$, so the intersection from the right-hand side of (5.2.48) reduces to an element. We have then from (5.2.48)

$$\partial_F(g \circ f)(\overline{x}) \subset \partial_F \langle \xi, f \rangle (\overline{x}).$$

Suppose $\chi \notin \partial_F(g \circ f)(\overline{x})$, and prove that $\chi \notin \partial_F \langle \xi, f \rangle (\overline{x})$. Observe, moreover, that $\eta \in \partial_F \langle \xi, f \rangle (\overline{x})$ is equivalent to $(\eta, -1) \in N_F(\text{epi} \langle \xi, f \rangle, (\overline{x}, \langle \xi, f \rangle (\overline{x})))$. By Proposition 5.2.8, it is sufficient to prove that $(\chi, -1) \notin N_F(\text{gr} \langle \xi, f \rangle, (\overline{x}, \langle \xi, f \rangle (\overline{x})))$. But since if $x \to \overline{x}$, due to the continuity of the function $x \mapsto \langle \xi, f \rangle (x)$, this is equivalent to $x \xrightarrow{\text{gr}\langle \xi, f\rangle} \overline{x}$, it means that we must prove that

$$\limsup_{x \to \overline{x}} \frac{\langle \chi, x - \overline{x} \rangle - \langle \xi, f(x) - f(\overline{x}) \rangle}{\|x - \overline{x}\| + |\langle \xi, f(x) - f(\overline{x}) \rangle|} > 0.$$

But then, since

$$\|x - \overline{x}\| + |\langle \xi, f(x) - f(\overline{x}) \rangle| \leq (1 + \|\xi\| \ell) \|x - \overline{x}\|,$$

it is sufficient to prove that

$$\limsup_{x \to \overline{x}} \frac{\langle \chi, x - \overline{x} \rangle - \langle \xi, f(x) - f(\overline{x}) \rangle}{\|x - \overline{x}\|} > 0. \tag{5.2.50}$$

But from $\chi \notin \partial_F(g \circ f)(\overline{x})$, we know that

$$\liminf_{x \to \overline{x}} \frac{\langle \xi, f \rangle (x) - \langle \xi, f \rangle (\overline{x}) - \langle \chi, x - \overline{x} \rangle}{\|x - \overline{x}\|} < 0,$$

which means exactly (5.2.50). The theorem is completely proved. □

The next result is an approximate mean value theorem for lower semicontinuous functions.

Theorem 5.2.37. *Let $\varphi : \mathbb{R}^p \to \mathbb{R}$ be a lower semicontinuous function and $a \neq b$. Consider any point $c \in [a, b)$ at which the function*

$$\psi(x) := \varphi(x) - \frac{\varphi(b) - \varphi(a)}{\|b - a\|} \|x - a\|$$

attains its minimum on $[a, b]$; *such a point always exists. Then there are sequences* $(x_n) \xrightarrow{\varphi} c$ *and* $\xi_n \in \partial_F \varphi(x_n)$ *such that*

$$\liminf_{n \to \infty} \langle \xi_n, b - x_n \rangle \geq \frac{\varphi(b) - \varphi(a)}{\|b - a\|} \|c - a\|, \tag{5.2.51}$$

$$\liminf_{n \to \infty} \langle \xi_n, b - a \rangle \geq \varphi(b) - \varphi(a). \tag{5.2.52}$$

Moreover, when $c \neq a$ *one has*

$$\lim_{n \to \infty} \langle \xi_n, b - a \rangle = \varphi(b) - \varphi(a).$$

Proof The function ψ is lower semicontinuous, and therefore attains its minimum on $[a, b]$ at some point c. Because $\psi(a) = \psi(b)$, we may suppose $c \in [a, b)$. Without loss of generality, suppose $\varphi(a) = \varphi(b)$, which means that $\psi(x) = \varphi(x)$ on $[a, b]$. Because φ is lower semicontinuous, there exists $r > 0$ and $y \in \mathbb{R}$ such that $\varphi(x) \geq y$ for every $x \in P := [a, b] + D(0, r)$. Then for every $n \in \mathbb{N}$, one can find $r_n \in (0, r)$ such that

$$\varphi(x) \geq \varphi(c) - \frac{1}{n^2}, \quad \text{for all } x \in [a, b] + D(0, r_n).$$

Moreover, choose $t_n \geq n$ such that $y + t_n r_n \geq \varphi(c) - \frac{1}{n^2}$. We have then

$$\varphi(c) \leq \inf_{x \in P} \varphi_n(x) + \frac{1}{n^2},$$

where $\varphi_n(x) := \varphi(x) + t_n d_{[a,b]}(x)$ is lower semicontinuous. By the use of Ekeland Variational Principle on the closed set P (see Remark 3.1.13), we get $x_n \in P$ such that

$$\|x_n - c\| \leq \frac{1}{n}, \quad \varphi_n(x_n) \leq \varphi(c), \quad \text{and}$$

$$\varphi_n(x_n) \leq \varphi_n(x) + \frac{1}{n} \|x - x_n\| \text{ for any } x \in P.$$

The last relation shows that the function $\varphi_n(x) + \frac{1}{n} \|x - x_n\|$ attains its minimum on P at $x = x_n$. Taking into account that $x_n \in \text{int } P$ for large n, we can conclude that this function attains a local minimum (without restrictions) at x_n. Using Lemma 5.2.31 for $\varepsilon = \varepsilon_n \downarrow 0$, we find sequences $(u_n) \xrightarrow{\varphi} c$, $(v_n) \to c$, $\xi_n \in \partial_F \varphi(u_n)$, $\eta_n \in \partial_F d_{[a,b]}(v_n)$, and $e_n \in D(0, 1)$ such that

$$\left\| \xi_n + t_n \eta_n + n^{-1} e_n \right\| \leq \varepsilon_n, \quad n \in \mathbb{N}. \tag{5.2.53}$$

Since $\eta_n \in \partial_F d_{[a,b]}(v_n) = \partial d_{[a,b]}(v_n)$, it follows that $\|\eta_n\| \leq 1$, and

$$\langle \eta_n, b - v_n \rangle \leq d_{[a,b]}(b) - d_{[a,b]}(v_n) \leq 0, \quad n \in \mathbb{N}.$$

Picking $w_n \in \text{pr}_{[a,b]}(v_n)$, we get

$$\langle \eta_n, b - w_n \rangle = \langle \eta_n, b - v_n \rangle + \langle \eta_n, v_n - w_n \rangle \leq d_{[a,b]}(b) - d_{[a,b]}(v_n) + \|\eta_n\| \cdot \|v_n - w_n\|$$

$$\leq -d_{[a,b]}(v_n) + \|v_n - w_n\| = 0.$$

It means that $\langle \eta_n, b - c \rangle \leq 0$ for large n, since $w_n \to c \neq b$, and hence, since a, b, c are colinear,

$$\langle \eta_n, b - a \rangle = \langle \eta_n, b - c \rangle \cdot \frac{\|a - b\|}{\|c - b\|} \leq 0 \text{ for large } n.$$

By using now (5.2.53), we get that there exists $b_n \in D(0, \varepsilon_n)$ such that

$$\xi_n = -t_n \eta_n - n^{-1} e_n - \varepsilon_n b_n,$$

hence,

$$\liminf_{n \to \infty} \langle \xi_n, b - u_n \rangle \geq 0, \quad \liminf_{n \to \infty} \langle \xi_n, b - a \rangle \geq 0,$$

which prove (5.2.51) and (5.2.52). Suppose that $c \neq a$. Then $v_n \neq a$ for large n, hence $\langle \eta_n, b - c \rangle = 0$. This implies that

$$\langle \xi_n, b - a \rangle = \langle \xi_n, b - c \rangle \cdot \frac{\|a - b\|}{\|c - b\|} = \left\langle -t_n \eta_n - n^{-1} e_n - \varepsilon_n b_n, b - c \right\rangle \cdot \frac{\|a - b\|}{\|c - b\|} \to 0,$$

where $b_n \in D(0, 1)$ exists from (5.2.53). This concludes the proof. $\qquad\square$

We end this subsection with a result which links the Clarke and Mordukhovich constructions presented throughout this chapter.

Theorem 5.2.38. *The following assertions hold:*
(i) Let $A \in \mathbb{R}^p$ be a closed set and $\overline{x} \in A$. Then

$$N_C(A, \overline{x}) = \text{cl conv } N_M(A, \overline{x}). \tag{5.2.54}$$

(ii) Let $f : \mathbb{R}^p \to \mathbb{R}$ be locally Lipschitz around $\overline{x}$. Then

$$\partial_C f(x) = \text{cl conv } \partial_M f(\overline{x}). \tag{5.2.55}$$

Proof We prove first that

$$\partial_M f(\overline{x}) \subset \partial_C f(\overline{x}), \tag{5.2.56}$$

which will show that cl conv $\partial_M f(\overline{x}) \subset \partial_C f(\overline{x})$, due to the fact that $\partial_C f(\overline{x})$ is a closed convex set. Take $\chi \in \partial_M f(\overline{x})$. Then there are $(x_n) \xrightarrow{f} \overline{x}$ and $(\chi_n) \to \chi$ such that $\chi_n \in \partial_F f(x_n)$ for every n. This means, due to the definition of the Fréchet subgradients, that for every n there is a neighborhood U_n of x_n such that

$$f(x) - f(x_n) - \langle \chi_n, x - x_n \rangle \geq -\frac{1}{n} \|x - x_n\|, \quad \forall x \in U_n.$$

But this shows that the function

$$\psi_n(x) := f(x + x_n) - \langle \chi_n, x \rangle + \frac{1}{n} \|x\|$$

attains a local minimum at 0. This proves, by Theorem 5.1.16, that

$$0 \in \partial_C \left[f(\cdot + x_n) - \langle \chi_n, \cdot \rangle + \frac{1}{n} \|\cdot\| \right] (0).$$

Since all the involved functions are locally Lipschitz, it follows from the sum rule for the Clarke subdifferential that

$$\chi_n \in \partial_C f(x_n) + D\left(0, \frac{1}{n}\right).$$

Using Proposition 5.1.6, we get that $\chi \in \partial_C f(\overline{x})$. Hence, the desired inclusion holds. For proving the reverse inclusion in (5.2.55), we will show the representation

$$f^\circ(\overline{x}, u) = \sup \left\{ \langle \xi, u \rangle \mid \xi \in \partial_M f(\overline{x}) \right\} \tag{5.2.57}$$
$$= \max \left\{ \langle \xi, u \rangle \mid \xi \in \mathrm{cl}\, \partial_M f(\overline{x}) \right\}.$$

This will prove that $f^\circ(\overline{x}, u)$ is the support function of the set $P := \partial_M f(\overline{x})$. Observe, by definition, that the support function of a set coincides with the support function of the closed convex hull of this set, hence

$$f^\circ(\overline{x}, u) = \sup \left\{ \langle \xi, u \rangle \mid \xi \in \mathrm{cl}\,\mathrm{conv}\, \partial_M f(\overline{x}) \right\}.$$

Moreover, we know that two convex and closed sets coincide if and only if their support functions coincide. Since $\partial_C f(x)$ and $\mathrm{cl}\,\mathrm{conv}\, \partial_M f(\overline{x})$ are both closed and convex, we get (5.2.55).

Take some realizing sequences for $f^\circ(\overline{x}, u)$. This means that there are $(t_n) \downarrow 0$ and $(x_n) \to \overline{x}$ such that

$$\frac{f(x_n + t_n u) - f(x_n)}{t_n} \to f^\circ(\overline{x}, u) \text{ for } n \to \infty.$$

By applying Theorem 5.2.37 to f on the interval $[x_n, x_n + t_n u]$ for every n, we get $(v_k) \to c_n \in [x_n, x_n + t_n u)$ for $k \to \infty$ and $\xi_k \in \partial_F f(v_k)$ such that

$$f(x_n + t_n u) - f(x_n) \le t_n \liminf_{k \to \infty} \langle \xi_k, u \rangle, \quad \forall n \in \mathbb{N}.$$

Since f is locally Lipschitz around $\overline{x}$ and $\xi_k \in \partial_F f(v_k)$, we know that (ξ_k) is bounded, hence without loss of generality we may suppose that (ξ_k) is convergent to some ξ. Passing to the limit above first for $k \to \infty$, and then for $n \to \infty$, we get that

$$f^\circ(\overline{x}, u) \le \langle \xi, u \rangle,$$

for some $\xi \in \partial_M f(\overline{x})$. Since $\langle \xi', u \rangle \le f^\circ(\overline{x}, u)$ for any $\xi' \in \partial_C f(\overline{x}) \supset \partial_M f(\overline{x})$, we get the representation (5.2.57). Hence, the proof of (5.2.55) is complete.

Let us prove now (5.2.54). Since due to (5.2.24) and (5.2.56) we have that

$$N_M(A, \overline{x}) = \bigcup_{\lambda > 0} \lambda \partial_M d_A(\overline{x}) \subset \bigcup_{\lambda > 0} \lambda \partial_C d_A(\overline{x}) = N_C(A, \overline{x}),$$

we also have cl conv $N_M(A, \overline{x}) = N_C(A, \overline{x})$, since $N_C(A, \overline{x})$ is closed and convex. For the reverse inclusion, we use (5.2.55) to get that

$$\bigcup_{\lambda>0} \lambda \partial_C d_A(\overline{x}) = \bigcup_{\lambda>0} \lambda \left[\text{cl conv } \partial_M d_A(\overline{x})\right] \subset \text{cl conv} \bigcup_{\lambda>0} \lambda \partial_M d_A(\overline{x}).$$

The theorem is now completely proved. □

5.2.5 Optimality Conditions

In this section, we make use of the exact version of the extremal principle in order to get necessary optimality conditions for the minimization problem considered in the previous section:

$$(MP) \quad \min f(x), \text{ subject to } g(x) \leq 0, \ h(x) = 0, \ x \in A, \tag{MP}$$

where the functions $f, g = (g_1, \ldots, g_n)$ and $h = (h_1, \ldots, h_m)$ are locally Lipschitz functions which map from $\mathbb{R}^p$ into $\mathbb{R}$, $\mathbb{R}^n$ and $\mathbb{R}^m$, respectively, and the set $A \subset \mathbb{R}^p$ is closed. The final aim is to get Fritz John necessary optimality conditions for problem (MP).

Theorem 5.2.39. *Let $\overline{x}$ be a solution of (MP), where f, g and h are Lipschitz around $\overline{x}$. Then there exist $\lambda_0 \geq 0$, $\lambda = (\lambda_1, \ldots, \lambda_n) \in \mathbb{R}^n$ and $\mu = (\mu_1, \ldots, \mu_m) \in \mathbb{R}^m$, with $\lambda_0 + \|\lambda\| + \|\mu\| \neq 0$, such that*

$$0 \in \lambda_0 \partial_M f(\overline{x}) + \sum_{i=1}^{n} \lambda_i \partial_M g_i(\overline{x}) + \sum_{j=1}^{m} \mu_j [\partial_M h_j(\overline{x}) \cup \partial_M(-h_j)(\overline{x})] + N_M(A, \overline{x}) \tag{5.2.58}$$

and

$$\lambda_i \geq 0, \ \lambda_i g_i(\overline{x}) = 0, \ \forall i \in \overline{1, n}, \tag{5.2.59}$$

$$\mu_j \geq 0, \ \forall j \in \overline{1, m}.$$

Proof Suppose, without loss of generality, that $f(\overline{x}) = 0$. Then it is easy to observe that the point $(\overline{x}, 0)$ is a local extremal point of the the following system of closed sets in $\mathbb{R}^{p+n+m+1}$:

$$A_0 := \left\{(x, \lambda_0, \lambda_1, \ldots, \lambda_n, \mu_1, \ldots, \mu_m) \mid \lambda_0 \geq f(x)\right\},$$
$$A_i := \left\{(x, \lambda_0, \lambda_1, \ldots, \lambda_n, \mu_1, \ldots, \mu_m) \mid \lambda_i \geq g_i(x)\right\}, \ i \in \overline{1, n},$$
$$A_{n+j} := \left\{(x, \lambda_0, \lambda_1, \ldots, \lambda_n, \mu_1, \ldots, \mu_m) \mid \mu_j = h_j(x)\right\}, \ j \in \overline{1, m},$$
$$A_{n+m+1} := A \times \{0\}.$$

Since by Theorem 5.2.28 the exact extremal principle holds in $\mathbb{R}^{p+n+m+1}$, we find elements

$$(\xi_0, -\lambda_0) \in N_M(\text{epi } f, (\overline{x}, 0)),$$

$$(\xi_i, -\lambda_i) \in N_M(\text{epi } g_i, (\overline{x}, 0)), \ i \in \overline{1, n}$$

$$(\xi_{n+j}, -\lambda_{n+j}) \in N_M(\text{gr } h_j, (\overline{x}, 0)), \ j \in \overline{1, m}$$

$$\widehat{\xi} \in N_M(A, \overline{x})$$

such that

$$\xi_0 + \ldots + \xi_{n+m} + \widehat{\xi} = 0, \tag{5.2.60}$$

$$\left\|(\xi_0, -\lambda_0)\right\| + \ldots + \left\|(\xi_{n+j}, -\lambda_n)\right\| + \left\|\widehat{\xi}\right\| = 1. \tag{5.2.61}$$

Using Proposition 5.2.15 on basic normals to epigraphs, we deduce that $\lambda_i \geq 0$ for $i \in \overline{0, n}$.

If $g_i(\overline{x}) < 0$ for some fixed $i \in \overline{1, n}$, then $g_i(x) < 0$ for all x around $\overline{x}$ due to the Lipschitz property of g_i. But this implies that $(\overline{x}, 0)$ is an interior point of epi g_i, which means by $N_M(\text{epi } g_i, (\overline{x}, 0)) = \{0\}$ that $\xi_i = 0$ and $\lambda_i = 0$.

Now, by Proposition 5.2.15 and Theorem 5.2.18, we know that

$$(\xi, -\lambda) \in N_M(\text{epi } \varphi, (\overline{x}, \varphi(\overline{x}))) \iff \xi \in \lambda \partial_M \varphi(\overline{x}), \ \lambda \geq 0$$

if φ is Lipschitz around $\overline{x}$. Also, by Proposition 5.2.8, for such function φ, one has that

$$(\xi, -\lambda) \in N_M(\text{gr } \varphi, (\overline{x}, \varphi(\overline{x}))) \iff \xi \in \partial_M(\lambda\varphi)(\overline{x}).$$

Taking into account also that

$$\partial_M(\lambda\varphi)(\overline{x}) \subset |\lambda| \left[\partial_M\varphi(\overline{x}) \cup \partial_M(-\varphi)(\overline{x})\right] \text{ for all } \lambda \in \mathbb{R},$$

we have, by denoting $\mu_j := \left|\lambda_{n+j}\right|$ for $j = \overline{1, m}$ and (5.2.60), that (5.2.58) holds, and also $\mu_j \geq 0$ for any $j \in \overline{1, m}$. This completes the proof of the theorem. □

Remark 5.2.40. *Observe that the set $\partial_M\varphi(\overline{x}) \cup \partial_M(-\varphi)(\overline{x})$ reduces to $\{\nabla\varphi(\overline{x}), -\nabla\varphi(\overline{x})\}$ when φ is C^1, so Theorem 5.2.39 is indeed a generalization to the nonsmooth case of the known results.*

Moreover, remark that, since $\partial_C\varphi(\overline{x})$ is always larger than $\partial_M\varphi(\overline{x})$, the previous theorem may provide more precise necessary optimality conditions than Theorem 5.1.34 in case that the equality constraints are not present. To observe this, consider for instance the minimization problem

$$\min f(x) := -|x|, \quad x \in \mathbb{R}.$$

Then $\overline{x} = 0$ is not a minimum point of this problem, while $0 \in \partial_C f(\overline{x}) = [-1, 1]$. On the other hand, $0 \notin \partial_M f(\overline{x}) = \{-1, 1\}$.

Another example is given by the minimization problem

$$\min x_1 \text{ subject to } \varphi(x_1, x_2) := |x_1| - |x_2| \leq 0.$$

Then $\partial_M \varphi(0,0) = \{(y_1, y_2) \mid y_1 \in [-1, 1], y_2 = \pm 1\}$. *It follows by Theorem 5.2.39 that the point* $(0,0)$ *cannot be optimal for the considered problem. Since*

$$\partial_C \varphi(0,0) = \text{cl conv } \partial_M \varphi(0,0) = [-1, 1] \times [-1, 1],$$

it means that we cannot decide that $(0,0)$ *is not an optimal candidate by Theorem 5.1.34.*

6 Basic Algorithms

The aim of this chapter is to present some fundamental ideas concerning algorithms for smooth optimization problems. We start by studying the Picard iterations (introduced and discussed in the second chapter) as a method to approximate the roots of certain nonlinear equations and this discussion naturally leads us to investigate some convergence acceleration techniques for the initial sequence of iterations. We then present Newton's method for solving nonlinear equations. In the convex framework, we present the proximal point algorithm. For optimization problems without restrictions we are interested in the line search method, which we discuss in detail, while for constrained problems, we give the sequential quadratic and the interior point methods. All the theoretical elements are then discussed and verified through some Matlab-based numerical simulations.

The situations we met in the majority of the concrete examples of optimization problems (see last chapter) are very hospitable, in the sense that we are able to find the solutions exactly, by solving the systems which give the solution candidates, and using the theoretical means we have previously developed. However, in many cases, the systems are not solvable, as in Section 3.4, when we considered the nonlinear case of the least square method, and the equation for finding the Lagrange multiplier for the problem of the computation of a projection on a generalized ellipsoid. For problems of this nature, it is necessary to develop methods called algorithms, in order to approximate the respective solutions. In general, there is a clear difference between the design of the algorithms for unconstrained optimization problems and those for constrained problems.

All the algorithms require a starting point, denoted x_0. Generally speaking, it is useful that this point is itself a good approximation of the solution we are looking for (especially if the solution is not unique). For instance, the function $f : \mathbb{R} \to \mathbb{R}$,

$$f(x) = \frac{x^4}{4} - \frac{x^3}{3} - x^2$$

has two minimal points: $x = -1$ is a local minimum and $x = 2$ is a global minimum. If we start with a value x_0 close to one of these points, then (roughly speaking) it is highly possible that the algorithm will converge to that point.

In general, after the choice of x_0, the algorithm generates a sequence of iterations $(x_k)_{k \in \mathbb{N}}$ with the aim of approaching the solution. The process of generating new iterations stops when no new progress can be made in the effort to come closer to the solution (according to the internal rule of the algorithm), or when an accuracy previously established was attained. Any algorithm should generate new iterations from the existing ones. In general, every new iteration should progress towards the solution. Some algorithms are called non-monotonic, and they do not necessarily progress at every step.

The study of efficient algorithms to detect (or to approximate) the solutions of optimization problems is a huge subject, several comprehensive monographs are fully dedicated to it. We just point out here the main ideas. Many of the very efficient algorithms are implemented into the functions of scientific software as Scilab or Matlab. We illustrate this at the end of the chapter.

There are at least two problems to be studied when an algorithm is designed: we are interested in knowing if the algorithm is global (i.e., it is convergent for any initial data), and to know its speed of convergence. Therefore, for a sequence $(x_k) \subset \mathbb{R}^p$ convergent to $\overline{x} \in \mathbb{R}^p$ with $x_k \neq \overline{x}$ for every $k \in \mathbb{N}^*$, one calls order of convergence the greatest natural number q for which

$$\lim_{k \to \infty} \frac{\|x_{k+1} - \overline{x}\|}{\|x_k - \overline{x}\|^q} \in [0, \infty).$$

If $q = 1$, then one has linear convergence, and if the above limit is 0, we have superlinear convergence. If $q = 2$, we have quadratic convergence. However, the aim is to design global algorithms which have a very good speed of convergence (at least quadratic).

We start with some methods to approximate the roots of some nonlinear equations. Some of the ideas from the next section make the link between the preceding results in the case of optimization algorithms. We have already seen a nonlinear equation that cannot be solved at the end of Chapter 3. Furthermore, let us remark that Theorem 3.2.6 transforms an optimization problem into the problem of solving the equation $L(x, (\lambda, \mu)) = 0$, which, in many cases, is highly nonlinear, and therefore not solvable.

6.1 Algorithms for Nonlinear Equations

6.1.1 Picard's Algorithm

To solve $g(x) = 0$ is equivalent to looking for the fixed points of the function $f(x) = g(x) + x$. The first part of our discussion, therefore, concerns the convergence of the Picard iteration, whose theoretical study was made in Section 2.3.2. Recall that if $f : [a, b] \to [a, b]$ is a differentiable function such that its derivative is bounded on $[a, b]$ by a positive constant strictly smaller than one, then for any initial data $x_0 \in [a, b]$ the Picard iteration defined by $x_{k+1} = f(x_k)$, $k \geq 0$ is convergent to the unique fixed point $\overline{x} \in [a, b]$ of f. Moreover, if the fixed point is not attained, then

$$\frac{x_{k+1} - \overline{x}}{x_k - \overline{x}} = \frac{f(x_k) - \overline{x}}{x_k - \overline{x}} \overset{k \to \infty}{\to} f'(\overline{x}).$$

Therefore, in general, we have a linear convergence: for k big enough, the error (the absolute value of the difference between the iteration and the real value of the fixed point) at the step $(k + 1)$ is proportional to the error at the step k. This kind of convergence is not very fast.

Let us consider the contraction $f : [0, \infty) \to [0, \infty)$ given by $f(x) = \frac{1}{1+x^2}$. According to Banach Principle, f has only one fixed point in $[0, \infty)$ and this is the unique real solution of the equation $x^3 + x - 1 = 0$, whose approximate value is $\bar{x} \approx 0.6823278$. Moreover, as before, for the Picard iterations, we have

$$\frac{x_{k+1} - \bar{x}}{x_k - \bar{x}} \overset{k \to \infty}{\to} f'(\bar{x}) = \frac{-2\bar{x}}{(1 + \bar{x}^2)^2} = -2\bar{x}^3 \approx -0.63534438165.$$

See Section 6.3 for the numerical implementation which gives the value above. Therefore, for k big enough, at every iterative step, the error is multiplied by the approximate value 0.6353. In contrast, if we study the restriction of $\sin x$ to the interval $[0, 1]$, which is a weak contraction with $\bar{x} = 0$ as fixed point, we deduce that

$$\frac{x_{k+1} - \bar{x}}{x_k - \bar{x}} \overset{k \to \infty}{\to} f'(\bar{x}) = 1,$$

and we do not expect a good speed of convergence. In the last section of the chapter, we illustrate these theoretical predictions.

Remark 6.1.1. *The actual speed of convergence for the Picard iterations is given by the value of $f'(\bar{x})$. In the best case where $f'(\bar{x}) = 0$ we can have better convergence than the linear case. In general, in the above framework, if $f'(\bar{x}) = 0$ and f is twice differentiable, then the double application of the L'Hôpital rule gives*

$$\lim_{x \to \bar{x}} \frac{f(x) - \bar{x}}{(x - \bar{x})^2} = \frac{f''(\bar{x})}{2}, \tag{6.1.1}$$

so, for every nonstationary Picard iteration,

$$\lim_{k \to \infty} \frac{x_{k+1} - \bar{x}}{(x_k - \bar{x})^2} = \frac{f''(\bar{x})}{2},$$

whence a quadratic convergence.

Remark 6.1.2. *In fact, for the quadratic convergence described before it is enough for $\lim_{x \to \bar{x}} \frac{f(x) - \bar{x}}{(x - \bar{x})^2}$ to exist or, more generally, $\lim_{x \to \bar{x}} \frac{|f(x) - \bar{x}|}{(x - \bar{x})^2}$, and this can happen without the twice differentiability of f : for instance, for $f : \mathbb{R} \to \mathbb{R}$,*

$$f(x) = \begin{cases} x^2, & x \geq 0 \\ -x^2, & x < 0, \end{cases}$$

there exists the limit

$$\lim_{x \to \bar{x}} \frac{|f(x) - \bar{x}|}{(x - \bar{x})^2} = 1$$

at $\bar{x} = 0$, but the function is not twice differentiable at 0. This remark will be useful later.

If $f'(\overline{x}) \neq 0$, the speed of convergence of the Picard iterations is only linear. We present in the next section a method to overcome this difficulty. This technique is called the Aitken acceleration method, and was developed by the New Zealand-born mathematician Alexander Craig Aitken in 1926.

Consider $f : [a, b] \to [a, b]$ a differentiable function such that $|f'(x)| \in [0, 1)$ for every $x \in [a, b]$. For every $x_0 \in [a, b]$, the Picard iteration defined by $x_{k+1} = f(x_k)$, $k \geq 0$ is convergent to the unique fixed point of f in $[a, b]$, denoted $\overline{x}$. Suppose that $f'(\overline{x}) \neq 0$, so $|f'(\overline{x})| \in (0, 1)$. If (x_k) is nonstationary,

$$\frac{x_{k+1} - \overline{x}}{x_k - \overline{x}} = \frac{f(x_k) - \overline{x}}{x_k - \overline{x}} \overset{k \to \infty}{\to} f'(\overline{x}). \tag{6.1.2}$$

Therefore, we can write

$$f(x_k) - \overline{x} = \rho_k(x_k - \overline{x}),$$

where $\rho_k \overset{k \to \infty}{\to} f'(\overline{x})$, which means

$$\overline{x} = \frac{f(x_k) - \rho_k x_k}{1 - \rho_k},$$

that is

$$\overline{x} = x_k + \frac{f(x_k) - x_k}{1 - \rho_k}. \tag{6.1.3}$$

Aitken's initial idea was to find another sequence (μ_k) to approximate $f'(\overline{x})$. The result is given in the next result.

Proposition 6.1.3. *If (x_k) is a nonstationary Picard sequence, then the sequence defined by*

$$\mu_k = \frac{f(f(x_k)) - f(x_k)}{f(x_k) - x_k}, \ \forall k \in \mathbb{N}$$

has the limit $f'(\overline{x})$.

Proof Since $x_{k+1} = f(x_k)$, we have that $x_{k+2} = f(f(x_k))$ and from the definition of μ_k we deduce that

$$\mu_k = \frac{x_{k+2} - \overline{x} - (x_{k+1} - \overline{x})}{(x_{k+1} - \overline{x}) - (x_k - \overline{x})}$$

$$= \frac{\frac{x_{k+2} - \overline{x}}{x_{k+1} - \overline{x}} - 1}{1 - \frac{x_k - \overline{x}}{x_{k+1} - \overline{x}}},$$

and from (6.1.2), we get

$$\lim \mu_k = \frac{f'(\overline{x}) - 1}{1 - \frac{1}{f'(\overline{x})}} = f'(\overline{x}).$$

The proposition is proved. $\qquad\qquad\qquad\qquad\qquad\qquad\qquad\qquad\qquad\qquad\square$

Now, relation (6.1.3) and the above result suggest that we should consider the sequence:

$$y_k = x_k + \frac{f(x_k) - x_k}{1 - \frac{f(f(x_k)) - f(x_k)}{f(x_k) - x_k}}$$

$$= x_k - \frac{(f(x_k) - x_k)^2}{f(f(x_k)) - 2f(x_k) + x_k}$$

$$= \frac{x_k f(f(x_k)) - f(x_k)^2}{f(f(x_k)) - 2f(x_k) + x_k}.$$

We prove that the sequence (y_k) also converges to the fixed point $\overline{x}$, but faster than (x_k).

Theorem 6.1.4. *Let (x_k) be a Picard nonstationary sequence convergent to $\overline{x}$ such that*

$$\lim \frac{x_{k+1} - \overline{x}}{x_k - \overline{x}} = f'(\overline{x}) \in (-1, 1) \setminus \{0\}.$$

If the sequence (y_k) given by

$$y_k = x_k - \frac{(f(x_k) - x_k)^2}{f(f(x_k)) - 2f(x_k) + x_k}$$

is well defined, then it converges to $\overline{x}$ and, moreover,

$$\lim \frac{y_k - \overline{x}}{x_k - \overline{x}} = 0.$$

Proof We show first that (y_k) converges to $\overline{x}$. In the definition of (y_k), we add and subtract $\overline{x}$, and then we divide both numerator and denominator by $x_{k+1} - \overline{x}$. We obtain

$$y_k = x_k + \frac{\left(1 + \frac{\overline{x} - x_k}{x_{k+1} - \overline{x}}\right)(x_{k+1} - x_k)}{\left(1 + \frac{\overline{x} - x_k}{x_{k+1} - \overline{x}}\right) - \left(\frac{x_{k+2} - \overline{x}}{x_{k+1} - \overline{x}} - 1\right)}.$$

Passing to the limit (and denoting, for simplicity, $f'(\overline{x}) := \alpha$), we deduce:

$$\lim y_k = \overline{x} + \frac{1 - \alpha^{-1}}{1 - \alpha^{-1} - \alpha + 1} \lim(x_{k+1} - x_k) = \overline{x}.$$

We show now that (y_k) converges faster than (x_k). We have

$$\frac{y_k - \overline{x}}{x_k - \overline{x}} = \frac{x_k - \overline{x} - \frac{(f(x_k) - x_k)^2}{f(f(x_k)) - 2f(x_k) + x_k}}{x_k - \overline{x}}$$

$$= 1 + \frac{\frac{(f(x_k) - x_k)^2}{(x_{k+1} - x_k) - (x_{k+2} - x_{k+1})}}{x_k - \overline{x}}.$$

By the same method as above, we get

$$\frac{y_k - \overline{x}}{x_k - \overline{x}} = 1 + \frac{\left(1 + \frac{\overline{x} - x_k}{x_{k+1} - \overline{x}}\right)\left(-1 + \frac{x_{k+1} - \overline{x}}{x_k - \overline{x}}\right)}{\left(1 + \frac{\overline{x} - x_k}{x_{k+1} - \overline{x}}\right) - \left(-1 + \frac{x_{k+2} - \overline{x}}{x_{k+1} - \overline{x}}\right)},$$

so,

$$\lim \frac{y_k - \overline{x}}{x_k - \overline{x}} = 1 + \frac{(1 - \alpha^{-1})(-1 + \alpha)}{2 - \alpha - \alpha^{-1}} = 0.$$

Consequently, (y_k) converges faster than (x_k). □

Despite the fact that (y_k) produces an acceleration of the speed of convergence, the order of convergence does not change: similar calculations show that

$$\lim \frac{y_{k+1} - \overline{x}}{y_k - \overline{x}} = f'(\overline{x}),$$

that is as well linear convergence. We call this the weak Aitken acceleration method.

However, starting from this idea, we consider a Picard iteration, but for the function

$$h(x) = \frac{xf(f(x)) - f(x)^2}{f(f(x)) - 2f(x) + x}$$

(this time, $f(x)^2$ means $f(x) \cdot f(x)$). The function h cannot be formally defined at $\overline{x}$, but one can extend its definition at that point by continuity, since

$$\lim_{x \to \overline{x}} h(x) = \frac{f(f(\overline{x})) + \overline{x}f'(f(\overline{x}))f'(\overline{x}) - 2f(\overline{x})f'(\overline{x})}{f'(f(\overline{x}))f'(\overline{x}) - 2f'(\overline{x}) + 1}$$

$$= \frac{\overline{x} + \overline{x}(f'(\overline{x}))^2 - 2\overline{x}f'(\overline{x})}{(f'(\overline{x}))^2 - 2f'(\overline{x}) + 1} = \overline{x},$$

(recall that $|f'(\overline{x})| < 1$). So, with this extension $(h(\overline{x}) = \overline{x})$, the fixed point of f is a fixed point of h. Suppose that there exists a neighborhood $V := [\overline{x} - \mu, \overline{x} + \mu]$ of $\overline{x}$ $(\mu > 0)$ with the property that for every $x \in (V \setminus \{\overline{x}\}) \cap [a, b]$, $f(f(x)) - 2f(x) + x \neq 0$. Then the converse holds as well: if u is a fixed point of h from $V \cap [a, b]$, then the equality $h(u) = u$ leads to $(f(u) - u)^2 = 0$. Therefore, the sole fixed point of h in $V \cap [a, b]$ is $\overline{x}$. It is also clear that h is derivable on $(V \setminus \{\overline{x}\}) \cap [a, b]$.

Moreover, we suppose that f is of class C^2 on $V \cap [a, b]$. We can show that h is derivable at $\overline{x}$ and its derivative at $\overline{x}$ is $h'(\overline{x}) = 0$. For simplicity, suppose that $\overline{x} \in (a, b)$ so that we can think that $V \subset (a, b)$. However, this is not essential. Write down Taylor's Formula for f around $\overline{x}$: for every ε with $|\varepsilon| < \mu$, there exists $\theta_\varepsilon \in (0, 1)$ such that

$$f(\overline{x} + \varepsilon) = f(\overline{x}) + f'(\overline{x})\varepsilon + \frac{f''(\overline{x} + \theta_\varepsilon\varepsilon)}{2}\varepsilon^2$$

$$= \overline{x} + f'(\overline{x})\varepsilon + \frac{f''(\overline{x} + \theta_\varepsilon\varepsilon)}{2}\varepsilon^2.$$

We fix ε. For ease of computation, we denote

$$\frac{f''(\overline{x} + \theta_\varepsilon\varepsilon)}{2} =: A_\varepsilon \text{ and } f'(\overline{x})\varepsilon + \frac{f''(\overline{x} + \theta_\varepsilon\varepsilon)}{2}\varepsilon^2 =: \delta_\varepsilon.$$

It is obvious that for ε small enough, $|\delta_\varepsilon| < \mu$, so

$$f(\overline{x} + \delta_\varepsilon) = \overline{x} + f'(\overline{x})\delta_\varepsilon + A_{\delta_\varepsilon}\delta_\varepsilon^2.$$

From the expression for h and a few computations, we obtain

$$h(\overline{x} + \varepsilon) = \frac{(\overline{x} + \varepsilon)f(\overline{x} + \delta_\varepsilon) - (\overline{x} + \delta_\varepsilon)^2}{f(\overline{x} + \delta_\varepsilon) - 2(\overline{x} + \delta_\varepsilon) + (\overline{x} + \varepsilon)}$$

$$= \overline{x} - \frac{\delta_\varepsilon^2 - f'(\overline{x})\varepsilon\delta_\varepsilon - A_{\delta_\varepsilon}\varepsilon\delta_\varepsilon^2}{\varepsilon - 2\delta_\varepsilon + f'(\overline{x})\delta_\varepsilon + A_{\delta_\varepsilon}\delta_\varepsilon^2}$$

$$= \overline{x} - \varepsilon^2(f'(\overline{x}) + A_\varepsilon\varepsilon)\cdot$$

$$\frac{A_\varepsilon - A_{\delta_\varepsilon}f'(\overline{x}) - A_{\delta_\varepsilon}A_\varepsilon\varepsilon}{(1 - f'(\overline{x}))^2 - \varepsilon(2A_\varepsilon - A_\varepsilon f'(\overline{x}) - A_{\delta_\varepsilon}f'(\overline{x})^2 + 2A_\varepsilon A_{\delta_\varepsilon}f'(\overline{x})\varepsilon + A_{\delta_\varepsilon}A_\varepsilon^2\varepsilon^2)}\cdot$$

We write the quotient $\frac{h(\overline{x}+\varepsilon)-\overline{x}}{\varepsilon}$, and pass to the limit as $\varepsilon \to 0$. Then, since $f'(\overline{x}) \neq 1$ and

$$A_\varepsilon \overset{\varepsilon\to 0}{\to} \frac{f''(\overline{x})}{2}, \quad A_{\delta_\varepsilon} \overset{\varepsilon\to 0}{\to} \frac{f''(\overline{x})}{2},$$

we deduce that there exists

$$\lim_{\varepsilon\to 0} \frac{h(\overline{x} + \varepsilon) - \overline{x}}{\varepsilon} = 0.$$

The claim is proved. Furthermore,

$$\lim_{x\to\overline{x}} \frac{h(x) - \overline{x}}{(x - \overline{x})^2} = \lim_{\varepsilon\to 0} \frac{h(\overline{x} + \varepsilon) - \overline{x}}{\varepsilon^2} = -\frac{f''(\overline{x})}{2} \cdot \frac{f'(\overline{x})}{1 - f'(\overline{x})} \in \mathbb{R}.$$

In particular, if one changes the neighborhood V, if needed, h is a contraction from V to V. By the use of this result and Remark 6.1.1, the sequence (x_k) defined by

$$x_{k+1} = x_k - \frac{(f(x_k) - x_k)^2}{f(f(x_k)) - 2f(x_k) + x_k}$$

with well chosen initial data is convergent towards $\overline{x}$ quadratically, i.e.,

$$\lim \frac{|x_{k+1} - \overline{x}|}{(x_k - \overline{x})^2} \in [0, \infty).$$

We call this method the strong Aitken acceleration method.

The drawback is that x_0 should be chosen from V, so it should be close enough to $\overline{x}$ (such that the equation $f(f(x)) - 2f(x) + x = 0$ should not have another root in V, except $\overline{x}$). Another supplementary assumption was linked to the order of smoothness of f. In fact, this is the price which must be paid in order to have such a good speed of convergence. Let us remark that, at first sight, the former requirement looks pretty heavy: it is unnatural to ask for an initial data close to the point $\overline{x}$ we want to approximate. A possible solution to this would be to generate, for some steps, the Picard iterations in order to get close to $\overline{x}$ and then to apply the strong Aitken method.

6.1.2 Newton's Method

The celebrated Newton's method is one of the most well-known iterative procedures to approximate the roots of a function with sufficient differentiability properties. We shall see that this is a local algorithm (since in order to have the desired convergence one should take as initial data a value sufficiently close to the solution), but it converges quadratically.

Let us consider a function $f : \mathbb{R}^p \to \mathbb{R}^p$ of class C^1 and let $\bar{x}$ be a nondegenerate root of f (i.e., $f(\bar{x}) = 0$, and $\nabla f(\bar{x})$ is nonsingular). We consider a value x_0 close enough to $\bar{x}$.

The sequence of Newton iterations starts from the equation

$$0 = f(x_k) + \nabla f(x_k)(x_{k+1} - x_k). \tag{6.1.4}$$

This equation, which gives the value of x_{k+1}, shows why it is necessary to have a simple solution and we should start from a point close to $\bar{x}$: we should put our initial point in a neighborhood of $\bar{x}$ where ∇f is invertible, and such neighborhood do exist exactly because ∇f is continuous and nonsingular at $\bar{x}$. In this way, we formally define the Newton iteration by:

$$x_{k+1}^t = x_k^t - \nabla f(x_k)^{-1} \cdot (f(x_k))^t. \tag{6.1.5}$$

Recall that $\nabla f(x_k)$ can be identified with the Jacobian matrix. As seen from the previous relations, as well as from the convergence result given in the sequel, the main drawbacks of the Newton's method can be summarized as follows:
- when the starting point x_0 is not close enough to the solution $\bar{x}$, the algorithm associated to (6.1.5) does not converge;
- if $\nabla f(x_k)$ is singular, one cannot define x_{k+1};
- it may be too expensive to compute exactly $\nabla f(x_k)^{-1}$ for large p;
- it may happen that $\nabla f(\bar{x})$ is singular.

Theorem 6.1.5. *Suppose f is Lipschitz continuously differentiable on an open convex set $D \subset \mathbb{R}^p$. Let $\bar{x}$ be a nondegenerate root of the equation $f(x) = 0$, and let (x_k) be a sequence of iterates generated by (6.1.5). Then when $x_0 \in D$ is sufficiently close to $\bar{x}$, one has*

$$\lim_{k \to \infty} \frac{\|x_{k+1} - \bar{x}\|}{\|x_k - \bar{x}\|^2} \in [0, \infty), \tag{6.1.6}$$

i.e., we have local quadratic convergence.

Proof Since $f(\bar{x}) = 0$, we have from Theorem 1.4.13 that

$$f(x_k) = f(x_k) - f(\bar{x}) = \nabla f(x_k)(x_k - \bar{x}) + w(x_k, \bar{x}), \tag{6.1.7}$$

where

$$w(x_k, \bar{x}) = \int_0^1 [\nabla f(\bar{x} + t(x_k - \bar{x})) - \nabla f(x_k)] (x_k - \bar{x}) \, dt.$$

We have then

$$\left\| w(x_k, \overline{x}) \right\| = \left\| \int_0^1 \left[\nabla f(\overline{x} + t(x_k - \overline{x})) - \nabla f(x_k) \right] (x_k - \overline{x}) \, dt \right\| \qquad (6.1.8)$$

$$\leq \int_0^1 \left\| \nabla f(\overline{x} + t(x_k - \overline{x})) - \nabla f(x_k) \right\| \left\| x_k - \overline{x} \right\| dt,$$

hence by Lagrange Theorem there is $c_k \in [0, 1]$ such that

$$\left\| w(x_k, \overline{x}) \right\| \leq \left\| \nabla f(\overline{x} + c_k(x_k - \overline{x})) - \nabla f(x_k) \right\| \left\| x_k - \overline{x} \right\| .$$

This gives, due to the Lipschitz continuity of ∇f, that

$$\left\| w(x_k, \overline{x}) \right\| \leq L \left\| x_k - \overline{x} \right\|^2 , \qquad (6.1.9)$$

where by L we have denoted the Lipschitz constant of ∇f.

Moreover, since $\nabla f(\overline{x})$ is nonsingular, there is a $\delta > 0$ sufficiently small and an $M > 0$ such that $\nabla f(x)$ is nonsingular on $D(\overline{x}, \delta)$ and

$$\left\| \nabla f(x)^{-1} \right\| \leq M, \quad \forall x \in D(\overline{x}, \delta).$$

One has, from (6.1.7) and (6.1.5), that

$$\begin{aligned}
x_{k+1} &= x_k - \nabla f(x_k)^{-1}(f(x_k)) \\
&= x_k - \nabla f(x_k)^{-1} \left(\nabla f(x_k)(x_k - \overline{x}) + w(x_k, \overline{x}) \right) \\
&= \overline{x} + \nabla f(x_k)^{-1}(w(x_k, \overline{x})),
\end{aligned}$$

hence

$$\left\| x_{k+1} - \overline{x} \right\| \leq \left\| \nabla f(x_k)^{-1} \right\| \cdot \left\| w(x_k, \overline{x}) \right\| \qquad (6.1.10)$$

$$\leq \left\| \nabla f(x_k)^{-1} \right\| \cdot L \left\| x_k - \overline{x} \right\|^2 .$$

Take x_0 such that $x_0 \in D(\overline{x}, \delta)$ and $ML \left\| x_0 - \overline{x} \right\| := \rho < 1$. It follows inductively from (6.1.10) that $\left\| x_k - \overline{x} \right\| \leq \rho^k \left\| x_0 - \overline{x} \right\|$ for every $k \geq 1$, hence $(x_k) \subset D(\overline{x}, \delta)$ and $x_k \to \overline{x}$. Moreover, since

$$\left\| x_{k+1} - \overline{x} \right\| \leq ML \left\| x_k - \overline{x} \right\|^2 , \ \forall k,$$

we get (6.1.6). □

In case $p = 1$, i.e., $f : \mathbb{R} \to \mathbb{R}$, the equation (6.1.5) becomes

$$x_{k+1} = x_k - \frac{f(x_k)}{f'(x_k)}.$$

In this case, the iterate x_{k+1} is exactly the one where the tangent to the graph of f at $(x_k, f(x_k))$ intersects Ox.

As in the case of Aitken method of acceleration, we discuss in this context ($p = 1$) some possibilities to choose the point x_0 sufficiently close to the solution such that the Newton iteration converges to it. An empirical possibility would be to study the graph of the function and to choose an x_0 value which seems to be close to the solution. Another possibility is to apply the method of halving interval. Let us suppose that we have a continuous function f and two real numbers $a < b$ for which $f(a)f(b) < 0$. Then f has a root in (a, b). We then generate two sequences (a_k) and (b_k) as follows: $a_0 = a$, $b_0 = b$. Let $x_0 = 2^{-1}(a_0 + b_0)$. If $f(x_0) = 0$ then x_0 is the solution we are looking for and the process stops. If $f(a_0)f(x_0) < 0$ then we choose $a_1 = a_0$ and $b_1 = x_0$, and if $f(x_0)f(b_0) < 0$ we choose $a_1 = x_0$ and $b_1 = b_0$. In the same way, we take $x_1 = 2^{-1}(a_1 + b_1)$. Going further, we get close to the solution with (x_k) by halving at every step the interval which contains the solution. In general, this convergence is not very rapid but is good enough to be used for some of the iterations in order to find initial data for the Newton method.

In the general case ($p \geq 1$), we remark that some methods, known as quasi-Newton methods, do not require the calculation of the Jacobian $\nabla f(x)$. Instead, they use an approximation of this matrix, updating it at each iteration in such a way that it mimics the behavior of the Jacobian over the current step. If we denote this approximation matrix by J_k, then equation (6.1.4) becomes

$$0 = (f(x_k))^t + J_k \cdot (x_{k+1} - x_k)^t,$$

which gives, when J_k is nonsingular, the explicit formula

$$x_{k+1}^t = x_k^t - J_k^{-1} \cdot (f(x_k))^t. \tag{6.1.11}$$

If we denote

$$s_k := x_{k+1} - x_k \quad \text{and} \quad y_k := f(x_{k+1}) - f(x_k),$$

then by using Theorem 1.4.13 we get that

$$y_k = \int_0^1 \nabla f(x_k + ts_k)s_k \, dt \approx \nabla f(x_{k+1})(s_k) + r(\|s_k\|),$$

where

$$\lim_{k \to \infty} \frac{r(\|s_k\|)}{\|s_k\|} = 0.$$

So, in order that J_k mimics the behavior of the Jacobian $\nabla f(x_k)$, one asks that J_{k+1} satisfies the so called secant equation:

$$y_k^t = J_{k+1}s_k^t, \tag{6.1.12}$$

which ensures that J_{k+1} and $\nabla f(x_{k+1})$ have similar behavior along s_k. In fact, (6.1.12) can be seen as a system of p equations with p^2 unknowns, where the unknowns are the elements of J_{k+1}, so for $p > 1$ the components of J_{k+1} are not uniquely determined.

One of the best ways to find J_{k+1} is described by Broyden's method, where the matrix is given by the recurrence

$$J_{k+1} = J_k + \frac{(y_k^t - J_k s_k^t) \cdot s_k}{\langle s_k, s_k \rangle}. \tag{6.1.13}$$

As shown by the next result, the Broyden update makes the smallest change to J_k (measured by the Euclidean norm) that is consistent to (6.1.12).

Proposition 6.1.6. *The matrix J_{k+1} given by (6.1.13) satisfies:*

$$\|J_{k+1} - J_k\| = \min \left\{ \|J - J_k\| \mid y_k = J s_k \right\}.$$

Proof Take J any matrix which satisfies $y_k^t = J s_k^t$. We have then

$$\|J_{k+1} - J_k\| = \left\| \frac{(y_k^t - J_k s_k^t) \cdot s_k}{\langle s_k, s_k \rangle} \right\| = \left\| \frac{((J - J_k)s_k^t) \cdot s_k}{\langle s_k, s_k \rangle} \right\|$$

$$\leq \|J - J_k\| \left\| \frac{s_k^t \cdot s_k}{s_k^t \cdot s_k} \right\| = \|J - J_k\|,$$

which finishes the proof. $\qquad\qquad\qquad\qquad\qquad\qquad\qquad\qquad\qquad\qquad\square$

6.2 Algorithms for Optimization Problems

6.2.1 The Case of Unconstrained Problems

There exist several general methods to design algorithms for unconstrained optimization, but we concentrate on the line search method. The aim of this algorithm is to realize at every step a decrease of the value of the objective function $f : \mathbb{R}^p \to \mathbb{R}$, which is considered to be of class C^2. One asks that $f(x_{k+1}) < f(x_k)$. The algorithm computes for every term k a direction p_k (a vector of norm 1) and a step $\alpha_k > 0$ to move on the direction p_k. Therefore, starting from x_k, the new iteration will be

$$x_{k+1} = x_k + \alpha_k p_k. \tag{6.2.1}$$

With this approach, the choice of both the direction and the step are very important. From Taylor's Formula, for fixed α, p, there exists $t \in (0, \alpha)$

$$f(x_k + \alpha p) = f(x_k) + \alpha \nabla f(x_k)(p) + \frac{1}{2} \alpha^2 \nabla^2 f(x_k + tp)(p, p). \tag{6.2.2}$$

Putting aside the second order term (which for small α is small), the direction on which f decreases most is in fact the solution of the minimization on the unit ball of the function (of p), $p \mapsto \nabla f(x_k)(p)$. Since

$$\nabla f(x_k)(p) = \|p\| \, \|\nabla f(x_k)\| \cos \theta = \|\nabla f(x_k)\| \cos \theta,$$

where θ is the angle between p and $\nabla f(x_k)$, it is clear that the minimum is attained for

$$p = -\frac{\nabla f(x_k)}{\|\nabla f(x_k)\|}$$

if $\|\nabla f(x_k)\| \neq 0$. So, if a critical point is attained, then we cannot go further. If is not the case, the choice of p_k as above is called the steepest descent method. On the other hand, every other direction for which the angle with $\nabla f(x_k)$ is greater that $\frac{\pi}{2}$ (i.e., $\cos \theta < 0$) produces a decrease of f if α is sufficiently small, since the second order term in (6.2.2) contains a factor of α^2. Such a direction (for which $\nabla f(x_k)(p_k) < 0$) is called a decrease direction. Now, one has the problem of the choice of α_k. The ideal choice would be the minimum for $\alpha > 0$, of the function $\alpha \mapsto f(x_k + \alpha p_k)$, but, again, this problem is not necessarily a simple one. Another possibility is to choose a number $\alpha > 0$ such that $f(x_k + \alpha p_k) < f(x_k)$, but this choice could be insufficient, since the function may have a not very important decrease. In order to avoid both the problem of exact solvability of the optimization problem and the latter mentioned difficulty, a compromise is to choose an α_k which satisfies

$$f(x_k + \alpha p_k) < f(x_k) + c_1 \alpha \nabla f(x_k)(p_k), \tag{6.2.3}$$

where $c_1 \in (0, 1)$. The above inequality is called the Armijo condition (and was introduced by the American mathematician Larry Armijo in 1966), and the possibility of choosing α to fulfill (6.2.3) is ensured by (6.2.2) and by $\nabla f(x_k)(p_k) < 0$. In general, in order that some values of α are sufficiently large to satisfy the condition (6.2.3), c_1 is taken to be small. Even so, there is a risk of choosing a value α that is too small, such that, usually, one needs a second condition of the type

$$c_2 \nabla f(x_k)(p_k) \leq \nabla f(x_k + \alpha p_k)(p_k), \tag{6.2.4}$$

where $c_2 \in (c_1, 1)$. This condition is called the curvature condition and says that the derivative of the mapping $\alpha \mapsto f(x_k + \alpha p_k)$ at α is bigger than the product between c_2 and the derivative of the same function at 0. It is clear that a lower value of $\nabla f(x_k)(p_k)$ implies a bigger decrease of f, so in condition (6.2.4) it is preferable that c_2 to be taken close to 1. The conditions (6.2.3) and (6.2.4) are called the Wolfe conditions (after the American mathematician Philip Wolfe who introduced them in 1968). If instead of (6.2.4) one takes

$$\left| \nabla f(x_k + \alpha p_k)(p_k) \right| \leq \left| c_2 \nabla f(x_k)(p_k) \right|,$$

then one talks about strong Wolfe conditions.

The consistency of these conditions is rigorously shown in the next sections.

Proposition 6.2.1. *Let $f : \mathbb{R}^p \to \mathbb{R}$ be of class C^1, p_k a decrease direction at x_k, and $0 < c_1 < c_2 < 1$. If f is lower bounded on the set $\{x_k + \lambda p_k \mid \lambda > 0\}$, then there exists $\alpha > 0$ which satisfies the Wolfe conditions and the strong Wolfe conditions.*

Proof According to the assumption, the function $\alpha \mapsto f(x_k + \alpha p_k)$ is lower bounded on $(0, \infty)$. Since $c_1 > 0$ and $\nabla f(x_k)(p_k) < 0$, for α small enough,

$$f(x_k + \alpha p_k) < f(x_k) + c_1 \alpha \nabla f(x_k)(p_k)$$

whence, taking into account the boundedness property, the equation (in α)

$$f(x_k + \alpha p_k) = f(x_k) + c_1 \alpha \nabla f(x_k)(p_k)$$

has at least a strictly positive solution. From the continuity (in α) of the functions involved, there exists a smallest strictly positive solution which we denote by α'. Obviously, for every $\alpha \in (0, \alpha')$, condition (6.2.3) holds. We apply again Taylor's Formula, and there exists $\alpha'' \in (0, \alpha')$ such that

$$f(x_k + \alpha' p_k) = f(x_k) + \alpha' \nabla f(x_k + \alpha'' p_k)(p_k),$$

hence

$$\nabla f(x_k + \alpha'' p_k)(p_k) = c_1 \nabla f(x_k)(p_k) > c_2 \nabla f(x_k)(p_k).$$

Therefore, for α'' condition (6.2.4) holds. Since for α'' the inequalities in both (6.2.3) and (6.2.4) are strict, there exists an interval around this point where these conditions are fulfilled. By the fact that $\nabla f(x_k + \alpha'' p_k)(p_k) < 0$, we infer that the strong Wolfe conditions hold in a whole interval around α''. $\qquad\square$

We discuss now the convergence of the algorithm of the line search method.

Theorem 6.2.2. *Let us consider the iteration (6.2.1), where (p_k) are decrease directions, and (α_k) satisfy the Wolfe conditions. Suppose that f is of class C^1 and lower bounded, and that ∇f is Lipschitz. Then the series*

$$\sum_{k=0}^{\infty} \cos^2 \theta_k \left\| \nabla f(x_k) \right\|^2$$

(where θ_k denotes the angle between $\nabla f(x_k)$ and p_k) is convergent.

Proof From the Wolfe conditions, for each $k \in \mathbb{N}^*$,

$$\nabla f(x_k + \alpha_k p_k)(p_k) - \nabla f(x_k)(p_k) \geq (c_2 - 1)\nabla f(x_k)(p_k),$$

that is

$$\nabla f(x_{k+1})(p_k) - \nabla f(x_k)(p_k) \geq (c_2 - 1)\nabla f(x_k)(p_k),$$

and the Lipschitz condition on the differential gives a positive constant L such that

$$\left\|\nabla f(x_{k+1}) - \nabla f(x_k)\right\| \le L \left\|x_{k+1} - x_k\right\|,$$

from where,

$$\left(\nabla f(x_{k+1}) - \nabla f(x_k)\right)(p_k) \le \alpha_k L \left\|p_k\right\|^2.$$

We infer that

$$\alpha_k \ge \frac{c_2 - 1}{L} \frac{\nabla f(x_k)(p_k)}{\left\|p_k\right\|^2}.$$

From (6.2.3), taking again into account the inequality $\nabla f(x_k)(p_k) < 0$, we get

$$f(x_{k+1}) \le f(x_k) + c_1 \frac{c_2 - 1}{L} \frac{\left(\nabla f(x_k)(p_k)\right)^2}{\left\|p_k\right\|^2}.$$

But

$$\frac{\left(\nabla f(x_k)(p_k)\right)^2}{\left\|p_k\right\|^2} = \cos^2 \theta_k \left\|\nabla f(x_k)\right\|^2,$$

so

$$f(x_{k+1}) - f(x_k) \le c_1 \frac{c_2 - 1}{L} \cos^2 \theta_k \left\|\nabla f(x_k)\right\|^2.$$

Summing up, we deduce

$$f(x_{k+1}) \le f(x_0) + c_1 \frac{c_2 - 1}{L} \sum_{i=0}^{k} \cos^2 \theta_i \left\|\nabla f(x_i)\right\|^2.$$

Since $c_2 - 1 < 0$, the lower boundedness of f yields the convergence of the series. $\qquad\square$

The above theorem ensures that

$$\cos^2 \theta_k \left\|\nabla f(x_k)\right\|^2 \to 0.$$

If the choice of p_k is made in such a way that $\cos \theta_k > \varepsilon$ for every k and for a fixed $\varepsilon > 0$, then $\nabla f(x_k) \to 0$. Such a situation is called the steepest descent method where $\cos \theta_k = -1$. The algorithm does not guarantee the convergence toward a minimum point. The fact that one gets a sequence of points where the norm of the gradient is smaller and smaller, however, gives us hope that we progress towards a critical point. Clearly, it can be a saddle point, hence not a minimum. Moreover, the speed of convergence is slow.

However, in particular situations, some versions of the line search algorithm can be analyzed more accurately, as the next example shows.

Example 6.2.3. *Let us consider the case of the function $f : \mathbb{R}^p \to \mathbb{R}$ given by $f(x) = \frac{1}{2} \left\langle (Ax^t)^t, x \right\rangle + \left\langle b, x \right\rangle$ where A is a symmetric, positive definite square matrix of dimension p, and $b \in \mathbb{R}^p$. Clearly, f is strictly convex and its level sets are bounded, so there exists* *a unique minimum point given by the equation* $\nabla f(x) = 0$. *Therefore,* $\overline{x} = -(A^{-1}b^t)^t$. *Let us study the behaviour of the algorithm given by the relations* $x_{k+1} = x_k + \alpha_k d_k$, *where* $d_k = -\nabla f(x_k) = -(Ax_k^t)^t - b$, *and* α_k *is the minimum of the function* $\alpha \mapsto f(x_k + \alpha d_k)$. *Suppose that the gradient does not vanish at this iteration points, which is equivalent (taking into account the convexity of the problem) to* $f(x_k) > \overline{f}$, *where* $\overline{f}$ *is the minimum value of the function, that is* $\overline{f} := f(\overline{x}) = -\frac{1}{2}\left\langle (A^{-1}b^t)^t, b \right\rangle$.

Thus,

$$f(x_k + \alpha d_k) = f(x_k) + \frac{1}{2}\alpha^2 \left\langle (Ad_k^t)^t, d_k \right\rangle + \alpha \left\langle (Ad_k^t)^t + b, d_k \right\rangle .$$

Since $d_k \neq 0$, *the minimum of this function is attained at*

$$\alpha_k = \frac{\|d_k\|^2}{\left\langle (Ad_k^t)^t, d_k \right\rangle},$$

and

$$d_{k+1} = -(Ax_{k+1}^t)^t - b = -(Ax_k^t)^t - \alpha_k(Ad_k^t)^t - b = d_k - \alpha_k(Ad_k^t)^t,$$

whence

$$\left\langle d_{k+1}, d_k \right\rangle = \left\langle d_k, d_k \right\rangle - \alpha_k \left\langle (Ad_k^t)^t, d_k \right\rangle = 0.$$

We infer that

$$f(x_{k+1}) = f(x_k) - \frac{1}{2}\frac{\|d_k\|^4}{\left\langle (Ad_k^t)^t, d_k \right\rangle},$$

so

$$f(x_{k+1}) - \overline{f} = \left(f(x_k) - \overline{f}\right)\left[1 - \frac{\|d_k\|^4}{2\left(f(x_k) - \overline{f}\right)\left\langle (Ad_k^t)^t, d_k \right\rangle}\right].$$

Therefore,

$$\left\langle (A^{-1}d_k^t)^t, d_k \right\rangle = \left\langle (A^{-1}((Ax_k^t)^t + b)^t)^t, (Ax_k^t)^t + b \right\rangle$$

$$= 2\left[\frac{1}{2}\left\langle (Ax_k^t)^t, x_k \right\rangle + \left\langle b, x_k \right\rangle + \left\langle (A^{-1}b^t)^t, b \right\rangle\right]$$

$$= 2\left(f(x_k) - \overline{f}\right).$$

So,

$$f(x_{k+1}) - \overline{f} = \left(f(x_k) - \overline{f}\right)\left[1 - \frac{\|d_k\|^4}{\left\langle (A^{-1}d_k^t)^t, d_k \right\rangle \cdot \left\langle (Ad_k^t)^t, d_k \right\rangle}\right].$$

From the inequality of Kantorovici (Theorem 2.2.31), we have

$$\frac{\|d_k\|^4}{\left\langle (A^{-1}d_k^t)^t, d_k \right\rangle \cdot \left\langle (Ad_k^t)^t, d_k \right\rangle} \geq 4\left[\sqrt{\frac{\lambda_1}{\lambda_p}} + \sqrt{\frac{\lambda_p}{\lambda_1}}\right]^{-2} = 4\frac{\lambda_1\lambda_p^{-1}}{(\lambda_1\lambda_p^{-1} + 1)^2},$$

where λ_1 and λ_p are the greatest and the smallest eigenvalue of A, respectively. We denote $c := \lambda_1 \lambda_p^{-1}$. We put together the above relations and get

$$f(x_{k+1}) - \bar{f} \leq \left(f(x_k) - \bar{f} \right) \left[1 - 4\frac{c}{(c+1)^2} \right].$$

It follows that

$$f(x_k) - \bar{f} \leq \left(f(x_0) - \bar{f} \right) \left(\frac{c-1}{c+1} \right)^{2k}, \ \forall k \in \mathbb{N}.$$

From Example 7.72,

$$
\begin{aligned}
f(x_k) - \bar{f} &= \frac{1}{2} \left\langle (Ax_k^t)^t, x_k \right\rangle + \langle b, x_k \rangle - \bar{f} \\
&= \frac{1}{2} \left\langle (A(x_k - \bar{x})^t)^t, x_k - \bar{x} \right\rangle \geq \frac{1}{2}\lambda_p \|x_k - \bar{x}\|^2,
\end{aligned}
$$

so

$$\|x_k - \bar{x}\| \leq \sqrt{\left[\frac{2(f(x_0) - \bar{f})}{\lambda_p} \right]} \left(\frac{c-1}{c+1} \right)^k, \ \forall k \in \mathbb{N},$$

and this relation allows us to conclude that the approximation of the minimum point depends on the value of c : if the difference between the greater and the smaller eigenvalues of A is small, then the convergence is rapid. At the limit, for $c = 1$, the first iteration already attains the minimum point.

On the other hand, there are several possible improvements of the general line search method and one of these possibilities would be to consider a second-order decrease direction (of Newton type). Therefore, let us suppose that we have a function $f : \mathbb{R} \to \mathbb{R}$ of class C^3. As above, the algorithm will search for critical points, i.e., to the solutions of the equation $f'(x) = 0$. If one supposes that $\bar{x}$ is a nondegenerate solution, and applies the Newton's method for this equation, they would be lead to consider the iterations

$$x_{k+1} = x_k - \frac{f'(x_k)}{f''(x_k)}.$$

For this algorithm we have a quadratic convergence (as shown in the previous section), the drawback being the same as we discussed for the Newton's method. On the other hand, if $f''(x_k)$ is not positive, then the direction $-\frac{f'(x_k)}{f''(x_k)}$ is not necessarily a decreasing one, so we can attain maximal points.

We close this section with a special look to the case of convex functions. Clearly, the above algorithm is well suited to these functions, but the particular form of convexity allows the design of some powerful specific algorithms. We now introduce the proximal point algorithm. An initial form of this was published by the French mathematician Bernard Martinet in 1970, and later generalized by the American mathematician Ralph Tyrrell Rockafellar in 1976.

Let $f : \mathbb{R}^p \to \mathbb{R}$ be a convex differentiable function. Thus, for every fixed $y \in \mathbb{R}^p$, we consider the function $g_y : \mathbb{R}^p \to \mathbb{R}$,

$$g_y(x) = f(x) + \frac{1}{2} \|x - y\|^2 .$$

This new application is convex and differentiable (as a sum of functions with these properties). Moreover, g_y is strictly convex because $x \mapsto \frac{1}{2} \|x - y\|^2$ has this property (see Theorem 2.2.15 (iv)).

Suppose that f satisfies the coercivity condition of Proposition 3.1.8, whence f attains its global minimum on $\mathbb{R}^p$. It is easy to see that the same condition is satisfied by g_y as well, therefore there exists a global minimum point of g_y on $\mathbb{R}^p$. By the fact that g_y is strictly convex, this minimum point is unique (Proposition 3.1.24), and we denote it by $\overline{x}_y$. Furthermore, according to Theorem 3.1.22 and the differentiation rules, $\overline{x}_y$ is characterized by the relation

$$\nabla f(\overline{x}_y) + \overline{x}_y - y = 0.$$

We generate now a sequence of iterations following the next rule: $x_0 \in \mathbb{R}^p$, and for every $k \geq 0$, $x_{k+1} = \overline{x}_{x_k}$, that is,

$$\nabla f(x_{k+1}) + x_{k+1} - x_k = 0.$$

Let $\overline{x}$ be a minimum point for f (it exists according to the above assumptions). The next relation holds:

$$\|x_{k+1} - \overline{x}\|^2 = \|x_k - \overline{x}\|^2 - \|x_{k+1} - x_k\|^2 + 2 \langle x_{k+1} - \overline{x}, x_{k+1} - x_k \rangle , \ \forall k.$$

But

$$\langle x_{k+1} - \overline{x}, x_{k+1} - x_k \rangle = -\nabla f(x_{k+1})(x_{k+1} - \overline{x}) \leq 0$$

(from Theorem 2.2.10), so

$$\|x_{k+1} - \overline{x}\|^2 \leq \|x_k - \overline{x}\|^2 - \|x_{k+1} - x_k\|^2 \leq \|x_k - \overline{x}\|^2 , \ \forall k.$$

We deduce the following facts: the sequence $(\|x_k - \overline{x}\|)_k$ is decreasing, whence convergent (being positive), while the sequence $(\|x_{k+1} - x_k\|)_k$ is convergent to 0. In particular, the sequence $(x_k)_k$ is bounded. We show that (x_k) is convergent to a minimum point of f. Let $x \in \mathbb{R}^p$ arbitrary but fixed. Then

$$\nabla f(x_{k+1})(x - x_{k+1}) = \langle x_k - x_{k+1}, x - x_{k+1} \rangle \geq - \|x_k - x_{k+1}\| \, \|x - x_{k+1}\| .$$

Let $z \in \mathbb{R}^p$ be a limit point of (x_k) (its existence is ensured by the boundedness of the sequence). Passing to the limit in the above relation, using that $\|x_{k+1} - x_k\| \to 0$ and that $(\|x - x_{k+1}\|)$ is bounded, we deduce

$$\nabla f(z)(x - z) \geq 0.$$

Since x is arbitrary, we get $\nabla f(z) = 0$, that is z is a critical point, whence a minimum point. Suppose that (x_k) has at least two different limit points z_1 and z_2. According to the above stage of the proof, both are minimum points of f. With the same reasoning as in the case of $\bar{x}$, we infer that the sequences $(\|x_k - z_1\|)_k$ and $(\|x_k - z_2\|)_k$ are convergent. But

$$\|x_k - z_2\|^2 = \|x_k - z_1\|^2 + 2\langle x_k - z_1, z_1 - z_2\rangle + \|z_1 - z_2\|^2, \ \forall k.$$

Therefore, there exists

$$2\lim_k \langle x_k - z_1, z_1 - z_2\rangle = \lim_k \|x_k - z_2\|^2 - \lim_k \|x_k - z_1\|^2 - \|z_1 - z_2\|^2.$$

By the fact that z_1 is a limit point of (x_k), $\lim_k \langle x_k - z_1, z_1 - z_2\rangle$ can be only 0, so

$$\lim_k \|x_k - z_2\|^2 - \lim_k \|x_k - z_1\|^2 = \|z_1 - z_2\|^2 > 0.$$

Changing the roles of z_1 and z_2,

$$\lim_k \|x_k - z_1\|^2 - \lim_k \|x_k - z_2\|^2 = \|z_1 - z_2\|^2 > 0,$$

so we arrive at a contradiction. Thus (x_k) is convergent to a minimum point of f.

6.2.2 The Case of Constraint Problems

For constrained problems, we adopt a slightly simplified framework which allows, nevertheless, the presentation of the main ideas of two important methods of searching for extrema, namely the sequential quadratic programing and the interior point methods.

6.2.2.1 Sequential quadratic programming

The sequential quadratic programming (SQP) is one of the most effective methods used to solve nonlinear optimization problems with constraints. We will restrict our approach to equality-constrained optimization, i.e., we consider for the problem (P) defined in the second section of Chapter 3 only equalities constraints. We take then the C^1 function $h : \mathbb{R}^p \to \mathbb{R}^m$ and

$$M := \left\{ x \in \mathbb{R}^p \mid h(x) = 0 \right\}.$$

The underlying idea of SQP is to model the problem (P) at the current iterate x_k by a quadratic programming subproblem, and then to construct the next iteration x_{k+1} by the use of the minimizer of this subproblem.

In order to continue our discussion in the general case of nonlinear optimization problems with constraints, we present some elements about quadratic programming.

An optimization problem where the objective function is quadratic and the constraints are linear is called a quadratic program (QP). As above, we limit our analyses to the case of equality constraints, i.e., we consider the problem

$$(QP) \quad \min_x \quad f(x) := \frac{1}{2}\langle (Qx^t)^t, x \rangle + \langle c, x \rangle,$$
$$\text{subject to} \quad Ax^t = b^t,$$

where Q is a symmetric $p \times p$ matrix, A is a $m \times p$ matrix with $m \leq p$, x, c are vectors in $\mathbb{R}^p$, and b is a vector in $\mathbb{R}^m$. If the Hessian matrix Q is positive definite, then we speak about convex QP, and in this case the analysis is similar to the case of linear programs. When Q is an indefinite matrix, more difficulties can arise, since in this case several stationary and local minima may appear.

In what follows, we will restrict to the case of convex QPs, and we suppose that the matrix A has full row rank (rank $A = m$). Then $\bar{x}$, the unique minimum point of (QP) (see Exercise 7.73), is fully characterized by the relations

$$A\bar{x}^t = b^t$$
$$\exists \mu \in \mathbb{R}^m \text{ such that } \nabla f(\bar{x}) + \mu A = 0,$$

which finally give

$$\mu^t = -(AQ^{-1}A^t)^{-1}(b + AQ^{-1}c^t)$$

and

$$\bar{x}^t = -Q^{-1}c^t + Q^{-1}A^t(AQ^{-1}A^t)^{-1}(b + AQ^{-1}c^t).$$

Remark also that the first-order optimality conditions for $\bar{x}$ can be written under matricial form as follows:

$$\begin{pmatrix} Q & A^t \\ A & 0 \end{pmatrix} \begin{pmatrix} \bar{x}^t \\ \bar{\mu}^t \end{pmatrix} = \begin{pmatrix} -c^t \\ b^t \end{pmatrix}. \tag{6.2.5}$$

Rewrite (6.2.5) in a form more useful for computation: take $\bar{x} = x + y$, where x is an estimate of the solution and y is the desired step. Then (6.2.5) becomes

$$\begin{pmatrix} Q & A^t \\ A & 0 \end{pmatrix} \begin{pmatrix} y^t \\ \bar{\mu}^t \end{pmatrix} = \begin{pmatrix} d^t \\ e^t \end{pmatrix}, \tag{6.2.6}$$

where

$$e^t = -Ax^t + b^t, \quad d^t = -c^t - Qx^t, \quad y = \bar{x} - x.$$

The previous comments show that, in order to find the unique global solution of (QP), we must solve the linear system (6.2.6). A first observation is that if $p \geq 1$, the Karush-Kuhn-Tucker matrix

$$K := \begin{pmatrix} Q & A^t \\ A & 0 \end{pmatrix}$$

is always indefinite. One option is to use a triangular factorization, as the QR (Householder) factorization, or to use the so-called Schur-complement method. For details see the book (Nocedal and Wrightm, 2006).

Coming back to the general case of constrained optimization problems with equality constraints, recall that the Lagrangian of (P) is the function

$$L(x, \mu) = f(x) + \sum_{j=1}^{m} \mu_j h_j(x).$$

Denote the Jacobian matrix of the constraints by $A(x)$, i.e.,

$$A(x) = \begin{pmatrix} \nabla h_1(x) \\ \cdots \\ \nabla h_m(x) \end{pmatrix}.$$

As shown by Theorem 3.2.6, the first order Karush-Kuhn-Tucker conditions for the problem (P) can be written as

$$F(x, \mu) := \begin{pmatrix} \nabla f(x) + \mu A(x) \\ h(x) \end{pmatrix} = 0. \tag{6.2.7}$$

One method is to solve the nonlinear equation (6.2.7) by the use of Newton's method. We have

$$\nabla F(x, \mu) = \begin{pmatrix} \nabla_{xx}^2 L(x, \mu) & A(x)^t \\ A(x) & 0 \end{pmatrix},$$

hence the Newton step, according to (6.1.5), is

$$\begin{pmatrix} x_{k+1}^t \\ \mu_{k+1}^t \end{pmatrix} = \begin{pmatrix} x_k^t \\ \mu_k^t \end{pmatrix} - \begin{pmatrix} \nabla_{xx}^2 L(x_k, \mu_k) & A(x_k)^t \\ A(x_k) & 0 \end{pmatrix}^{-1} \cdot \begin{pmatrix} (\nabla f(x_k) + \mu_k A(x_k))^t \\ (h(x_k))^t \end{pmatrix}. \tag{6.2.8}$$

Of course, in order that the Karush-Kuhn-Tucker matrix

$$K(x_k, \mu_k) := \begin{pmatrix} \nabla_{xx}^2 L(x_k, \mu_k) & A(x_k)^t \\ A(x_k) & 0 \end{pmatrix}$$

is nonsingular, we suppose that the constraint Jacobian $A(x)$ has full row rank, and that the matrix $\nabla_{xx}^2 L(x, \mu)$ is positive definite on the tangent space of the constraints, i.e.,

$$\langle (\nabla_{xx}^2 L(x, \mu) d^t)^t, d \rangle > 0, \ \forall d \neq 0 \text{ s.t. } A(x) d^t = 0. \tag{6.2.9}$$

Another way to view the iterations (6.2.8) is to consider the quadratic problem bellow at each iterate (x_k, μ_k) :

$$\begin{aligned} \min_y \quad & \tfrac{1}{2} \langle (\nabla_{xx}^2 L(x_k, \mu_k) y^t)^t, y \rangle + \langle \nabla f(x_k), y \rangle + f(x_k) \\ \text{subject to} \quad & A(x_k) y^t + (h(x_k))^t = 0. \end{aligned} \tag{6.2.10}$$

Under the assumptions made, we know by the comments before that this problem has a unique solution y_k for which there is a multiplier l_k such that:

$$(\nabla^2_{xx}L(x_k, \mu_k)y_k^t)^t + \nabla f(x_k) - l_k A(x_k) = 0,$$
$$A(x_k)y_k^t + (h(x_k))^t = 0. \qquad (6.2.11)$$

Moreover, the pair (y_k, l_k) can be identified with the one of (6.2.8). To see this, denote $y_k' := x_{k+1} - x_k$ and rewrite (6.2.8) as

$$\begin{pmatrix} \nabla^2_{xx}L(x_k, \mu_k) & A(x_k)^t \\ A(x_k) & 0 \end{pmatrix} \begin{pmatrix} (y_k')^t \\ \mu_{k+1}^t - \mu_k^t \end{pmatrix} = \begin{pmatrix} -(\nabla f(x_k) + \mu_k A(x_k))^t \\ -(h(x_k))^t \end{pmatrix}.$$

By subtracting $\mu_k A(x_k)$ in both sides of the previous relation, we get

$$\begin{pmatrix} \nabla^2_{xx}L(x_k, \mu_k) & A(x_k)^t \\ A(x_k) & 0 \end{pmatrix} \begin{pmatrix} (y_k')^t \\ \mu_{k+1}^t \end{pmatrix} = \begin{pmatrix} -(\nabla f(x_k))^t \\ -(h(x_k))^t \end{pmatrix}.$$

Hence, by the nonsingularity of the Karush-Kuhn-Tucker matrix $K(x_k, \mu_k)$ and by relations (6.2.11), we obtain that $y_k = y_k'$ and $\mu_{k+1} = l_k$.

Hence, the new iterate (x_{k+1}, μ_{k+1}) can be defined either as solution of the quadratic program (6.2.10), or as the Newton type iterate given by (6.2.8) applied to the optimality conditions of the problem.

We close our consideration with a result about the rate of convergence. Recall that the set of critical directions is in our case

$$C(\overline{x}, \overline{\mu}) = \left\{ u \in \mathbb{R}^p \mid \nabla h_j(\overline{x})(u) = 0, \text{ for every } j \in \overline{1, m}. \right\}$$

Theorem 6.2.4. *Suppose $\overline{x}$ is a local solution of the problem*

$$\min f(x), \quad \text{subject to } h(x) = 0,$$

such that f and h are twice continuously differentiable functions with Lipschitz continuous second derivatives. Moreover, suppose that the linear independence condition holds at $\overline{x}$, and that

$$\nabla^2_{xx}L(\overline{x}, (\overline{\lambda}, \overline{\mu}))(u, u) > 0, \quad \forall u \in C(\overline{x}, \overline{\mu}) \setminus \{0\}.$$

Then, if (x_0, μ_0) is sufficiently close to $(\overline{x}, \overline{\mu})$, then the sequence generated by (6.2.8) converges quadratically to $(\overline{x}, \overline{\mu})$.

Proof The proof follows from Theorem 6.1.5, because (6.2.8) is the Newton's method applied to the nonlinear system $F(x, \mu) = 0$, where F is given by (6.2.7). □

6.2.2.2 Interior-point methods

In the approach we present here, for the problem (P) defined in the second section of Chapter 3, we consider only inequality constraints. Therefore, we take $g : \mathbb{R}^p \to \mathbb{R}^n$ and

$$M := \{x \in \mathbb{R}^p \mid g(x) \leq 0\}.$$

We denote

$$\text{strict } M := \{x \in \mathbb{R}^p \mid g(x) < 0\}$$
$$= \{x \in \mathbb{R}^p \mid g_i(x) < 0, \ \forall i \in \overline{1, n}\}.$$

Let us observe that, in general, strict M does not coincide with the interior of M (it is enough to consider $g : \mathbb{R} \to \mathbb{R}^2$ with $g(x) = (-x^2, -x - 1)$, since in this case strict $M = \text{int } M \setminus \{0\}$).

The main idea is to transform the constrained problem into an unconstrained one through a penalization of the objective function f by an auxiliary function that contains the constraints. In fact, this also happens when one introduces the Lagrangian function, but then some parameters (λ, μ) depending on the solution were in force. At this moment, we consider the function (called the logarithmic barrier of (P)), $B(x, \mu) : \text{strict } M \times (0, \infty) \to \mathbb{R}$,

$$B(x, \mu) := f(x) - \mu \sum_{i=1}^{n} \ln(-g_i(x)).$$

It is clear that this function preserves the smoothness properties of the problem data. On the other hand, if a solution $\overline{x}$ lies in strict M, then for x close to $\overline{x}$, $\lim_{\mu \to 0} B(x, \mu) = f(x)$. If $\overline{x}$ lies in $M \setminus \text{strict } M$ then at least one constraint is active, so for a sequence $(x_k) \to \overline{x}$,

$$\lim_{k} \left(f(x_k) - \mu \sum_{i=1}^{n} \ln(-g_i(x_k)) \right) = \infty.$$

Consequently, a coercivity condition similar to that in Proposition 3.1.9 could be fulfilled (under some conditions). The idea (in both situations) is to ensure the existence of a unconstrained minimum for $B(\cdot, \mu)$ on strict M, denoted x_μ, then, for $\mu \to 0$, to show that x_μ converges towards a minimum of the problem (P). The below figure offers an intuitive image for this remark for the case of the function $f : \mathbb{R} \to \mathbb{R}, f(x) = e^x - x^3$ under the constrains $g_1(x) = 1 - x \leq 0, g_2(x) = x - 3 \leq 0$. The minimum is $\overline{x} = 3$, and the Figure 6.1 presents, besides the graph of f, the graphs of $B(x, 3^{-1})$ and $B(x, 7^{-1})$.

In order to prepare the main result, we need an additional preliminary discussion. We have already said that if we have minimum points that are close one to each other, then it is difficult to design algorithms which make the distinction between them, and in order to approach the desired point one should start from appropriate initial data. The most unpleasant situation occurs when a minimum point is not isolated in the set of local minima. Such an example for a C^2 function is given below. Let $f : \mathbb{R} \to \mathbb{R}$,

$$f(x) = \begin{cases} x^4 \left(2 + \cos \frac{1}{x}\right), & x \neq 0 \\ 0, & x = 0. \end{cases}$$

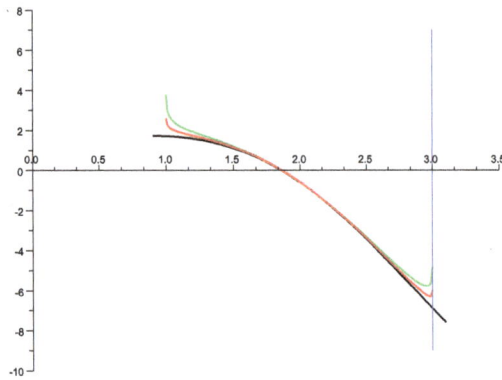

Figure 6.1: Barrier method illustration.

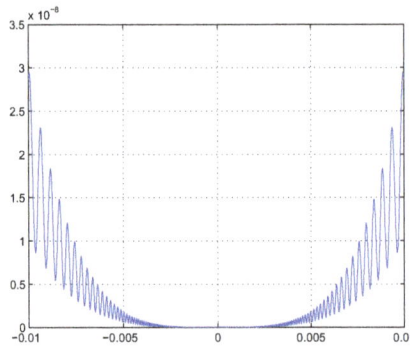

Figure 6.2: Non isolated minimum.

This function has a global minimum at 0, but there is a sequence of local minima which converges to 0 (see Figure 6.2).

It is now necessary to formulate some conditions concerning topological properties of the set of minima. We start with a definition.

Definition 6.2.5. *Let $A \subset B$. We say that A is an isolated subset of B if there exists a closed set E with $A \subset \text{int } E$ and $B \cap E = A$.*

The next result holds.

Proposition 6.2.6. *Let $M \subset \mathbb{R}^p$ and $\varphi : M \to \mathbb{R}$. We denote by N a set (that we suppose to be nonempty) of a local minima of φ on M, for which the value of the function is the* *same (denoted by $\overline{\varphi}$). Suppose that N^* is an isolated compact subset of N. Then there exists a compact set C such that $N^* \subset \text{int } C$ and $\varphi(x) > \overline{\varphi}$ for every $x \in (M \cap C) \setminus N^*$.*

Proof According to the preceding definition, there exists a closed set E with $N^* \subset \text{int } E$ and $N \cap E = N^*$. Since N^* consists only of minima, for every $x \in N^*$ there exists an open neighborhood V_x of x with

$$\varphi(x) = \overline{\varphi} \le f(u), \quad \forall u \in M \cap V_x.$$

Then the set $G := \bigcup_{x \in N^*} V_x$ is open and includes N^*, and $\varphi(u) \ge \overline{\varphi}$ for every $u \in M \cap G$.

Then, by the compactness of N^* and the openness of $G \cap \text{int } E$, there exists a compact set C with

$$N^* \subset \text{int } C \subset C \subset G \cap \text{int } E \subset G \cap E$$

(it can be shown quite simply that one can take C as $\{x \in \mathbb{R}^p \mid d(x, C) \le n^{-1}\}$ for sufficiently large $n \in \mathbb{N}^*$). Clearly, $N^* \subset \text{int } C \cap M$. Take now $x \in (M \cap C) \setminus N^*$. Since $x \in M \cap G$, one has $\varphi(x) \ge \overline{\varphi}$. On the other hand, since $C \subset E$, one has $x \in E \setminus N^*$, whence $x \notin N$. Therefore, $\varphi(x) \ne \overline{\varphi}$. Consequently, $\varphi(x) > \overline{\varphi}$ is the only possibility. $\square$

We present now the main result of this section.

Theorem 6.2.7. *Suppose that f and g are continuous. Let N be the set (supposed to be nonempty) of local minima of f on M for which the value of the function is the same (denoted by $\overline{f}$), and let $(\mu_k) \subset (0, \infty)$ be a strictly decreasing sequence convergent to 0. We suppose that:*
   (a) there exists an isolated compact subset N^* of N;
   (b) $N^* \cap \text{cl}(\text{strict } M) \ne \emptyset$.
   Then:
   (i) there exists a compact set C such that $N^* \subset \text{int } C$, and for every $x \in (M \cap C) \setminus N^*$, $f(x) > \overline{f}$;
   (ii) there exist an infinity of numbers k for which there exists a global minimum $y_k \in \text{strict } M \cap \text{int } C$ without restrictions of $B(\cdot, \mu)$ on strict $M \cap \text{int } C$ with

$$B(y_k, \mu_k) = \min\{B(x, \mu_k) \mid x \in \text{strict } M \cap C\};$$

   (iii) every limit point of (y_k) is in N^*, and if (x_l) is a subsequence of (y_k) convergent to the underlying limit point, then

$$\lim_l f(x_l) = \overline{f} = \lim_l B(x_l, \mu_l).$$

Proof The first item, (i), follows easily from (a) and from Proposition 6.2.6. So, there exists a compact set C such that $N^* \subset \text{int } C \cap M$, and the value of f at the points of N^* is the smallest one in $C \cap M$. Since the function $B(\cdot, \mu_k)$ verifies the assumptions of Proposition 3.1.9 for $D = \text{strict } M$, there exists a global minimum for $B(\cdot, \mu_k)$ on strict $M \cap C$.

At this moment, this minimum point, denoted by y_k, is not without restrictions (note that C is closed). Since (y_k) is bounded, there exists a subsequence of it, denoted (x_l), which has a limit, denoted by x_∞, lying in $M \cap C$. Therefore x_∞ is a feasible point. We show now that $x_\infty \in N^*$. Otherwise, from the preceding part, $f(x_\infty) > \bar{f}$. In order to arrive at a contradiction, we use (b). Take $x^* \in N^* \cap \mathrm{cl}(\mathrm{strict}\,M)$. We distinguish two situations, and in both cases we show that there exists $x_{\mathrm{int}} \in C \cap \mathrm{strict}\,M$ with $f(x_\infty) > f(x_{\mathrm{int}})$. Firstly, we consider that $x^* \in \mathrm{strict}\,M$. Then $f(x_\infty) > f(x^*)$, so we can take $x_{\mathrm{int}} = x^*$. Suppose that $x^* \in \mathrm{cl}(\mathrm{strict}\,M)\backslash\mathrm{strict}\,M$. From $N^* \subset \mathrm{int}\,C$, we deduce that $x^* \in \mathrm{int}\,C$. Since $f(x_\infty) > \bar{f} = f(x^*)$ and f is continuous, there exists a neighborhood V of x^* such that $f(x_\infty) > f(x)$ for every $x \in V$. In particular, there exists $x_{\mathrm{int}} \in C \cap \mathrm{strict}\,M$ with the desired property. Thus, in every situation, there exists $x_{\mathrm{int}} \in C \cap \mathrm{strict}\,M$ with $f(x_\infty) > f(x_{\mathrm{int}})$. Again, from the continuity of f, for every l big enough, $f(x_l) > f(x_{\mathrm{int}})$. But x_l is a global minimum on $C \cap \mathrm{strict}\,M$ for $B(\cdot, \mu_l)$, so

$$f(x_l) - \mu_l \sum_{i=1}^{n} \ln(-g_i(x_l)) \leq f(x_{\mathrm{int}}) - \mu_l \sum_{i=1}^{n} \ln(-g_i(x_{\mathrm{int}})).$$

Since $\sum_{i=1}^{n} \ln(-g_i(x_{\mathrm{int}})) \in \mathbb{R}$, passing to the limit as $l \to \infty$,

$$\lim_{l} \left(f(x_{\mathrm{int}}) - \mu_l \sum_{i=1}^{n} \ln(-g_i(x_{\mathrm{int}})) \right) = f(x_{\mathrm{int}}).$$

If $x_\infty \in \mathrm{strict}\,M$, as above,

$$\lim_{l} \left(f(x_l) - \mu_l \sum_{i=1}^{n} \ln(-g_i(x_l)) \right) = f(x_\infty),$$

whence

$$f(x_\infty) \leq f(x_{\mathrm{int}}),$$

in contradiction to the step before.

If $x_\infty \in (M \cap C) \backslash \mathrm{strict}\,M$, adding $-\mu_l \sum_{i=1}^{n} \ln(-g_i(x_{\mathrm{int}}))$ in the inequality $f(x_l) > f(x_{\mathrm{int}})$, we infer, by the use of the relation before,

$$f(x_l) - \mu_l \sum_{i=1}^{n} \ln(-g_i(x_{\mathrm{int}})) > f(x_{\mathrm{int}}) - \mu_l \sum_{i=1}^{n} \ln(-g_i(x_{\mathrm{int}}))$$

$$\geq f(x_l) - \mu_l \sum_{i=1}^{n} \ln(-g_i(x_l)),$$

whence

$$-\sum_{i=1}^{n} \ln(-g_i(x_{\mathrm{int}})) > -\sum_{i=1}^{n} \ln(-g_i(x_l)).$$

For $l \to \infty$, in the left-hand side we get a real number, and in the right-hand side we get $+\infty$, which is a contradiction. We conclude that the assumption made was false, so $f(x_\infty) = \bar{f}$, that is $x_\infty \in N^*$. We conclude that every limit point of (y_k) satisfies this property. From the fact that $x_\infty \in N^*$, we infer that $x_\infty \in$ int C, so, eventually, x_l belongs to int C. This means that the geometric restriction $x \in C$ is not active, whence x_l is a minimum without constraints for $B(\cdot, \mu_l)$ on strict $M \cap$ int C. The second item, (ii), is proved.

The first part of (iii) is obvious, since $\lim_l f(x_l) = f(x_\infty) = \bar{f}$. It remains to show that $\bar{f} = \lim_l B(x_l, \mu_l)$.

If $x_\infty \in$ strict M, then the sum $\sum_{i=1}^n \ln(-g_i(x_l))$ is finite for all big l. Then

$$\lim_l B(x_l, \mu_l) = f(x_\infty) = \bar{f}.$$

Now, suppose that $x_\infty \notin$ strict M, so at least a constraint goes to 0 on (x_l) for $l \to \infty$. Since for every l, x_l is the minimum point for $B(\cdot, \mu_l)$ on strict $M \cap C$, we have

$$B(x_l, \mu_l) \le B(x_{l+1}, \mu_l) \text{ and } B(x_{l+1}, \mu_{l+1}) \le B(x_l, \mu_{l+1}).$$

We multiply the first inequality by $\mu_l^{-1}\mu_{l+1} \in (0, 1)$ and we add the second inequality. We get

$$f(x_{l+1})\left(1 - \frac{\mu_{l+1}}{\mu_l}\right) \le f(x_l)\left(1 - \frac{\mu_{l+1}}{\mu_l}\right),$$

so $f(x_{l+1}) \le f(x_l)$.

For every l big enough, $\sum_{i=1}^n \ln(-g_i(x_l)) < 0$, whence $B(x_l, \mu_l) > f(x_l)$. Since $f(x_{l+1}) \le f(x_l)$ and $\lim_l f(x_l) = f(x_\infty) = \bar{f}$, we infer that the sequence $(B(x_l, \mu_l))$ is lower bounded. On the other hand,

$$0 < \mu_{l+1} < \mu_l \text{ and } \sum_{i=1}^n \ln(-g_i(x_l)) < 0,$$

so for l big enough,

$$-\mu_{l+1}\sum_{i=1}^n \ln(-g_i(x_l)) < -\mu_l\sum_{i=1}^n \ln(-g_i(x_l)),$$

hence

$$B(x_l, \mu_{l+1}) < B(x_l, \mu_l).$$

Therefore,

$$B(x_{l+1}, \mu_{l+1}) \le B(x_l, \mu_{l+1}) \le B(x_l, \mu_l).$$

We deduce that the sequence $(B(x_l, \mu_l))$ is monotone, and converges towards a limit denoted by $\bar{B}$. The inequality $\bar{B} \ge \bar{f}$ is ensured by the above considerations. We suppose, by way of contradiction, that $\bar{B} > \bar{f}$, and we take $\varepsilon := 2^{-1}(\bar{B} - \bar{f}) > 0$. From the continuity of f, there exists a neighborhood V of x_∞ for which

$$f(x) < f(x_\infty) + \varepsilon = \bar{B} - \varepsilon, \ \forall x \in V.$$

There exists at least one point of $x' \in$ strict M in V. So,

$$B(x_l, \mu_l) \le B(x', \mu_l) = f(x') - \mu_l \sum_{i=1}^{n} \ln(-g_i(x')).$$

But $\sum_{i=1}^{n} \ln(-g_i(x')) \in \mathbb{R}$, and for every l big enough,

$$-\mu_l \sum_{i=1}^{n} \ln(-g_i(x')) < 2^{-1}\varepsilon.$$

But

$$f(x') < \overline{B} - \varepsilon,$$

so

$$B(x_l, \mu_l) < \overline{B} - \varepsilon + 2^{-1}\varepsilon = \overline{B} - 2^{-1}\varepsilon,$$

which contradicts $B(x_l, \mu_l) \to \overline{B}$. So $\overline{B} = \overline{f}$ and the proof is complete. $\qquad\square$

The hypotheses of the above result are quite weak. It is remarkable that the problem data are supposed to be only continuous. We can say that the assumption (b) is the most demanding one, but it is quite natural. However, there exist situations when this assumption is not fulfilled. To see this, it is sufficient to consider the problem of minimizing the function $f : \mathbb{R} \to \mathbb{R}$, $f(x) = (x + 1)^2$ under the constraint $g(x) \le 0$, where $g : \mathbb{R} \to \mathbb{R}^2$, $g(x) = (x(1 - x), -x)$. Then $M = \{0\} \cup [1, \infty)$, and the only solution is $\overline{x} = 0$. But strict $M = (1, \infty)$, hence $N^* \cap \mathrm{cl}(\text{strict } M) = \emptyset$.

The good part of the conclusion is that we established convergence to the solutions of the problem (P) without qualification conditions. However, the sequence (y_k) can be divergent: we know that it has convergent subsequences. Let us remark that

$$\nabla_x B(x, \mu) = \nabla f(x) + \sum_{i=1}^{n} \frac{\mu}{g_i(x)} \nabla g_i(x).$$

If $\overline{x}$ is a minimum point without restrictions for $B(\cdot, \mu)$, then $\nabla_x B(\overline{x}, \mu) = 0$, and if μ is small enough and $\overline{x}$ is close to the solution of the problem (P), then the Lagrange multipliers λ_i can be approximated by $\mu g_i^{-1}(\overline{x})$.

Subsequently, Theorem 6.2.7 can be combined with other hypotheses and techniques in order to obtain several enhancements of the conclusions and in order to include the equality constraints in the discussion.

6.3 Scientific Calculus Implementations

In this section we aim at illustrating the theoretical discussions above through their numerical implementation in the scientific calculus software Matlab. Thus we verify, from a practical point of view, the results we have proved before, and we split our exemplifications into two categories of codes: on one hand, there are codes which use some default functions of Matlab (which, in turn, encapsulate various numerical algorithms) and, on the other hand, we directly implement many of the studied algorithms.

1. (least squares method - linear dependence) The system obtained by the modelling the least squares method for the affine dependence $v = at + b$, has, as established before (Section 3.4), a unique solution:

$$\begin{pmatrix} a \\ b \end{pmatrix} = \begin{pmatrix} \sum_{i=1}^{N} t_i^2 & \sum_{i=1}^{N} t_i \\ \sum_{i=1}^{N} t_i & N \end{pmatrix}^{-1} \begin{pmatrix} \sum_{i=1}^{N} t_i v_i \\ \sum_{i=1}^{N} v_i \end{pmatrix}.$$

Let us take the concrete example: $N = 5$, $t_1 = 0$, $t_2 = 1$, $t_3 = 2$, $t_4 = 3$, $t_5 = 4$ and $v_1 = 1$, $v_2 = 2.5$, $v_3 = 5.1$, $v_4 = 6.7$, $v_5 = 8.3$.

For the calculus of the parameters a, b we implement the following Matlab code:

```
t=[0,1,2,3,4];
v=[1,2.5,5.1,6.7,8.3];
A=[sum(t.^2) sum(t)
sum(t) 5]
B=[sum(t.*v)
sum(v) ]
U=A^(-1)*B
x=linspace(0,4.5,90);
y=U(1)*x+U(2);
plot(t,v,'ro','Linewidth',2);
hold on;
plot(x,y);
```

Then we obtain the values 1.88 and 0.96 as well as the figure below (Figure 6.3).

2. (least squares method - nonlinear dependence) A dedicated Matlab function for solving nonlinear problems coming from the application of the least squares method is the function `lsqnonlin` which can be used with the syntax `[x,resnorm]=lsqnonlin(@fun, x0)`, where `fun` is a function which depends upon the parameters of the model. Therefore, `lsqnonlin` takes as objective function the sum of the squares of the components of the vector `fun` and returns both the minimum point and the minimal value.

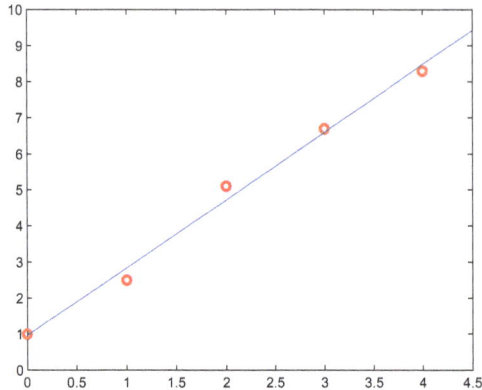

Figure 6.3: Least squares method: linear case.

Let us suppose that from the direct observation of a specific physical phenomenon at the moments t_i we get the m_i data, as in the table below.

i	1	2	3	4	5	6
t_i	0.1	0.3	0.5	0.7	0.8	0.9
m_i	0.7	1.5	4.5	22.3	94	387.9

Moreover, we suppose that one can observe a behaviour of type $(1 - t)^{x_1}$ around 0, and a behaviour of type t^{-x_2} around 1. The suggested continuous model at every moment t is $f : \mathbb{R}^3 \to \mathbb{R}$,

$$f(x) = x_3 (1 - t)^{x_1} t^{-x_2}.$$

The function lsqnonlin will minimize the objective function

$$x \to \sum_{i=1}^{6} \left[m_i - x_3 (1 - t)^{x_1} t^{-x_2} \right]^2.$$

The program consists of a function file with the code

```
function z=fun(p)
measures=[0.1 0.7;0.3 1.5;0.5 4.5;0.7 22.3; 0.8 94; 0.9 387.9];
z=measures(:,2)-p(3)*(1-measures(:,1)).^p(1)./measures(:,1).^p(2)
```

and the main file as follows:

```
p0=[1 1 1];
[p,difference]=lsqnonlin(@fun,p0)
measures=[0.1 0.7;0.3 1.5;0.5 4.5;0.7 22.3; 0.8 94; 0.9 387.9];
```

which generates

```
p = -0.9368 -6.7935 91.8653
difference =27.7334
```

while the additional lines

```
plot(measures(:,1),measures(:,2),'o','Linewidth',2)
hold on;
model= '91.8653*(1-x)^(-0.9368)/x^(-6.7935)';
fplot(model,[0 0.95],'-r')
```

give, on the same figure, the discrete (measured) model and the continuous model (see the figure below).

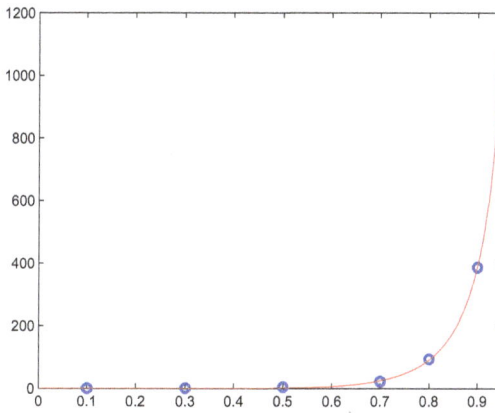

Figure 6.4: Least squares method: nonlinear case.

3. (fixed points - basic approximations) We want to approximate the solution of the equation $\cos x = x$, $x \in [0, 1]$. Clearly, a solution of this equation is a fixed point of the restriction of the function $\cos$ to the interval $[0, 1]$. The $\cos$ function is a contraction, since $\sup_{x \in [0,1]} |\cos' x| = \sin 1 < 1$. Moreover, every calculator gives us the approximate value of the contraction constant: $\sin 1 \simeq 0.84147$, so

$$|\cos x - \cos y| \le 0.8415 \, |x - y| \, .$$

Thus, according to the Banach Principle, there exists a unique solution (denoted $\bar{x}$) of the mentioned equation which can be approximated arbitrarily well by the sequence of the Picard iterations starting from every initial date $x_0 \in [0, 1]$. We intend to investigate how accurately the solution has been approximated after a given number of iterations. One may ask how many iterations are needed to obtain the value of $\bar{x}$ with an error smaller than $\frac{1}{1000}$. Answering this question is now possible in view of the estimations concerning the speed of convergence of Picard approximations in the Banach Principle. We recall that:

$$|x_n - \overline{x}| \leq |x_1 - x_0| \frac{\lambda^n}{1 - \lambda} \tag{6.3.1}$$

$$|x_n - \overline{x}| \leq \frac{\lambda}{1 - \lambda} |x_n - x_{n-1}| \tag{6.3.2}$$

This discussion helps us to understand these inequalities and to obtain the desired approximation of $\overline{x}$ starting from $x_0 = 0$. Taking into account (6.3.1) and (6.3.2), we have

$$|x_n - \overline{x}| \leq \frac{0.8415^n}{1 - 0.8415} \text{ and } |x_n - \overline{x}| \leq \frac{0.8415}{1 - 0.8415} |x_n - x_{n-1}|.$$

In the Matlab programs given below, one sees that the second estimation is better than the first one, since the value $\frac{0.8415^n}{1-0.8415}$ is less than 0.001 starting from $n = 51$, while the right-side member in the second part is under 0.001 faster (for $n = 22$). For instance, for $n = 22$ in the second relation, we get

$$|x_{22} - \overline{x}| < 0.0009,$$

so $\overline{x} \in (x_{22} - 0.0009, x_{22} + 0.0009)$, that is $\overline{x}$ is between 0.7381 and 0.7399. The Matlab programs are as follow :

```
lambda=0.8415;
c=1/(1-lambda)
i=0;
while c*lambda>0.001
i=i+1; c=c*lambda;
end
```

which gives

```
-> disp(i+1); disp(c*lambda);
51.
0.0009500
```

and, respectively,

```
u=0; i=0;
while (0.8415/(1-0.8415))*abs((u-cos(u)))>0.001
    i=i+1; u=cos(u);
end
disp(i+1); disp((0.8415/(1-0.8415))*abs((u-cos(u))))
```

which gives

```
-> disp(i+1); disp((0.8415/(1-0.8415))*abs((u-cos(u))))
22.
0.0008820.
```

Again, the approximation process is faster if one starts form an initial value close to $\bar{x}$. For instance, taking $x_0 = 0.7$, the estimations

$$|x_n - \bar{x}| \le \frac{0.8415^n}{1 - 0.8415} |\cos(0.7) - 0.7| \text{ and } |x_n - \bar{x}| \le \frac{0.8415}{1 - 0.8415} |x_n - x_{n-1}|$$

give an error less than 0.001 for $n = 35$, and $n = 16$, respectively. These new values can be verified easily, by making the obvious modifications in the above programs.

4. (speed of convergence of Picard iterations: the case $f'(\bar{x}) \in (0, 1)$) Let us consider the function $f : [0, \infty) \to [0, \infty)$ given by $f(x) = \frac{1}{1+x^2}$. We have seen that this is a contraction, has a unique fixed point which is the unique positive solution of the equation $x^3 + x - 1 = 0$, and the sequence of the Picard iterations satisfies:

$$\frac{x_{n+1} - \bar{x}}{x_n - \bar{x}} \overset{n \to \infty}{\to} f'(\bar{x}) = \frac{-2\bar{x}}{(1 + \bar{x}^2)^2} = -2\bar{x}^3 \in (0, 1).$$

Let us study the speed of convergence of this sequence by means of a Matlab program. The stopping criterion is the attainment of a maximum number of iterations (1000), or the situation where the absolute value of the difference between two consecutive iterations is under an admissible tolerance (10^{-7}).

```
funct='1/(1+x^2)'
tol=1e-7; maxiter=1000;
n=0; x=1; x_old=0;
%Picard
while abs(x-x_old)>tol & n<maxiter
      x_old=x; x=eval(funct); n=n+1;
end
%endPicard
disp(x);
disp(n);
```

The displayed results are:

```
->disp(u); 0.6823278 ->disp(n); 35
```

so the algorithm stopped after 35 iterations and found the approximate value of the solution as being 0.6823278, starting from the initial data $x_0 = 1$. The speed of convergence, which is a relatively good one, is due to the fact that $|f'(\bar{x})|$ is smaller than 1.

5. (speed of convergence of Picard iterations: the case $f'(\bar{x}) = 1$) For the restriction of the function $\sin x$ to the interval $[0, 1]$, which satisfies the assumptions in the Picard Theorem with $\bar{x} = 0$ as the unique fixed point, the speed of convergence dramatically changes, as every Picard iteration (x_k) satisfies

$$\frac{x_{k+1} - \bar{x}}{x_k - \bar{x}} \overset{k \to \infty}{\to} f'(\bar{x}) = 1$$

and we expect that the progress made in approaching the solution from a iteration to another is very small. In the above program, we change f and we get the results:

```
->disp(u);
0.0545930
->disp(n);
1000
```

which means that after 1000 iterations we get a quite unsatisfactory approximation of the fixed point. The situation changes insignificantly if one starts with the value $x_0 = 0.1$, closer to $\overline{x}$:

```
->disp(u); 0.0480222 ->disp(n); 1000.
```

6. (speed of convergence of Picard iterations: the case $f'(\overline{x}) = 0$) Consider the case of the function $f : [\sqrt{2}, \infty) \to [\sqrt{2}, \infty)$ given by

$$f(x) = \frac{x}{2} + \frac{1}{x}.$$

It is easy to see that f is well defined (the means inequality). Moreover,

$$\left|f'(x)\right| = \left|\frac{1}{2} - \frac{1}{x^2}\right| \leq \frac{1}{2},$$

so f is a contraction and its unique fixed point is $\overline{x} = \sqrt{2}$. One observes that $f'(\overline{x}) = 0$, so, for every nonstationary Picard iteration,

$$\frac{x_{k+1} - \overline{x}}{(x_k - \overline{x})^2} = \frac{1}{2x_k} \to \frac{1}{2\sqrt{2}} = f''(\overline{x}),$$

so we have quadratic convergence, and we expect a very good speed of convergence. We repeat the above program for the new function and we get:

```
->disp(u); 1.4142136 ->disp(n); 5
```

so the algorithm stops after only 5 iterations and we have a very good approximation of the fixed point $\overline{x} = \sqrt{2}$.

7. (the Aitken acceleration methods) We come back to the function sin (restricted to the interval $[0, 1]$) for which the convergence of the Picard iterations sequence is slow. We want to test the two Aitken acceleration methods: the weak and the strong ones (which actually work as well for the case $\left|f'(\overline{x})\right| = 1$). Briefly, for the weak method, besides a Picard sequence associated to f, one considers the sequence

$$y_k = x_k - \frac{(f(x_k) - x_k)^2}{f(f(x_k)) - 2f(x_k) + x_k}$$

which converges faster, however, without a modification of the order of convergence. For the strong method, we work directly with the sequence

$$x_{k+1} = x_k - \frac{(f(x_k) - x_k)^2}{f(f(x_k)) - 2f(x_k) + x_k}.$$

We have the code below:

```
funct='sin(x)'
functcomp='sin(sin(x))'
tol=1e-7;
maxiter=20000;
y=1;x=1;y_old=0;n=0;
%Aitken (weak)
while abs(y-y_old)>tol & n<maxiter
 y_old=y; x=eval(funct);
 y=x-((eval(funct)-x)^2)/(eval(functcomp)-2*eval(funct)+x);
n=n+1;
end
disp(x); disp(y); disp(n);
```

The result is the following:

```
0.0122
0.0082
20000
```

which means that for the initial data $x_0 = 1$, after 20000 Picard iterations we are not able to approximate the fixed point $\bar{x} = 0$ with an error smaller that 10^{-2} (we get 0.0122449), while this happens after 20000 weak Aitken iterations (the value is 0.0081631).

For the strong Aitken method, we do not change the Picard method, and for the code we made the following modifications: we introduce a new constant `tol1=1e-4`, and eliminate the variable `w`, and the strong Aitken method looks like:

```
funct='sin(x)'
functcomp='sin(sin(x))'
tol1=1e-4; maxiter=1000;
n=0; x=1; x_old=0;
while abs(x-x_old)>tol1 & n<maxiter
 x_old=x;
 x=x-((eval(funct)-x)^2)/(eval(functcomp)-2*eval(funct)+x);
 n=n+1;
end
disp(x);
disp(n);
```

with the result:

```
1.8779e-004
21
```

which is clear indication of the huge progress for the speed of convergence.

8. (particular acceleration of the Picard iterations) There are particular cases where the Picard iterations can be accelerated by means of an auxiliary function. For instance, consider $f : \mathbb{R} \to \mathbb{R}$, $f(x) = x^3 + 4x^2 + x - 10$. By the basic methods of mathematical analysis, the equation $f(x) = x$ has only one real solution situated in the interval $[1, 2]$. Clearly, f is not a contraction. However, the application $g : [1, 2] \to \mathbb{R}$,

$$g(x) = \frac{2x^3 + 4x^2 + 10}{3x^2 + 8x}$$

satisfies the fact that the equality $g(x) = x$ is equivalent to $f(x) = x$. But

$$g'(x) = \frac{(6x^2 + 8x)(3x^2 + 8x) - (6x + 8)(2x^3 + 4x^2 + 10)}{(3x^2 + 8x)^2}$$

$$= \frac{(6x + 8)(x^3 + 4x^2 - 10)}{(3x^2 + 8x)^2}.$$

Since $\overline{x}$ is the solution of the equation $x^3 + 4x^2 - 10 = 0$, we conclude that $g'(\overline{x}) = 0$. This means that around $\overline{x}$, g is a contraction and, moreover, the convergence of the associated Picard iterations is quadratic. The next program shows that after only 9 steps, the Picard iterations of f blow up, while using g we get a good approximation of the solution after only 5 iterations.

```
tol=1e-7;maxiter=1000;
u=1; u_old=0; n=0; t=1; t_old=0; p=0;
%Picard g
while abs(u-u_old)>tol & n<maxiter
        u_old=u; u=(2*u^3+4*u^2+10)/(3*u^2+8*u); n=n+1;
end
%Picard f
while abs(t-t_old)>tol & p<maxiter
        t_old=t; t=t^3+4*t^2+t-10; p=p+1;
end
disp(u); disp(n); disp(t); disp(p);
```

The result is:

```
-> disp(u); disp(n); disp(t); disp(p);
1.36523
5.
Nan
9.
```

9. (Newton method - one root) We test the Newton method for the function $f : \mathbb{R} \to \mathbb{R}$ given by the relation

$$f(x) = x + e^x + \frac{10}{1 + x^2} - 5$$

which has a solution of order 1 in $(-2, 0)$, as one can also observe in the figure below:

Starting with the initial data $x_0 = 1.5$ we converge rapidly to this solution, as shown in the next code:

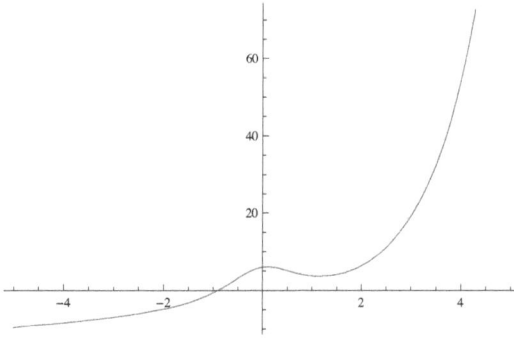

Figure 6.5: The graph of $x + e^x + \frac{10}{1+x^2} - 5$.

```
f_el='x+exp(x)+10/(1+x^2)-5';
syms x
g=diff(f_el);tol=1e-5;maxiter=1000;
x=1.5;x_old=0;n=1;
%Newton
while abs(x-x_old)>tol & n<maxiter
    x_old=x;x=x-eval(f_el)/eval(g);n=n+1;
end
x
n
eval(f_el)
```

which returns:

```
x = -0.9046
n = 35
ans = 8.8818e-016
```

If one starts with the initial date $u = -1.5$, then one gets the approximation after only 5 iterations.

10. (distance to a generalized ellipsoid) Let have a look again at the nonlinear equation whose solution gives the number needed to compute the projection of a point on a generalized ellipsoid. Let us implement the Newton method for approximation of the square of the equation

$$\sum_{i=1}^{p} \frac{a_i^2 v_i^2}{\left(a_i^2 + \lambda\right)^2} = 1,$$

for $p = 5$, $a_i = 6 - i$, $v_i = 9$, $i \in \overline{1, 5}$. The code:

```
f_el='2500/((25+x)^2)+1600/((16+x)^2)+900/((9+x)^2)
+400/((4+x)^2)+100/((1+x)^2)-1';
syms x
```

```
g=diff(f_el);tol=1e-5;maxiter=1000;
x=97;x_old=0;n=1;
%Newton
while abs(x-x_old)>tol & n<maxiter
x_old=x;x=x-eval(f_el)/eval(g);n=n+1;
end
x
n
eval(f_el)
```

returns:

```
x =57.5719; n =9; ans =-1.1102e-016
```

11. (Newton method- several roots) Let us consider $f : \mathbb{R} \to \mathbb{R}, f(x) = e^x - 2x^2$. It is not very difficult to see that f has three real roots, among which one is negative and two are positive. We apply the Newton method (as in the previous examples) for various initial data.

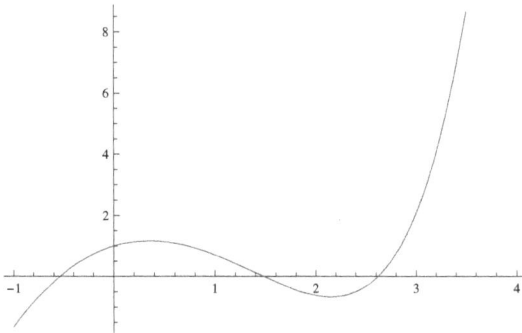

Figure 6.6: The graph of $e^x - 2x^2$ on $[-1, 4]$.

If we start with $u = 0$ we obtain:

```
x = -0.5398; n = 7.
```

Therefore, starting from 0 we find the approximation of the negative root after 7 iterations.

Starting from $u = 1$, we get:

```
x = 1.4880; n = 5
```

and for $u = 3$, we get:

```
x = 2.6179; n = 6.
```

12. (Newton method for fixed points) Besides Picard iterations which, in general, have linear convergence, we can use the Newton's method for approximating fixed points with quadratic orders of convergence. For instance, in the above case of the fixed point of cos in $[0, 1]$, according to the theory, it is sufficient to approximate the solution of the equation $\cos x = x$, and this means that we should consider the Picard iterations associated to

$$g(x) = x + \frac{\cos x - x}{1 + \sin x}.$$

The next Matlab program proves the advantages of this new approach:

```
f_el='x+(cos(x)-x)/(1+sin(x))';
syms x
tol=1e-5;maxiter=1000;
x=1;x_old=0;n=1;
%Newton
while abs(x-x_old)>tol & n<maxiter
    x_old=x;x=eval(f_el);n=n+1;
end
%Picard
t=1;t_old=0; p=0;
while abs(t-t_old)>tol & p<maxiter
    t_old=t; t=cos(t); p=p+1;
end
x
n
t
p
```

and the results are:

```
x = 0.7391
n = 5
t = 0.7391
p = 29
```

13. (proximal point algorithm) We implement now the proximal point algorithm. Let $f : \mathbb{R} \to \mathbb{R}$,

$$f(x) = \frac{x^4}{4} + \frac{x^2}{2} - 3x + 1.$$

This is a convex function since $f''(x) = 3x^2 + 1 > 0$ for every $x \in \mathbb{R}$. Its graph is below (Figure 6.7). In order to approximate the minimum point we use the proximal point algorithm in the following code:

we define in a function file:

```
function F = prox_fun(x,y)
    F = x.^3+2.*x-3-y;
end
```

and in a M-file we use the default function `fsolve` in its parametric version:

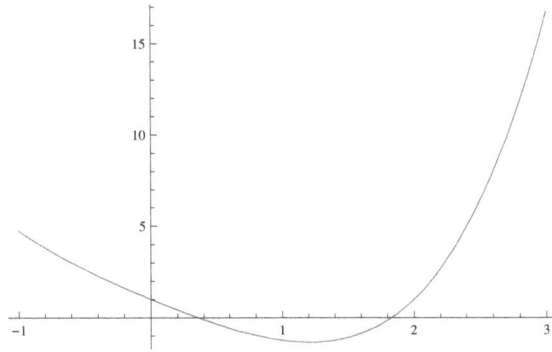

Figure 6.7: The graph of $\frac{x^4}{4} + \frac{x^2}{2} - 3x + 1$.

```
maxiter=100;n=0;y=-10;
while n<maxiter
    y = fsolve(@(x) prox_fun(x,y),[-1;y(1)]);
    n=n+1;
end
disp(y(1));
```
We get the approximate value $\bar{x}$ = 1.213411662762243.

14. (the algorithm of steepest descent direction) Let us consider the steepest descent method. Usually, the direction is chosen as

$$p_k = -\frac{\nabla f(x_k)}{\|\nabla f(x_k)\|}$$

if $\|\nabla f(x_k)\| \neq 0$ and, concerning the step α_k, we use only the Armijo condition

$$f(x_k + \alpha_k p_k) < f(x_k) + c_1 \alpha_k \nabla f(x_k)(p_k).$$

Therefore, we test, at every step, if the condition is fulfilled and, contrary, we multiply α_k by a factor less than 1, in order to obtain a smaller step.

For illustration, we chose the function $f : \mathbb{R} \to \mathbb{R}$, $f(x) = e^x - 2x^2$. This function has a local minimum point close to 2.

We have the code:

```
funct=@(x) exp(x)-2*x^2;
derivative=@ (x)exp(x)-4*x;
tol=1e-7; maxiter=20;u=5; u_old=1; n=0; factor=0.5;
c1=0.01; alpha=1;
 while abs(derivative(u))>tol
 u_old=u;
 u=u-alpha*derivative(u)/abs(derivative(u));
 while funct(u)>=funct(u_old)+c1*alpha*abs(derivative(u_old))
```

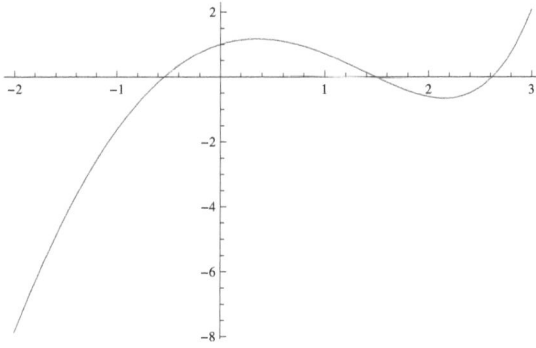

Figure 6.8: The graph of $e^x - 2x^2$ on $[-2, 3]$.

```
alpha=factor*alpha;
u=u-alpha*derivative(u)/abs(derivative(u));
end
n=n+1;
end
u
n
alpha
derivative(u)
```

and the results are

```
u = 2.153292357921600
n = 15
alpha = 5.960464477539063e - 008
ans = -2.854974923138798e - 008
```

So, the algorithm stops after it obtains a convenient approximation of the minimum. The evolution of the decreasing of the gradient absolute value is also interesting.

15. (the algorithm of steepest descent direction: several variables) The next example refers to the approximation of the minimum point for the Rosenbrock function (see Exercise 7.51). In code below, we implement the steepest descent method for this function, with the chosen step following an heuristic rule. The graphic representations we get give us informations about the construction if the iterates. We have the code:

```
t1=linspace(-0.6,1.3,20);
t2=linspace(-0.4,1.3,20);
function z=r(x, y)
    z=100.0*(y-x^2)^2 + (1-x)^2
z=feval(t1,t2,r);contour(t1,t2,z,40,flag=[2, 2 0]);
function z=g_r(x)
```

```
    z=[-2*(1-x(1))-400*x(1)*(x(2)-x(1)^2),200*(x(2)-x(1))]
function z=r(x)
    z=100.0*(x(2)-x(1)^2)^2 + (1-x(1))^2
maxiter=50;u=[-0.4 0.6];n=1;v=u;alpha=0.25;
for i=1:maxiter
    n=n+1;
    u=u-alpha*g_r(u)/norm(g_r(u));v=[v;u];
    alpha=0.25/log(n);
end
plot(v(:,1),v(:,2),'-');plot(1,1,'r*')
```

We get the next picture:

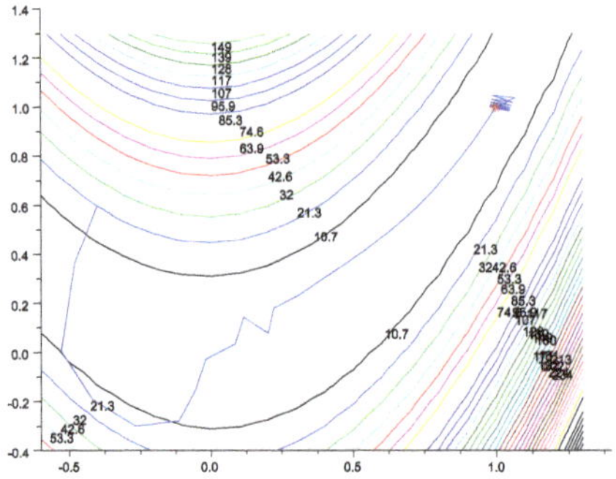

Figure 6.9: The descent method for Rosenbrock function.

Observe that the iterations oscillate around the point $(1, 1)$ which, due to Exercise 7.51, is the minimum point of the function. A detail of the previous picture convinces us of this.

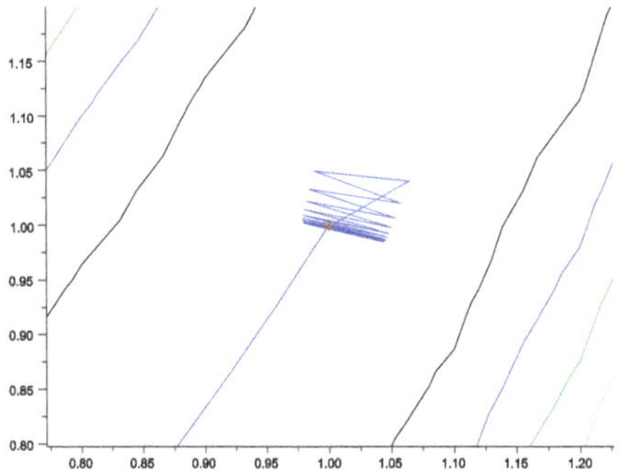

Figure 6.10: Detail: the oscillation of the iterations.

Supplemental details can be obtained by displaying at every step the value of the iteration, the distance to the minimum and the norm of the gradient.

16. (the QP method) We consider the algorithm for the function $f(x) = \frac{1}{2}\langle(Ax^t)^t, x\rangle + \langle b, x\rangle$ from Example 6.2.3. We saw that the speed of this algorithm depends of the ratio of the biggest and the smallest eigenvalue of A. We give a generic code for the method described in Example 6.2.3 for the situation of a matrix A of the form $\begin{pmatrix} \lambda_1 & 0 \\ 0 & \lambda_2 \end{pmatrix}$ (having hence the eigenvalues λ_1, λ_2), and for $b = 0$.

```
t1=linspace(-0.3,0.3,20);t2=linspace(-0.3,0.3,20);
a=20;b=1;
function z=r(x, y)
    z=a*x^2 + b*y^2
z=feval(t1,t2,r);
contour(t1,t2,z,10,flag=[2, 2 0]);
function z=g_r(x)
    z=[2*a*x(1),2*b*x(2)]
function z=r(x)
    z=a*x(1)^2 +b* x(2)^2
maxiter=10;u=[-0.1 0.3];v=u;
for i=1:maxiter
    u=u-g_r(u)*norm(g_r(u))^2/(2*r(g_r(u)));v=[v;u];
end
plot(v(:,1),v(:,2),'-o');plot(0,0,'r*')
```

In the case considered here we have $c = 20$, the level lines of the function being ellipses with relatively big eccentricity.

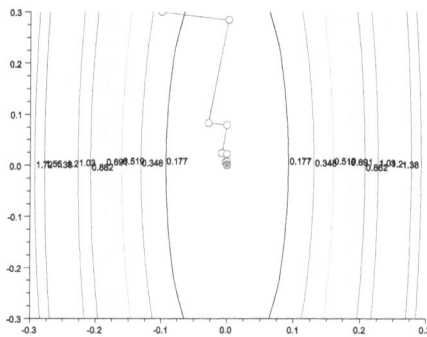

Figure 6.11: The QP method.

We obtain the same phenomenon of oscillation of the iterations, according to the next picture.

Figure 6.12: The QP method: detail.

17. (the SQP method) Consider the functions $f, h : \mathbb{R}^2 \to \mathbb{R}$ given by

$$f(x) = e^{x_1 x_2} - \frac{1}{2}\left(x_1^3 + x_2^3 + 1\right)^2, \quad h(x) = x_1^2 + x_2^2 - 5.$$

We implement the algorithm described in (6.2.8) to find the solution of the problem of minimizing f with the restriction $h(x) = 0$. Following the steps described at the SQP method, we generate the code:

```
f='exp(x*y)-1/2*(x^3+y^3+1)^2'
h='x^2+y^2-5'
L='exp(x*y)-1/2*(x^3+y^3+1)+l*(x^2+y^2-5)'
syms x y l
df=[diff(f,x);diff(f,y)]
dL=[diff(diff(L,x),x), diff(diff(L,x),y); diff(diff(L,x),y),
diff(diff(L,y),y)]
A=[diff(h,x) diff(h,y)]
At=[diff(h,x);diff(h,y)]
F=[df-l*At;h]
dF=[dL -At;A 0]
x=1;y=2;l=1;
v=[x;y;l];
i=0;
while(norm(eval(F))>10^-7)
  i=i+1
  v=v-inv(eval(dF))*eval(F)
```

```
x=v(1);y=v(2);l=v(3);
end
v
```

Taking as initial value $(x_{10}, x_{20}, \mu_0) = (1, 2, 1)$, the code generates after 39 iterations the solution $(\bar{x}_1, \bar{x}_2) = (1.5811388, 1.5811388)$, $\bar{\mu} = -15.0304612$. Starting from $(x_{10}, x_{20}, \mu_0) = (-\sqrt{2}, \sqrt{3}, 1)$, the algorithm generates the same solution after 49 steps.

16. (the barrier method) Let $f, g : \mathbb{R}^2 \to \mathbb{R}$,

$$f(x) = x_1^2 + \frac{x_2^2}{3} + x_1 x_2 + x_1, \ g(x) = x_1 + x_2 - 1.$$

Consider the optimization problem of minimizing f with the restriction $g(x) \leq 0$. We have shown that this problem has the solution $\bar{x} = (-2, 3)$. We verify that the minimal points of the barrier functions obtained for different values of the parameter μ approximate this solution. This can be done by considering the unconstrained minimization problems given by the barrier functions.

A selection of the obtained values in this way is given in the table below:

μ	1	0.0625	0.0123457	0.0004165	0.0001
$\bar{x}_1$	−1.2928932	−1.8232233	−1.9214326	−1.9855692	−1.9929289
$\bar{x}_2$	0.8786797	2.4696699	2.7642977	−0.9983152	2.9787868

Therefore, we have a good approximation of the actual solution since μ is small.

7 Exercises and Problems, and their Solutions

This chapter is dedicated to the concrete application of the methods and techniques from the previous chapters. All these applications are given as exercises, together with their solutions. At the same time, there are problems with complete solutions that highlight different theoretical aspects which were not included in the previous chapters. This means that in what follows, there appear several theoretical conclusions important on their own. The organization of the material in this chapter follows the main topics of the book.

7.1 Analysis of Real Functions of One Variable

Exercise 7.1. *Determine the extreme points of the functions below:*

(i) $f : \mathbb{R} \to \mathbb{R}$, $f(x) = (x + 2)^2(x - 1)^3$;

(ii) $f : \mathbb{R} \to \mathbb{R}$, $f(x) = \sin^3 x + \cos^3 x$;

(iii) $f : \mathbb{R} \to \mathbb{R}$, $f(x) = \sqrt[3]{x^2} - \sqrt[3]{x^2 - 1}$;

(iv) $f : \mathbb{R} \to \mathbb{R}$, $f(x) = \frac{x^2 - 5x + 6}{x^2 + 1}$;

(v) $f : (0, \infty) \to \mathbb{R}$, $f(x) = \frac{|\ln x|}{\sqrt{x}}$;

(vi) $f : \mathbb{R} \setminus \{-\sqrt{2}, 0\} \to \mathbb{R}$, $f(x) = \frac{x^2 e^{\frac{1}{x}}}{x + \sqrt{2}}$;

(vii) $f : \mathbb{R} \to \mathbb{R}$, $f(x) = x \arcsin \frac{x - 1}{\sqrt{2(x^2 + 1)}}$.

Solution (i) The derivative of the function is

$$f'(x) = (x + 2)(x - 1)^2(5x + 4),$$

so the critical (stationary) points are $-2; 1; -\frac{4}{5}$. From the interval of monotonicity of the function (according to the sign of derivative) one immediately infers that -2 is a local maximum point, $-\frac{4}{5}$ is a local minimum point, while 1 is not an extremum point.

(ii) It is enough to study the function on the interval $[0, 2\pi)$ (taking into account its periodicity). The derivative is

$$f'(x) = 3 \sin x \cos x(\sin x - \cos x),$$

with the roots

$$0, \frac{\pi}{4}, \frac{\pi}{2}, \pi, \frac{5\pi}{4}, \frac{3\pi}{2}.$$

From the analysis of the sign of the derivative on the corresponding intervals, and extending the argument to the entire real line, we obtain that $0 + 2k\pi$, $\frac{\pi}{2} + 2k\pi$, $\frac{5\pi}{4} + 2k\pi$ ($k \in \mathbb{Z}$) are local maxima, and $\frac{\pi}{4} + 2k\pi$, $\pi + 2k\pi$, $\frac{3\pi}{2} + 2k\pi$ are local minima.

Another possible solution is to compute the second order derivative of f and to use the following result: for a critical point $\overline{x}$, if $f''(\overline{x}) > 0$, then $\overline{x}$ is a local minimum, and if $f''(\overline{x}) < 0$, then $\overline{x}$ is a local maximum.

(iii) The function is differentiable on $\mathbb{R} \setminus \{-1, 0, 1\}$, and the derivative is

$$f'(x) = \frac{2}{3} \frac{(x^2 - 1)^{\frac{2}{3}} - x^{\frac{4}{3}}}{x^{\frac{1}{3}}(x^2 - 1)^{\frac{2}{3}}}.$$

The critical points are $-\frac{1}{\sqrt{2}}, \frac{1}{\sqrt{2}}$, so the candidates for extrema are

$$-1, 0, 1, -\frac{1}{\sqrt{2}}, \frac{1}{\sqrt{2}}.$$

From the analysis of the derivative sign around those points, we infer that $-\frac{1}{\sqrt{2}}, \frac{1}{\sqrt{2}}$ are local maxima and 0 is a local minimum.

(iv) The discussion is similar to that from item (ii). One gets that $1 - \sqrt{2}$ is a local maximum and $1 + \sqrt{2}$ is a local minimum.

(v) The function is differentiable on $(0, \infty) \setminus \{1\}$, and

$$f'(x) = \begin{cases} -\frac{2 - \ln x}{2x\sqrt{x}}, & x \in (0, 1) \\ \frac{2 - \ln x}{2x\sqrt{x}}, & x \in (1, \infty). \end{cases}$$

The candidates for extrema are 1 and e^2. From the variation of f, we decide that 1 is a local minimum and e^2 is a local maximum.

(vi) Similar arguments show that $-1 - \sqrt{2}$ is a local maximum and $-\sqrt{2} + 2$ is a local minimum. We remark as well that for $|f|$ both points are local minima, but no one is global minimum, since $\inf_{x \in \mathbb{R}} |f(x)| = 0$ is obtained for $x \to 0+$.

(vii) In order to decide the behaviour of the first derivative, we should compute the second one as well. Then

$$f'(x) = \begin{cases} \arcsin \frac{x-1}{\sqrt{2(x^2+1)}} + \frac{x}{x^2+1}, & x > -1 \\ \arcsin \frac{x-1}{\sqrt{2(x^2+1)}} - \frac{x}{x^2+1}, & x < -1, \end{cases}$$

and

$$f''(x) = \begin{cases} \frac{2}{(x^2+1)^2}, & x > -1 \\ -\frac{2}{(x^2+1)^2}, & x < -1. \end{cases}$$

Finally, the function has neither critical points nor extrema (the only possible candidate is the point where f is not differentiable, i.e., -1). □

Exercise 7.2. *Decide if $x = 0$ is an extreme point for $f : \mathbb{R} \to \mathbb{R}$, $f(x) = e^x + e^{-x} + 2\cos x$.*

Solution We successively compute the derivative at this point, until we find a nonvanishing one. We obtain

$$f'(0) = 0; \ f''(0) = 0; \ f'''(0) = 0; \ f^{iv}(0) = 4.$$

So, $x = 0$ is a local minimum point, according to Theorem 1.3.14. □

Problem 7.3. *Let $a, b \in \mathbb{R}$, $a < b$, and $f, g : [a, b] \to \mathbb{R}$ be a continuous functions. One defines $h : \mathbb{R} \to \mathbb{R}$,*

$$h(t) = \sup\{f(x) + tg(x) \mid x \in [a, b]\}.$$

Show that h is a Lipschitz function.

Solution Let $s, t \in \mathbb{R}$. Since the applications $x \mapsto f(x) + tg(x)$ and $x \mapsto f(x) + sg(x)$ are continuous, by virtue of Weierstrass Theorem, on the compact interval $[a, b]$, there exist $x_t, x_s \in [a, b]$ such that

$$h(t) = f(x_t) + tg(x_t)$$
$$h(s) = f(x_s) + sg(x_s).$$

Then

$$
\begin{aligned}
h(t) - h(s) &= f(x_t) + tg(x_t) - (f(x_s) + sg(x_s))\\
&\leq f(x_t) + tg(x_t) - f(x_t) - sg(x_t)\\
&= g(x_t)(t - s).
\end{aligned}
$$

Similarly,

$$
\begin{aligned}
h(t) - h(s) &= f(x_t) + tg(x_t) - (f(x_s) + sg(x_s))\\
&\geq f(x_s) + tg(x_s) - f(x_s) - sg(x_s)\\
&= g(x_s)(t - s).
\end{aligned}
$$

We denote $M := \max_{x \in [a,b]} |g(x)|$ and observe that $M \in \mathbb{R}$ (using again Weierstrass Theorem). Then we get

$$\left| h(t) - h(s) \right| \leq M |t - s|,$$

that is the conclusion. □

Exercise 7.4. *Let $f : \mathbb{R} \to \mathbb{R}$, $f(x) = \min\{x + 1, 0, 1 - x\}$. Show that $\overline{x} = 0$ is a local minimum point of f, but it is not global maximum for f.*

Solution The easy-to-draw graph of f proves the assertions. Clearly, a more rigorous proof could be given as well, also starting from the investigation of the graph. □

Exercise 7.5. *Let $x_1 < x_3 < x_2$ be real numbers, and $f : \mathbb{R} \to \mathbb{R}$ be a C^1 function on $[x_1, x_2]$. Suppose that*

$$\max\{f(x_1), f(x_2)\} < f(x_3).$$

Then there exists $\overline{x} \in (x_1, x_2)$ a critical point of f with

$$f(\overline{x}) = \max_{x \in [x_1, x_2]} f(x).$$

Solution On $[x_1, x_2]$, the function f admits a maximum point (Weierstrass' Theorem). From the assumption, this point cannot be x_1 or x_2. So, this maximum point (denoted by $\bar{x}$) lies in the interior of the interval, so it is a local maximum of f. Fermat Theorem gives the conclusion. □

Exercise 7.6. *Let $a_1, \ldots, a_n$ be strictly positive real numbers such that*

$$a_1^x + a_2^x + \ldots + a_n^x \geq n, \ \forall x \in \mathbb{R}.$$

Show that $a_1 a_2 \ldots a_n = 1$.

Solution Let $f : \mathbb{R} \to \mathbb{R}$, $f(x) = a_1^x + a_2^x + \ldots + a_n^x$. Since $f(0) = n$, from the hypothesis we infer that $x = 0$ is a global minimum point for f so, from Fermat Theorem, $f'(0) = 0$, which leads to the conclusion.

Another solution could be given using the well-known (fundamental) limit: $\lim_{x \to 0} \frac{a^x - 1}{x} = \ln a$ for $a > 0$. □

Problem 7.7. *Let $a, b \in \mathbb{R}$, $a < b$, and $f : [a, b] \to \mathbb{R}$ be a continuous function, derivable at a and b, with $f'(a)f'(b) < 0$. Show that f admits a local extremum in (a, b).*

Solution From Weierstrass Theorem, f admits a minimum and a maximum on $[a, b]$. If the conclusion would not hold, then these points should be a and b. Without loss of generality, suppose that $f'(a) < 0$ (whence, $f'(b) > 0$). Since

$$f'(a) = \lim_{x \to a+} \frac{f(x) - f(a)}{x - a} < 0,$$

for x sufficiently close to a, $f(x) < f(a)$, so a should be the maximum point. Whence b is the minimum point, so $f(x) \geq f(b)$ for every $x \in [a, b]$. Then

$$\frac{f(x) - f(b)}{x - b} \leq 0, \ \forall x \in [a, b].$$

We infer that $f'(b) \leq 0$, which is absurd. This contradiction can be resolved only if the conclusion holds. □

Problem 7.8. *Let $f : \mathbb{R} \to \mathbb{R}$ be a derivable function such that $\lim_{x \to \infty} \frac{f(x)}{x} = \infty$ and $\lim_{x \to -\infty} \frac{f(x)}{x} = -\infty$. Show that $\operatorname{Im} f' = \mathbb{R}$ (i.e., f' is surjective).*

Solution Let $r \in \mathbb{R}$ and $g : \mathbb{R} \to \mathbb{R}$, $g(x) := f(x) - rx$. It is clear that $\lim_{|x| \to \infty} g(x) = \infty$ and, therefore, g attains its (global) minimum at a point $\bar{x} \in \mathbb{R}$. Then, from Fermat Theorem, $0 = g'(\bar{x}) = f'(\bar{x}) - r$. □

Exercise 7.9. *Show that $f : (1, \infty) \to \mathbb{R}$, $f(x) = -\ln(\ln x)$ is convex. Deduce that for every $a, b > 1$, one has*

$$\sqrt{\ln a \ln b} \leq \ln \left(\frac{a + b}{2} \right).$$

Solution The second-order derivative on the definition interval is

$$f''(x) = \frac{1}{x^2 \ln x} + \frac{1}{x^2 \ln^2 x}.$$

Since $x > 1$, this function has only positive values, so f is convex.

Now, using this property, one gets

$$-\ln\left(\ln\frac{a+b}{2}\right) \le -\frac{1}{2}\left(\ln(\ln a) + \ln(\ln b)\right) = -\ln(\sqrt{\ln a \ln b}),$$

and the desired inequality follows. $\qquad\square$

Problem 7.10. *Let $g : \mathbb{R} \to \mathbb{R}$ be a continuous function. Show that g is convex if and only if for every integrable (in the Riemann sense) function $f : [0, 1] \to \mathbb{R}$, the following inequality holds*

$$g\left(\int_0^1 f(u)du\right) \le \int_0^1 g(f(u))du.$$

Solution If g is convex then for any partition of $[0, 1]$ and for every system of intermediate points (with the notation from Definition 1.4.1) it holds that

$$g\left(\sum_{i=1}^n f(\xi_i)(x_i - x_{i-1})\right) \le \sum_{i=1}^n g(f(\xi_i))(x_i - x_{i-1}).$$

Passing to the limit for the norm of the partition going to 0, we get the inequality from the conclusion.

For the converse implication, we fix $x, y \in \mathbb{R}$, $\alpha \in (0, 1)$, and we consider $f : [0, 1] \to \mathbb{R}$,

$$f(u) = \begin{cases} x, & \text{if } u \in [0, \alpha] \\ y, & \text{if } u \in (\alpha, 1]. \end{cases}$$

Then f is Riemann integrable on $[0, 1]$ (see Theorems 1.4.9, 1.4.10). Applying the assumption of this stage, we find $g(\alpha x + (1 - \alpha)y) \le \alpha g(x) + (1 - \alpha)g(y)$, whence the convexity of g. $\qquad\square$

Problem 7.11. *(i) Let $f : [1, \infty) \to \mathbb{R}$ given by $f(x) = x\sqrt[x]{x}$. Show that f is increasing and concave.*

(ii) Show that the sequence $a_n := (n+1)\sqrt[n+1]{n+1} - n\sqrt[n]{n}$, $n \in \mathbb{N}\setminus\{0, 1\}$ is monotone and bounded. Find $\lim a_n$.

Solution (i) The function f could be written as

$$f(x) = e^{\left(1+\frac{1}{x}\right)\ln x},$$

and

$$f'(x) = f(x)\frac{x + 1 - \ln x}{x^2}, \quad f''(x) = f(x)\frac{(1 - \ln x)^2 - x}{x^4}.$$

Obviously, f' has positive values, so f is increasing. In order to establish the sign of f'' we compare, by use of some auxiliary functions, the expressions $|1 - \ln x|$ and $\sqrt{x}$ considering, separately, the cases $x \le e$ and $x \ge e$. In both cases, we deduce that $f''(x) \le 0$, whence f is concave.

(ii) We can write, for every $n \in \mathbb{N} \setminus \{0, 1\}$, $a_n = f(n+1) - f(n)$, and by the concavity of f we get

$$f(n + 1) = f\left(\frac{n+2}{2} + \frac{n}{2}\right) \ge \frac{f(n+2)}{2} + \frac{f(n)}{2},$$

which leads to $a_n \ge a_{n+1}$. Therefore, (a_n) is a decreasing sequence. The monotonicity of f proves that the terms of (a_n) are positive, whence (a_n) is convergent. In order to find its limit, we can apply the Stolz-Cesàro Criterion (Proposition 1.1.24) to the sequence

$$b_n = \sqrt[n]{n} = \frac{n\sqrt[n]{n}}{n}, \quad n \in \mathbb{N} \setminus \{0, 1\}.$$

We have that

$$\lim b_n = 1 = \lim \frac{(n + 1)\sqrt[n+1]{n+1} - n\sqrt[n]{n}}{n + 1 - n} = \lim a_n. \qquad \square$$

Problem 7.12. *Let $a, b \in \mathbb{R}$, $a < b$, and $f : [a, b] \to \mathbb{R}$ be a C^2 function with $f(a) = f(b) = 0$. Let $M := \sup_{x \in [a,b]} |f''(x)|$ and*

$$g(x) := f(x) - M\frac{(x - a)(b - x)}{2}; \quad h(x) := -f(x) - M\frac{(x - a)(b - x)}{2}.$$

Show that g, h are convex and deduce the inequality

$$|f(x)| \le M\frac{(x - a)(b - x)}{2}, \quad \forall x \in [a, b].$$

Solution We have

$$g''(x) = f''(x) + M \ge 0, \quad \forall x \in [a, b]$$
$$h''(x) = -f''(x) + M \ge 0, \quad \forall x \in [a, b],$$

which shows that both functions are convex.

The convexity of g and the fact that $g(a) = g(b) = 0$ lead to the conclusion that $g(x) \le 0$ for every $x \in [a, b]$. Hence,

$$f(x) \le M\frac{(x - a)(b - x)}{2}.$$

Analogously, arguing for h,

$$-f(x) \le M\frac{(x - a)(b - x)}{2}.$$

These two relations lead to the conclusion. $\qquad \square$

Problem 7.13. *Let $f : \mathbb{R} \to \mathbb{R}$ be a convex function.*

(i) Let $a, b \in \mathbb{R}$, $a < b$. Study the position of the graph of f with respect to the line joining the points $(a, f(a))$ and $(b, f(b))$.

(ii) Deduce that if f is bounded, then it is constant.

Solution (i) It is clear (from the geometrical interpretation of convexity) that for $x \in [a, b]$, the graph of f is under the line joining $(a, f(a))$ and $(b, f(b))$. Notice that this line has the equation

$$y = \frac{f(b) - f(a)}{b - a}(x - a) + f(a).$$

For $x > b$, from convexity and from $a < b < x$ we deduce

$$\frac{f(b) - f(a)}{b - a} \leq \frac{f(x) - f(a)}{x - a},$$

that is

$$f(x) \geq \frac{f(b) - f(a)}{b - a}(x - a) + f(a),$$

whence the graph of f is above the mentioned line. The same conclusion holds analogously for $x < a$.

(ii) Suppose that f would not be constant. Then it would exist $a, b \in \mathbb{R}$ with $a < b$ and $f(a) \neq f(b)$. We can consider, without loss of generality, that $f(b) > f(a)$. Since for $x > b$,

$$f(x) \geq \frac{f(b) - f(a)}{b - a}(x - a) + f(a),$$

we get that $\lim_{x \to \infty} f(x) = +\infty$, that is f is not bounded. This is a contradiction, so f is constant. $\square$

Problem 7.14. *Let $a, b \in \mathbb{R}$, $a < b$, and $f : (a, b) \to \mathbb{R}$ be convex. Show that f is bounded from below. It is true that f is bounded?*

Solution Let $x_0 < x_1 < x_2$ be three points in (a, b). For $x < x_1$, we have

$$\frac{f(x_1) - f(x)}{x_1 - x} \leq \frac{f(x_2) - f(x_1)}{x_2 - x_1},$$

which gives

$$(x_1 - x)\frac{f(x_2) - f(x_1)}{x_2 - x_1} + f(x_1) \leq f(x),$$

so f is bounded from below on $(a, x_1]$. Similarly, for $x > x_1$,

$$\frac{f(x) - f(x_1)}{x - x_1} \geq \frac{f(x_1) - f(x_0)}{x_1 - x_0}$$

that is

$$(x - x_1)\frac{f(x_1) - f(x_0)}{x_1 - x_0} + f(x_0) \leq f(x).$$

We obtain that f is bounded from below on $[x_1, b)$, so it is also bounded from below on (a, b).

Generally, the upper boundedness is not ensured by the convexity. As an example, consider the function $f : (-\frac{\pi}{2}, \frac{\pi}{2}) \to \mathbb{R}$, $f(x) = |\text{tg } x|$. □

Exercise 7.15. *Let $f : \mathbb{R} \to \mathbb{R}$ be a convex, increasing function. Show that f is constant or $\lim_{x \to \infty} f(x) = \infty$.*

Solution If f is not constant, then there exist $a, b \in \mathbb{R}$, $a < b$ and $f(a) < f(b)$. Let $y = mx + n$ be the line joining $(a, f(a))$ and $(b, f(b))$. Clearly, $m > 0$. For $x > b$, we have, as seen before, $f(x) \geq mx + n$, whence $\lim_{x \to \infty} f(x) = \infty$. □

Problem 7.16. *Let $f : \mathbb{R} \to \mathbb{R}$ be a convex function.*
(i) Show that if $\lim_{x \to \infty} f(x) = 0$, then $f(x) \geq 0$, for every $x \in \mathbb{R}$.
(ii) Show that if f admits asymptote to $+\infty$, then its graph is above the asymptote.

Solution (i) Suppose, by way of contradiction, that exists $x_0 \in \mathbb{R}$ with $f(x_0) < 0$. By hypothesis, there exists $x_1 > x_0$ with $f(x_1) > f(x_0)$. For $x > x_1$, by means of convexity

$$\frac{f(x_1) - f(x_0)}{x_1 - x_0} \leq \frac{f(x) - f(x_1)}{x - x_1},$$

whence

$$f(x_1) + (x - x_1)\frac{f(x_1) - f(x_0)}{x_1 - x_0} \leq f(x).$$

We arrive at the contradiction $\lim_{x \to \infty} f(x) = \infty$.

(ii) Let $y = ax + b$ be the equation of the asymptote. The function $g : \mathbb{R} \to \mathbb{R}$, $g(x) = f(x) - ax - b$ is convex (as a sum between a convex function and an affine one) and, moreover, $\lim_{x \to \infty} g(x) = 0$. The application of the above step gives us the conclusion. □

Problem 7.17. *(i) Let $a, b \in \mathbb{R}$, $a < b$, $M > 0$, and $(f_n) : [a, b] \to \mathbb{R}$ be a sequence of M–Lipschitz functions on $[a, b]$. Show that if (f_n) is pointwise convergent on $[a, b]$, then (f_n) is uniformly convergent on $[a, b]$.*
(ii) Let $a, b \in \mathbb{R}$, $a < b$, and $(f_n) : (a, b) \to \mathbb{R}$ be a sequence of convex functions which is pointwise convergent on (a, b). Show that (f_n) is uniformly convergent on every closed subinterval of (a, b).

Solution (i) It is clear that the pointwise limit of (f_n) is itself an M–Lipschitz function, which we denote by f. Take $\varepsilon > 0$. We fix as well a partition of the form

$$a = \alpha_0 < \alpha_1 < \ldots < \alpha_p = b$$

of the interval (a, b), with the norm smaller than $\frac{\varepsilon}{M}$. The pointwise convergence gives a natural number n_0 sufficiently large such that for every $i \in \overline{0, p}$,

$$\left| f_n(\alpha_i) - f(\alpha_i) \right| < \varepsilon.$$

Let $x \in [a, b]$. There exists $i \in \overline{0, p-1}$ with $x \in [\alpha_i, \alpha_{i+1}]$. We have

$$|f_n(x) - f(x)| \leq |f_n(x) - f_n(\alpha_i)| + |f_n(\alpha_i) - f(\alpha_i)| + |f(\alpha_i) - f(x)|$$
$$< M\,|x - \alpha_i| + \varepsilon + M\,|x - \alpha_i| \leq 3\varepsilon.$$

So (f_n) is uniformly convergent towards f.

(ii) We reduce this situation to the preceding one. Let $[\alpha, \beta] \subset (a, b)$ and $\alpha' \in (a, \alpha), \beta' \in (\beta, b)$. Since f_n is convex, for every $x, y \in [\alpha, \beta]$ with $x \neq y$,

$$\frac{f_n(\alpha) - f_n(\alpha')}{\alpha - \alpha'} \leq \frac{f_n(x) - f_n(y)}{x - y} \leq \frac{f_n(\beta) - f_n(\beta')}{\beta - \beta'}.$$

The outer members of this inequality are bounded (by the pointwise convergence), so there exists $M > 0$ with

$$\left| \frac{f_n(x) - f_n(y)}{x - y} \right| \leq M, \ \forall x, y \in [\alpha, \beta], \ x \neq y, \ \forall n \in \mathbb{N}.$$

Therefore we can apply (i) to the functions (f_n) on $[\alpha, \beta]$. $\qquad\square$

Problem 7.18. (i) Let $f, g : \mathbb{R} \to \mathbb{R}$. Show that if f is convex and increasing and g is convex, then $f \circ g$ is convex.

(ii) Let $f : \mathbb{R} \to (0, \infty)$. Show that $\ln f$ is convex if and only if for every $\alpha > 0$, the function f^α is convex.

Solution (i) Let $x, y \in \mathbb{R}$ and $\lambda \in (0, 1)$. Then using the properties of f and g, we have

$$(f \circ g)(\lambda x + (1 - \lambda)y) \leq f(\lambda g(x) + (1 - \lambda)g(y))$$
$$\leq \lambda f(g(x)) + (1 - \lambda)f(g(y)).$$

(ii) Suppose first that $\ln f$ is convex. For every $\alpha > 0$,

$$f^\alpha = e^{\alpha \ln f}.$$

Since the mapping $x \mapsto e^{\alpha x}$ is convex and increasing, the preceding item applies.

Conversely, suppose that for every $\alpha > 0$, the function f^α is convex. Hence, for every $x, y \in \mathbb{R}$ and $\lambda \in [0, 1]$, $u(\alpha) \leq v(\alpha)$ where

$$u(\alpha) = e^{\alpha \ln f(\lambda x + (1 - \lambda)y)}$$
$$v(\alpha) = \lambda e^{\alpha \ln f(x)} + (1 - \lambda)e^{\alpha \ln f(y)}.$$

But $u(0) = v(0)$. From $u(\alpha) \leq v(\alpha)$ and $u(0) = v(0)$ we obtain $u'(0) \leq v'(0)$, which leads to

$$\ln f(\lambda x + (1 - \lambda)y) \leq \lambda \ln f(x) + (1 - \lambda)\ln f(y),$$

that is the desired relation. $\qquad\square$

Definition 7.1.1. *Let $I \subset \mathbb{R}$ be an interval, and $f : I \to \mathbb{R}$ be a function. We say that f admits a support functional at $x \in I$ if there exists an affine function of the form $s(u) = f(x) + m(u - x)$ (where $u \in \mathbb{R}$) such that $s(u) \leq f(u)$ for every $u \in I$.*

Problem 7.19. *Let $a, b \in \mathbb{R}$, $a < b$. Show that $f : (a, b) \to \mathbb{R}$ is convex if and only if it admits a support functional at every $x \in (a, b)$.*

Solution Suppose first that f is convex. We know already that f has lateral derivatives at any point of the interval. Let $\bar{x} \in (a, b)$ be fixed and $m \in [f'_-(\bar{x}), f'_+(\bar{x})]$. Then, for every $x \in (a, b)$, $x > \bar{x}$ we have

$$\frac{f(x) - f(\bar{x})}{x - \bar{x}} \geq m,$$

and for every $x \in (a, b)$, $x < \bar{x}$ the inverse inequality holds. In both cases,

$$m(x - \bar{x}) \leq f(x) - f(\bar{x}),$$

hence f admits a support functional at $\bar{x}$.

Conversely, suppose that f admits a support functional at every $x \in (a, b)$. Let $x, y \in (a, b)$ and $\lambda \in (0, 1)$. Then $\bar{x} = \lambda x + (1 - \lambda)y \in (a, b)$, so there exists $m \in \mathbb{R}$ such that

$$s(u) := f(\bar{x}) + m(u - \bar{x}) \leq f(u), \ \forall u \in (a, b).$$

Consequently,

$$f(\bar{x}) = s(\bar{x}) = \lambda s(x) + (1 - \lambda)s(y) \leq \lambda f(x) + (1 - \lambda)f(y).$$

Therefore, f is convex. □

From the above problem and its solution one infers the next result.

Theorem 7.1.2. *Let $a, b \in \mathbb{R}$, $a < b$. A function $f : (a, b) \to \mathbb{R}$ is convex if and only if for every $c \in (a, b)$ there exists $y \in \mathbb{R}$ such that*

$$f(x) \geq f(c) + y(x - c), \ \forall x \in (a, b).$$

Moreover, y can be arbitrarily chosen in the interval $[f'_-(c), f'_+(c)] = \partial f(c)$.

Problem 7.20. *Let $a, b \in \mathbb{R}$, $a < b$ and $f : [a, b] \to \mathbb{R}$ be a convex, continuous function. Show that for every $c \in (a, b)$, there exists $y \in \mathbb{R}$ such that*

$$f(c) + y\frac{a + b - 2c}{2} \leq \frac{1}{b - a} \int_a^b f(x)dx.$$

Show that the equality holds if and only if f is affine, that is has the form

$$f(x) = f(c) + y(x - c), \ \forall x \in (a, b).$$

Infer that the first inequality in the Hermite-Hadamard Inequality holds as equality if and only if f is affine.

Solution Using the preceding theorem and applying the Riemann integral, we get the desired inequality. For the equality, if there exists $x \in (a, b)$ with

$$f(c) + y(x - c) < f(x),$$

then from the continuity of the involved functions we arrive at a contradiction. So, in order to have equality one must have an affine function.

The obtained inequality is a refinement of the first inequality in the Hermite-Hadamard Inequality: it is enough to take $c = \frac{a+b}{2}$. □

From the above facts and the already known theory, we get that, under the continuity assumption, in both parts of the Hermite-Hadamard Inequality, equality holds if and only if f is affine.

Exercise 7.21. *Show that the following functions are convex and write Hermite-Hadamard Inequality in every case on mentioned intervals:*
   (i) $f : [0, \infty) \to \mathbb{R}$, $f(x) = (x + 1)^{-1}$ on $[0, x]$ and $[n - 1, n]$;
   (ii) $f : \mathbb{R} \to \mathbb{R}$, $f(x) = e^x$ on $[a, b]$;
   (iii) $f : [0, \pi] \to \mathbb{R}$, $f(x) = -\sin x$ on $[a, b]$.

Solution In all three cases, the convexity is easy to verify using the second-derivative criterion. No one of these function is affine, so both inequalities in Hermite-Hadamard Inequality are strict.

For the first function, Hermite-Hadamard Inequality on an interval of the form $[0, x]$, $x \in \mathbb{R}$ (for instance) leads to

$$x - \frac{x^2}{x + 2} < \ln(x + 1) < x - \frac{x^2}{2(x + 1)},$$

and on an interval of the form $[n - 1, n]$, $n \in \mathbb{N}^*$ to

$$\frac{2}{2n + 1} < \ln(n + 1) - \ln n < \frac{1}{2}\left(\frac{1}{n} + \frac{1}{n + 1}\right).$$

For the second function we get

$$e^{\frac{a+b}{2}} < \frac{e^b - e^a}{b - a} < \frac{e^a + e^b}{2}, \quad \forall a, b \in \mathbb{R}, \ a \neq b.$$

In particular,

$$\sqrt{xy} < \frac{x - y}{\ln x - \ln y} < \frac{x + y}{2}, \quad \forall x, y \in (0, \infty), \ x \neq y.$$

For the third function we find

$$\frac{\sin a + \sin b}{2} < \frac{\cos a - \cos b}{b - a} < \sin\left(\frac{a + b}{2}\right), \quad \forall a, b \in \mathbb{R}, \ a \neq b.$$

From here one can deduce the well-known inequalities

$$\sin x < x < \operatorname{tg} x, \ \forall x \in \left[0, \frac{\pi}{2}\right].$$ □

Problem 7.22. *Let $a, b \in \mathbb{R}$, $a < b$, and $f : [a, b] \to \mathbb{R}$ be a convex, continuous function. Show that for every $c \in [0, \frac{b-a}{4}]$ one has*

$$\frac{1}{2}\left(f\left(\frac{a+b}{2}-c\right)+f\left(\frac{a+b}{2}+c\right)\right) \leq \frac{1}{b-a}\int_a^b f(x)dx.$$

Solution We apply the inequality from Problem 7.20 for f on $\left[a, \frac{a+b}{2}\right]$ and $\left[\frac{a+b}{2}, b\right]$, and take into account that one can choose y like in Theorem 7.1.2. Therefore, for $c \in \left[0, \frac{b-a}{4}\right]$ we find

$$\frac{2}{b-a}\int_a^{\frac{a+b}{2}} f(x)dx \geq f\left(\frac{b+a}{2}-c\right)+f'_-\left(\frac{b+a}{2}-c\right)\frac{\frac{a+b}{2}+a-2\left(\frac{a+b}{2}-c\right)}{2}$$

and

$$\frac{2}{b-a}\int_{\frac{a+b}{2}}^b f(x)dx \geq f\left(\frac{b+a}{2}+c\right)+f'_-\left(\frac{b+a}{2}+c\right)\frac{\frac{a+b}{2}+b-2\left(\frac{a+b}{2}+c\right)}{2}.$$

After computation and considering the fact that

$$f'_-\left(\frac{b+a}{2}+c\right) \geq f'_-\left(\frac{b+a}{2}-c\right)$$

we get the conclusion. □

Problem 7.23. *Let $a, b \in \mathbb{R}$, $a < b$ and $f : [a, b] \to \mathbb{R}$ be a C^2 function. Take $m, M \in \mathbb{R}$ such that $m \leq f''(x) \leq M$ for every $x \in [a, b]$ (note that such constants do exist according to Weierstrass' Theorem). Show that*

$$m\frac{(b-a)^2}{24} \leq \frac{1}{b-a}\int_a^b f(x)dx - f\left(\frac{a+b}{2}\right) \leq M\frac{(b-a)^2}{24},$$

and

$$m\frac{(b-a)^2}{12} \leq \frac{f(a)+f(b)}{2} - \frac{1}{b-a}\int_a^b f(x)dx \leq M\frac{(b-a)^2}{12}.$$

Solution It is easy to observe that the mappings

$$x \mapsto f(x) - \frac{mx^2}{2}$$

$$x \mapsto \frac{Mx^2}{2} - f(x)$$

are convex and continuous. Both inequalities follow from the application of Hermite-Hadamard Inequality to these two functions. □

Problem 7.24. *Let $a, b \in \mathbb{R}$, $a < b$, and $f : [a, b] \to \mathbb{R}$ be a L-Lipschitz ($L > 0$) function. Show that*

$$\left| f(x) - \frac{1}{b-a} \int_a^b f(t)dt \right| \leq \left[\frac{1}{4} + \left(\frac{x - \frac{a+b}{2}}{b-a} \right)^2 \right] L(b-a).$$

Solution We can write successively,

$$\left| f(x) - \frac{1}{b-a} \int_a^b f(t)dt \right| = \left| \frac{1}{b-a} \int_a^b (f(x) - f(t))dt \right|$$

$$\leq \frac{L}{b-a} \int_a^b |x - t|\, dt$$

$$= \left[\frac{1}{4} + \left(\frac{x - \frac{a+b}{2}}{b-a} \right)^2 \right] L(b-a),$$

and the inequality is proved. □

7.2 Nonlinear Analysis

Problem 7.25. *Show that a nonempty set $D \subset \mathbb{R}^p$ is convex if and only if it contains all the convex combinations of its elements.*

Solution Clearly, if D contains all the convex combinations of its elements it contains as well all the convex combinations with two of its elements and this is the definition of convexity.

Conversely, if D is convex then, by definition, contains all the convex combinations of any two of its elements. Consider three elements $x_1, x_2, x_3 \in D$, $\alpha_1, \alpha_2, \alpha_3 \in (0, 1)$ with $\alpha_1 + \alpha_2 + \alpha_3 = 1$. Then

$$\alpha_1 x_1 + \alpha_2 x_2 + \alpha_3 x_3 = (\alpha_1 + \alpha_2) \left(\frac{\alpha_1}{\alpha_1 + \alpha_2} x_1 + \frac{\alpha_2}{\alpha_1 + \alpha_2} x_2 \right) + \alpha_3 x_3.$$

Now $z := \frac{\alpha_1}{\alpha_1 + \alpha_2} x_1 + \frac{\alpha_2}{\alpha_1 + \alpha_2} x_2$ is a convex combination of two elements of D, so $z \in D$. Furthermore, $(\alpha_1 + \alpha_2)z + \alpha_3 x_3$ is again a convex combination of two elements of D, so, finally, $\alpha_1 x_1 + \alpha_2 x_2 + \alpha_3 x_3 \in D$. Of course, the complete proof needs a mathematical induction argument which uses the idea from above: suppose that for a $n \in \mathbb{N}\setminus\{1, 2\}$, D contains the convex combinations of n of its elements. Take $x_1, x_2, \ldots, x_n, x_{n+1} \in D$, $\alpha_1, \alpha_2, \ldots, \alpha_n, \alpha_{n+1} \in (0, 1)$ with $\alpha_1 + \ldots + \alpha_{n+1} = 1$. Then

$$\sum_{i=1}^{n+1} \alpha_i x_i = \sum_{i=1}^{n} \alpha_i x_i + \alpha_{n+1} x_{n+1} = \sum_{i=1}^{n} \alpha_i \sum_{i=1}^{n} \frac{\alpha_i}{\sum_{i=1}^{n} \alpha_i} x_i + \alpha_{n+1} x_{n+1}.$$

Now $z := \sum_{i=1}^{n} \frac{\alpha_i}{\sum_{i=1}^{n} \alpha_i} x_i$ is a convex combinations of n elements of D, whence, from the assumption, $z \in D$. Next, $\sum_{i=1}^{n} \alpha_i z + \alpha_{n+1} x_{n+1}$ is a convex combination of two elements of D, so $\sum_{i=1}^{n+1} \alpha_i x_i \in D$ and the proof is complete. $\qquad \square$

Problem 7.26. *Let $D \subset \mathbb{R}^p$ be a nonempty set. Show that $\operatorname{conv} D$ is the smallest convex set (with respect to the set inclusion order relation) which contains D.*

Solution The inclusion $D \subset \operatorname{conv} D$ is obvious (take the elements with $n = 1$ in the definition of $\operatorname{conv} D$). On the other hand, if C is a convex set which includes D, from the above problem, is contains, in particular, all the convex combinations of the elements of D, that is it contains $\operatorname{conv} D$. It remains to show that $\operatorname{conv} D$ is a convex set. Take $x, y \in \operatorname{conv} D$ and $\mu_1, \mu_2 \in (0, 1)$ with $\mu_1 + \mu_2 = 1$. Since $x \in \operatorname{conv} D$, there exist $n \in \mathbb{N}^*$, $(\alpha_i)_{i \in \overline{1,n}} \subset (0, \infty)$, $\sum_{i=1}^{n} \alpha_i = 1$, $(x_i)_{i \in \overline{1,n}} \subset D$ with $x = \sum_{i=1}^{n} \alpha_i x_i$. Similarly, for y, there exist $m \in \mathbb{N}^*$, $(\beta_i)_{i \in \overline{1,m}} \subset (0, \infty)$, $\sum_{i=1}^{m} \beta_i = 1$, $(y_i)_{i \in \overline{1,m}} \subset D$ with $y = \sum_{i=1}^{m} \beta_i y_i$. Then

$$\mu_1 x + \mu_2 y = \mu_1 \sum_{i=1}^{n} \alpha_i x_i + \mu_2 \sum_{i=1}^{m} \beta_i y_i = \sum_{i=1}^{n} \mu_1 \alpha_i x_i + \sum_{i=1}^{m} \mu_2 \beta_i y_i.$$

It is enough to observe that

$$\sum_{i=1}^{n} \mu_1 \alpha_i + \sum_{i=1}^{m} \mu_2 \beta_i = \mu_1 \sum_{i=1}^{n} \alpha_i + \mu_2 \sum_{i=1}^{m} \beta_i = \mu_1 + \mu_2 = 1$$

in order to deduce that $\mu_1 x + \mu_2 y$ is a convex combination of $n + m$ elements of D, so $\mu_1 x + \mu_2 y \in \operatorname{conv} D$. $\qquad \square$

Problem 7.27. *Show that the convex hull of a closed set is not necessarily closed. Show that the convex hull of a compact set is compact.*

Solution For the first part, let us consider the following closed subset of $\mathbb{R}^2$:

$$D = \left([0, \infty) \times \{0\}\right) \cup \left(\{0\} \times [0, 1]\right).$$

It is easy to observe that $\operatorname{conv} D = \left([0, \infty) \times [0, 1)\right) \cup \{(0, 1)\}$. This set is not closed: every point $(a, 1)$ with $a > 1$ is in $\operatorname{cl} \operatorname{conv} D \setminus \operatorname{conv} D$.

For the second part, consider $D \subset \mathbb{R}^p$ a compact convex set. We use the Carathéodory Theorem (Theorem 2.1.17). Since

$$\operatorname{conv} D = \left\{ \sum_{i=1}^{p+1} \alpha_i x_i \mid (\alpha_i)_{i \in \overline{1,p+1}} \subset [0, \infty), \ \sum_{i=1}^{p+1} \alpha_i = 1, \ (x_i)_{i \in \overline{1,p+1}} \subset D \right\},$$

we define the function $f : [0, 1]^{p+1} \times D^{p+1} \to \mathbb{R}^p$ given as

$$f(\alpha_1, \ldots, \alpha_{p+1}, x_1, \ldots, x_{p+1}) = \sum_{i=1}^{p+1} \alpha_i x_i$$

and observe that conv $D = f(M \times D^{p+1})$ where M is the unit simplex of $\mathbb{R}^{p+1}$. Since M and D are compact, we deduce that $M \times D^{p+1}$ is also compact. Moreover, f is continuous, whence conv D is the image of a compact set through a continuous function. Therefore, conv D is compact. $\qquad\square$

Problem 7.28. *Let $A \subset \mathbb{R}^p$ be a nonempty set. Recall that the conic hull of A is* cone $A := [0, \infty)A$. *Show that* cone A *is the smallest cone (with respect to the set inclusion order relation) which contains A. Show that* cone(conv A) = conv(cone A) *and* $A^- = (\text{cl cone } A)^-$.

Solution All the affirmations are easy to prove by the use of the definitions of the involved objects. $\qquad\square$

Problem 7.29. *Let $f, g : \mathbb{R}^p \to \mathbb{R}$ be convex functions. Show that the function* max(f, g) *is convex. Is it true for* min(f, g)?

Solution Denote $h : \mathbb{R}^p \to \mathbb{R}$, $h(x) = \max(f, g)$. For every $x, y \in \mathbb{R}^p$ and $\alpha \in (0, 1)$,

$$f(\alpha x + (1 - \alpha)y) \le \alpha f(x) + (1 - \alpha)f(y) \le \alpha h(x) + (1 - \alpha)h(y)$$
$$g(\alpha x + (1 - \alpha)y) \le \alpha g(x) + (1 - \alpha)g(y) \le \alpha h(x) + (1 - \alpha)h(y),$$

so $h(\alpha x + (1 - \alpha)y) \le \alpha h(x) + (1 - \alpha)h(y)$, whence h is convex. For the min(f, g) it is not longer true: it is enough to look at the convex functions $f, g : \mathbb{R} \to \mathbb{R}$, $f(x) = x^2$ and $g(x) = (x - 1)^2$. $\qquad\square$

Problem 7.30. *Let $K \subset \mathbb{R}^p$ be a closed convex cone with nonempty interior which does not coincide with $\mathbb{R}^p$, and let $e \in \text{int } K$. Show that:*
   (i) $K + [0, \infty)e \subset K$;
   (ii) $K + (0, \infty)e = \text{int } K$;
   (iii) $\mathbb{R}e - K = \mathbb{R}^p$;
   (iv) for any $x \in \mathbb{R}^p$, $x + \mathbb{R}e \not\subset K$;

Solution Observe first that $e \ne 0$ since otherwise $0 \in \text{int } K$ implies $K = \mathbb{R}^p$ which is impossible.
   (i) Since K is a convex cone, $\alpha K \subset K$ for any $\alpha \ge 0$ and $K + K \subset K$. Since $e \in K$, we have successively that $[0, \infty)e \subset K$ and then $K + [0, \infty)e \subset K$.
   (ii) We have that $e \in \text{int } K$, so there exists $\varepsilon > 0$ such that $e + B(0, \varepsilon) \subset K$. Then for any $\alpha > 0$, $\alpha e + B(0, \alpha\varepsilon) \subset K$, whence $\alpha e \in \text{int } K$ for any $\alpha > 0$. Fix now $\alpha > 0$ and take $k \in K$. Since $B(0, \alpha\varepsilon)$ is absorbing, there exists $t > 0$ such that $tk \in B(0, \alpha\varepsilon)$.

But since $B(0, \alpha\varepsilon)$ is open, it is a neighborhood of tk, so there exists $\delta > 0$ such that $tk + B(0, \delta) \subset B(0, \alpha\varepsilon)$. Henceforth, $\alpha e + tk + B(0, \delta) \subset \alpha e + B(0, \alpha\varepsilon) \subset K$, so $\alpha e + tk \in$ int K. The inclusion $K + (0, \infty)e \subset$ int K is proved. Let us prove the converse. Take $v \in$ int K, that is there exists $\varepsilon > 0$ such that $v - D(0, \varepsilon) \subset K$. But $\varepsilon \|e\|^{-1} e \in D(0, \varepsilon)$, whence $\varepsilon \|e\|^{-1} e - v \in -K$, so there exists $k \in K$ with $\varepsilon \|e\|^{-1} e - v = -k$, i.e., $v = k + \varepsilon \|e\|^{-1} e \subset K + (0, \infty)e$. The equality is proved.

(iii) Take $x \in \mathbb{R}^p$. As above, there exists $\varepsilon > 0$ such that $e + B(0, \varepsilon) \subset K$ and $t > 0$ such that $tx \in B(0, \varepsilon)$. Of course, we also have that $-tx \in B(0, \varepsilon)$, and we get that $e - tx \in K$, which means that $x \in t^{-1}e - K \subset \mathbb{R}e - K$.

(iv) Assume that there exists $x \in \mathbb{R}^p$ with $x + \mathbb{R}e \subset K$. Take $k \in K$ and $t \in \mathbb{R}$. The convexity of K allows us to deduce that

$$\frac{n-1}{n}k + \frac{1}{n}(x + tne) \in K$$

for every $n \in \mathbb{N} \setminus \{0\}$. Passing to the limit as $n \to \infty$ and taking into account the closedness of K, we get $k + te \subset K$, so $K + \mathbb{R}e \subset K$. Consequently, $\mathbb{R}e - K \subset -K$. From (iii) we infer that $\mathbb{R}^p \subset -K$, whence $\mathbb{R}^p = K$, which is a contradiction. Therefore, the conclusion holds. $\qquad\square$

Problem 7.31. *Show that f is sublinear if and only if its epigraph is a convex cone.*

Solution Clearly the property $f(\alpha x) = \alpha f(x)$ for all $\alpha \geq 0$ and $x \in \mathbb{R}^p$ is equivalent to the fact that the epigraph of f is a cone, while the property $f(x + y) \leq f(x) + f(y)$ for all $x, y \in \mathbb{R}^p$ is equivalent to the fact that this cone is convex. $\qquad\square$

Problem 7.32. *Let $K \subset \mathbb{R}^p$ be a closed convex cone with nonempty interior, and let $e \in$ int K. Show that for every $v \in K^- \setminus \{0\}$, $\langle v, e \rangle < 0$.*

Solution The fact that $\langle v, e \rangle \leq 0$ follows from the definition of K^-. Suppose that $\langle v, e \rangle = 0$. Since $e \in$ int K, there exists $\varepsilon > 0$ such that $e + B(0, \varepsilon) \subset K$. Then one obtains that $\langle v, u \rangle \leq 0$ for all $u \in B(0, \varepsilon)$. If follows that $v = 0$, a contradiction. $\qquad\square$

Problem 7.33. *Let $K \subset \mathbb{R}^p$ be a closed convex cone with nonempty interior, and let $\emptyset \neq A \subset \mathbb{R}^p$. Show that the following assertions are equivalent:*
(i) there exists $v \in K^- \setminus \{0\}$, $\langle v, a \rangle \leq 0$ for all $a \in A$;
(ii) conv $A \cap -$ int $K = \emptyset$.

Solution Firstly, we prove that (i) implies (ii). If conv $A \cap -$ int $K \neq \emptyset$, then there exists $u \in$ conv A with $u \in -$ int K. Then, from the preceding proposition, $\langle v, u \rangle > 0$, which contradicts (i). Whence (ii) holds.

Suppose that (ii) holds. Then, from the convex sets separation theorem, there exists $v \in \mathbb{R}^p \setminus \{0\}$ such that

$$\langle v, a \rangle \leq \langle v, u \rangle, \quad \forall a \in \text{conv } A, \ \forall u \in -\text{int } K.$$

Easy arguments show that $v \in K^- \setminus \{0\}$ and $\langle v, a \rangle \leq 0$, for all $a \in A$. $\qquad\square$

Problem 7.34. *Let $K \subset \mathbb{R}^p$ be a closed convex cone with nonempty interior, and let $\emptyset \neq A \subset \mathbb{R}^p$. Show that the following assertions are equivalent:*
(i) $A \cap -\operatorname{int} K = \emptyset$;
(ii) $\operatorname{cl} A \cap -\operatorname{int} K = \emptyset$;
(iii) $(A + K) \cap -\operatorname{int} K = \emptyset$;
(iv) $\operatorname{cl}(\operatorname{cone}(A + K)) \cap -\operatorname{int} K = \emptyset$.

Solution These assertions are simple applications of the definitions. $\qquad\square$

The next result is named after the American mathematician James Caristi who published it in 1976. A multifunction from $\mathbb{R}^p$ to $\mathbb{R}^q$ is an applications which maps every point from $\mathbb{R}^p$ into a subset of $\mathbb{R}^q$.

Theorem 7.2.1 (Caristi Fixed Point Theorem). *Let $\varphi : \mathbb{R}^p \to \mathbb{R}$ be a lower semicontinuous and lower bounded function. Let $T : \mathbb{R}^p \rightrightarrows \mathbb{R}^p$ be a multifunction (with nonempty values) with the property that*

$$\varphi(y) \leq \varphi(x) - \|x - y\|, \ \forall x \in \mathbb{R}^p, \ \forall y \in T(x).$$

Then there exists $\overline{x} \in \mathbb{R}^p$ with $\overline{x} \in T(\overline{x})$.

Let us shed light on the links between this result and Ekeland Variational Principle.

Problem 7.35. *Show that Caristi Fixed Point Theorem and the third conclusion from Ekeland Variational Principle are equivalent.*

Solution The proof of Caristi Fixed Point Theorem using Ekeland Variational Principle. For function φ and $\varepsilon > 0$, $\delta := \varepsilon + 1$, we apply Ekeland Variational Principle, and from its third conclusion we deduce that there exists $\overline{x} \in \mathbb{R}^p$ with

$$\varphi(\overline{x}) \leq \varphi(x) + \frac{\varepsilon}{\varepsilon + 1} \|x - \overline{x}\|, \ \forall x \in \mathbb{R}^p,$$

from where

$$\varphi(\overline{x}) < \varphi(x) + \|x - \overline{x}\|, \ \forall x \in \mathbb{R}^p \setminus \{\overline{x}\}.$$

If, ab absurdum, $\overline{x} \notin T(\overline{x})$, then for every $y \in T(\overline{x})$ we have $y \neq \overline{x}$, whence

$$\varphi(\overline{x}) < \varphi(y) + \|y - \overline{x}\|,$$

in contradiction to the hypothesis of Caristi Theorem.

The proof of the third conclusion of Ekeland Variational Principle by the use of Caristi Fixed Point Theorem. Suppose that the conclusion does not hold. Then for every $x \in \mathbb{R}^p$, we consider the nonempty set

$$T(x) = \left\{ y \in \mathbb{R}^p \mid y \neq x, \ \frac{\delta}{\varepsilon} f(x) \geq \frac{\delta}{\varepsilon} f(y) + \|y - x\| \right\}.$$

For $\varphi(\cdot) = \frac{\delta}{\varepsilon} f(\cdot)$ we can apply Caristi Theorem, so there exists $\overline{x} \in \mathbb{R}^p$ with $\overline{x} \in T(\overline{x})$, which is, obviously, impossible. $\qquad\square$

Problem 7.36. *Show that Caristi Fixed Point Theorem implies the existence part from Banach Fixed Point Principle.*

Solution Let $f : \mathbb{R}^p \to \mathbb{R}^p$ be a contraction of constant $\lambda \in (0, 1)$, which we identify to the mapping T from Caristi Fixed Point Theorem. Let $\varphi : \mathbb{R}^p \to \mathbb{R}$,

$$\varphi(x) = \frac{1}{1-\lambda} \left\| x - f(x) \right\| .$$

Clearly, φ is continuous and lower bounded (by 0). Moreover, the condition from Caristi Fixed Point Theorem is automatically fulfilled since $y \in T(x)$ (which here means $y = f(x)$). So T has a fixed point. $\qquad\square$

Definition 7.2.2. *One says that $f : \mathbb{R}^p \to \mathbb{R}^p$ is a directional contraction if it is continuous and there exists $\lambda \in (0, 1)$ such that for every $x \in \mathbb{R}^p$ with $x \neq f(x)$ there exists*

$$y \in (x, f(x)) = \{u \in \mathbb{R}^p \mid \exists t \in (0, 1), \ u = tx + (1 - t)f(x)\}$$

with the property

$$\left\| f(x) - f(y) \right\| \leq \lambda \left\| x - y \right\| .$$

Exercise 7.37. *Using the function $f : \mathbb{R}^2 \to \mathbb{R}^2$,*

$$f(x, y) = \left(\frac{3x}{2} - \frac{y}{3}, x + \frac{y}{3} \right) ,$$

deduce that every contraction is a directional contraction, while the converse is false.

Solution From $\left\| f(x, y) - f(z, y) \right\| = \frac{\sqrt{13}}{2} |x - z|$ for every $x, y, z \in \mathbb{R}$, we deduce that f is not a contraction. Let $(x, y) \in \mathbb{R}^2$ with $f(x, y) \neq (x, y)$. Denote $f(x, y) = (a, b)$. Notice that $f(x, y) = (x, y) \iff a = x \iff b = y$. Consider $x \neq a$ and observe, after calculations, that the point $(u, v) = 2^{-1} (a + x, b + y)$ satisfies the required property with $\lambda = 5/24$, whence f is a directional contraction. $\qquad\square$

Problem 7.38. *Let $f : \mathbb{R}^p \to \mathbb{R}^p$ be a directional contraction of constant λ. Then f has a fixed point.*

Solution Let $g : \mathbb{R}^p \to \mathbb{R}$,

$$g(x) = \left\| x - f(x) \right\| .$$

Clearly, g is continuous and lower bounded. We apply the last conclusion of Ekeland Variational Principle for g and for $\varepsilon := \frac{1-\lambda}{2}$, $\delta := 1$. We infer the existence of an element $x \in \mathbb{R}^p$ such that for every $y \in \mathbb{R}^p$

$$\left\| x - f(x) \right\| \leq \left\| y - f(y) \right\| + \frac{1 - \lambda}{2} \left\| x - y \right\| .$$

If $x = f(x)$ the proof is over. We want to show that this is the only possible situation. Suppose, by way of contradiction, that $f(x) \neq x$. From the directional contraction condition there exists $y \neq x$ with

$$\|x - y\| + \|y - f(x)\| = \|x - f(x)\|$$

and

$$\|f(x) - f(y)\| \leq \lambda \|x - y\|.$$

Therefore, putting together all the relations, we get

$$
\begin{aligned}
0 &\leq \lambda \|x - y\| - \|f(x) - f(y)\| \\
&\leq \lambda \|x - y\| - \|f(y) - y\| + \|y - f(x)\| \\
&= (\lambda - 1) \|x - y\| - \|f(y) - y\| + \|x - f(x)\| \\
&\leq \frac{\lambda - 1}{2} \|x - y\|.
\end{aligned}
$$

Since $\lambda < 1$, we obtain that $x = y$, which is a contradiction. Therefore, $f(x) = x$ is the sole possibility, and this ends the proof. $\square$

Exercise 7.39. *Decide if the directional contractions imply the uniqueness of the fixed point.*

Solution It is sufficient to analyze the case of the function from Exercise 7.37, where all the elements of the form $(x, \frac{3x}{2})$, $x \in \mathbb{R}$ are fixed points. So, the uniqueness property of the fixed point do not hold. $\square$

Exercise 7.40. *Let $f, g : \mathbb{R}_+ \to \mathbb{R}$ be defined by*

$$f(x) = \begin{cases} \dfrac{x}{e^x - 1}, & x \neq 0 \\ 1, & x = 0. \end{cases} \quad , \quad g(x) = (x - 2)e^{2x} + (x + 2)e^x.$$

(i) Show that $g(x) \geq 0$, for every $x \in \mathbb{R}_+$.
(ii) Show that f is of class C^1 on $\mathbb{R}_+$.
(iii) Show that

$$f''(x) = \frac{g(x)}{(e^x - 1)^3}, \quad \forall x \in (0, \infty)$$

and $|f'(x)| \leq 2^{-1}$ for every $x \in \mathbb{R}_+$.
(iv) One defines the sequence (x_n) by $x_0 = 0$, and $x_{n+1} = f(x_n)$ for every $n \in \mathbb{N}$. Show that

$$|x_n - \ln 2| \leq 2^{-n} \ln 2, \quad \forall n \in \mathbb{N}.$$

Solution (i) It is enough to prove that $(x - 2)e^x + x + 2 \geq 0$, for every $x \in \mathbb{R}_+$. This can be done easily by studying the variation of this expression through its derivatives (up to the order two).

(ii) Clearly, f is derivable on $(0, \infty)$ and, on this interval,

$$f'(x) = \frac{e^x - 1 - xe^x}{(e^x - 1)^2}.$$

We compute the limit of this derivative at 0 (with a combination between a fundamental limit and the L'Hôpital Rule)

$$\lim_{x \to 0+} f'(x) = \lim_{x \to 0+} \frac{e^x - 1 - xe^x}{(e^x - 1)^2} = \lim_{x \to 0+} \frac{e^x - 1 - xe^x}{x^2} \frac{x^2}{(e^x - 1)^2}$$
$$= \lim_{x \to 0+} \frac{e^x - 1 - xe^x}{x^2} = \lim_{x \to 0+} \frac{-e^x}{2} = -\frac{1}{2}.$$

By the use of some of Lagrange Theorem consequences, we deduce that f is differentiable at 0, and its derivative is continuous at 0. Moreover, $f'(0) = -2^{-1}$.

(iii) The function f is twice derivable on $(0, \infty)$, and the announced relation can be shown by direct calculation. According to (i), f' is increasing on $(0, \infty)$, and the continuity of f', relation $f'(0) = -2^{-1}$ and the remark

$$\lim_{x \to \infty} f'(x) = 0$$

show that $f'(x) \in [-2^{-1}, 0)$ for every $x \in \mathbb{R}_+$, whence the conclusion.

(iv) The preceding step shows that f is a 2^{-1}-contraction on $\mathbb{R}_+$ which takes values in $\mathbb{R}_+$. Since (x_n) is a Picard iteration with the initial data 0, we infer, by the Banach Principle, that (x_n) converges to the unique fixed point of f from $\mathbb{R}_+$, which proves to be, by direct calculus, $\overline{x} = \ln 2$. The estimation follows by induction as

$$|x_n - \overline{x}| = |f(x_{n-1}) - f(\overline{x})| \le 2^{-1} |x_{n-1} - \overline{x}| \le \dots \le 2^{-n} |x_0 - \overline{x}|.$$

Consequently, the inequality holds. $\qquad\square$

Exercise 7.41. *Let consider the function $f : \mathbb{R} \setminus \{0\} \to \mathbb{R}$,*

$$f(x) = 1 + \frac{1}{4} \sin \frac{1}{x}.$$

For initial data $x_0 \in \mathbb{R} \setminus \{0\}$ consider the Picard iteration associated to (x_n). Study the convergence of this sequence.

Solution The image of f is the interval $I := \left[\frac{3}{4}, \frac{5}{4}\right]$. We consider the restriction of f to this interval and we show that this is a contraction from I to I. To this end, we compute (for $x \in I$)

$$f'(x) = -\frac{1}{4x^2} \cos \frac{1}{x},$$

from where

$$|f'(x)| \le \frac{4}{9} < 1, \ \forall x \in I.$$

Since $x_1 = f(x_0) \in I$, we can apply Banach Principle in order to obtain that (x_n) is convergent towards the unique fixed point of f from I.

Notice that the approximate value of the fixed point of f can be found by using the Matlab code given in the previous chapter. $\qquad\square$

Exercise 7.42. *Let consider the equation $x^3 - x - 1 = 0$ for $x \in I := [1, 2]$.*

(i) Transform this equation into a problem of finding a fixed point for a suitable contraction.

(ii) Deduce the existence and the uniqueness of the solution of the initial equation and indicate a sequence (x_n) convergent towards this solution. Determine a sufficient number of terms to be computed in order to approximate the solution with a less that 10^{-5} error.

Solution (i) The equation is equivalent to

$$x = x^3 - 1,$$

but in this formulation we should take $g(x) := x^3 - 1$, but $g([1, 2]) \not\subset [1, 2]$, and g is not a contraction. So, we write the initial equation equivalently as

$$x^3 = x + 1 \Leftrightarrow x = \sqrt[3]{x + 1}.$$

Consider $f : I \to \mathbb{R}$, $f(x) = \sqrt[3]{x + 1}$. It is easy to observe that $f(I) \subset I$ and

$$f'(x) = \frac{1}{3\sqrt[3]{(x + 1)^2}} \le \frac{1}{3\sqrt[3]{4}} < 1,$$

so f is a contraction from I to I.

(ii) From the Banach Principle and the above formulation we infer the existence and the uniqueness of the solution (denoted by $\overline{x}$) of the initial equation. Every Picard iteration associated to f is convergent to the solution. We take $x_0 = 1$ and $x_{n+1} = f(x_n)$ for every $n \in \mathbb{N}$. Furthermore, for every n,

$$|x_n - \overline{x}| \le \left(\frac{1}{3\sqrt[3]{4}} \right)^n |x_0 - \overline{x}| \le \left(\frac{1}{3\sqrt[3]{4}} \right)^n.$$

It is then sufficient to estimate n for which

$$\left(\frac{1}{3\sqrt[3]{4}} \right)^n \le 10^{-5}.$$

Moreover, it is sufficient to have

$$\frac{1}{4^n} \le 10^{-5},$$

that is $n \ge 5 \log_4 10$. Therefore, $n = 9$ satisfies the requirement. $\qquad\square$

Problem 7.43. *Show that every weak contraction defined on a compact set K and taking values into K has a unique fixed point.*

Solution The function f is continuous (being Lipschitz). Further, the function $x \in K \mapsto \|f(x) - x\|$ is continuous on K, hence admits a global minimum point. Consequently, there exists $\bar{x} \in K$ such that

$$\|f(\bar{x}) - \bar{x}\| \leq \|f(x) - x\|, \ \forall x \in K.$$

If $f(\bar{x}) \neq \bar{x}$, then, from weak contraction property,

$$\|f(\bar{x}) - f(f(\bar{x}))\| < \|f(\bar{x}) - \bar{x}\|.$$

Since $f(\bar{x}) \in K$, these two relations are in contradiction. So $f(\bar{x}) = \bar{x}$. The uniqueness is obvious. □

Problem 7.44. *Let $a, b \in \mathbb{R}$, $a < b$, and $f : [a, b] \to \mathbb{R}$ be a continuous function. Show that the next assertions hold:*

(i) If $[a, b] \subset f([a, b])$, then f has at least a fixed point.

(ii) Every closed interval from $f([a, b])$ is the image of a closed interval from $[a, b]$.

(iii) Let $n \in \mathbb{N}^*$. If there exist n closed intervals $I_0, I_1, \ldots, I_{n-1}$ contained in $[a, b]$, such that for every $k \in \overline{0, n-2}$, $I_{k+1} \subset f(I_k)$ and $I_0 \subset f(I_{n-1})$, then f^n has at least one fixed point.

Solution (i) By continuity, $f([a, b])$ is a compact interval which we denote by $[m, M]$, where $m, M \in \mathbb{R}$. By hypothesis $[a, b] \subset f([a, b])$, we deduce that

$$m \leq a < b \leq M.$$

Since $m, M \in f([a, b])$, there exist $x_m, x_M \in [a, b]$ such that $m = f(x_m)$ and $M = f(x_M)$. But

$$f(x_m) - x_m = m - x_m \leq a - x_m \leq 0$$

and

$$f(x_M) - x_M = M - x_M \geq b - x_M \geq 0,$$

and the function $g(\cdot) = f(\cdot) - \cdot$ vanishes in $[a, b]$, that is f has a fixed point in this interval.

(ii) Let $I = [c, d] \subset f([a, b])$. Obviously, there exist $u, v \in [a, b]$ such that $f(u) = c$ and $f(v) = d$. We can suppose that $u \leq v$. We consider the set $A := \{x \in [u, v] \mid f(x) = c\}$. This set is bounded (as a subset of $[a, b]$), nonempty ($u \in A$) and closed (since f is continuous and $A = [u, v] \cap f^{-1}(c)$) so, there exist $\alpha = \max A \in A$. Similarly, the set $B := \{x \in [\alpha, v] \mid f(x) = d\}$ has a minimum point, denoted by β. Then, $f(\alpha) = c, f(\beta) = d$ and for every $x \in (\alpha, \beta)$ one has $f(x) \neq c$ and $f(x) \neq d$. From the Darboux property, $(c, d) \subset f((\alpha, \beta))$, and the interval $f((\alpha, \beta))$ does not contain the points c and d. Consequently, $[c, d] = f([\alpha, \beta])$.

(iii) From the inclusion $I_0 \subset f(I_{n-1})$ and from *(ii)*, there exists a closed interval $J_{n-1} \subset I_{n-1}$ such that $I_0 = f(J_{n-1})$. But $J_{n-1} \subset I_{n-1} \subset f(I_{n-2})$. Again, using *(ii)*, there exists a closed interval $J_{n-2} \subset I_{n-2}$ such that $J_{n-1} = f(J_{n-2})$. We repeat this argument, and we infer the existence of n closed interval $J_0, J_1, ..., J_{n-1}$ such that

$$J_k \subset I_k, \ \forall k \in \overline{0, n-1},$$

and

$$J_{k+1} = f(J_k), \forall k \in \overline{0, n-2} \text{ and } I_0 = f(J_{n-1}).$$

So,

$$J_0 \subset I_0 = f(J_{n-1}) = f(f(J_{n-2})) = ... = f^n(J_0).$$

We now apply *(i)* for the continuous function f^n and for interval J_0, and we get the conclusion. □

Problem 7.45. *Let $f : [0, 1] \to [0, 1]$ be a continuous function with $f(0) = 0$ and $f(1) = 1$. Show that there exist $m \in \mathbb{N}^*$ such that $f^m(x) = x$ for every $x \in [0, 1]$, then $f(x) = x$ for every $x \in [0, 1]$.*

Solution Since f^m is a bijection, it follows that f itself is a bijection, from a well-known result concerning the injectivity and surjectivity of the compositions. The continuity ensures through Theorem 1.2.27 that f is strictly monotone, and since $f(0) = 0$, $f(1) = 1$, we deduce that f is strictly increasing. Suppose, by way of contradiction, that there exists $x \in (0, 1)$ such that $f(x) > x$ (the case $f(x) < x$ is similar). Then from the monotonicity, for every $n \in \mathbb{N}$,

$$f^n(x) > f^{n-1}(x) > ... > f(x) > x.$$

In particular, for $n = m$, we get a contradiction, hence the assumption made was false. □

Problem 7.46. *Let $f : \mathbb{R} \to \mathbb{R}$ be a continuous function such that $f \circ f$ has a fixed point. Show that f has a fixed point.*

Solution If we suppose that f has no fixed point, from continuity, it follows that either $f(x) > x$, for every $x \in \mathbb{R}$, or $f(x) < x$, for every $x \in \mathbb{R}$. In the first situation, passing x into $f(x)$ we get: $f(f(x)) > f(x) > x$, for every $x \in \mathbb{R}$, so $(f \circ f)(x) > x$, for every $x \in \mathbb{R}$. Therefore, $f \circ f$ has no fixed point, which contradicts the assumptions. The second case is similar. □

Problem 7.47. *Let $a, b \in \mathbb{R}$, $a < b$, and $f : [a, b] \to [a, b]$ be a Lipschitz function of constant $L > 0$. Define the sequence $(x_n)_{n \in \mathbb{N}^*}$ by $x_0 \in [a, b]$, and for every $n \geq 0$,*

$$x_{n+1} = (1 - \lambda)x_n + \lambda f(x_n),$$

where $\lambda := \frac{1}{L+1}$. Show that (x_n) is monotone and convergent towards a fixed point of f.

Solution We observe that if one of the terms x_n is a fixed point for f, then starting from n, the sequence is stationary, and the conclusion follows. Suppose that $f(x_n) \neq x_n$ for every $n \in \mathbb{N}^*$. Without loss of generality, suppose that $f(x_0) > x_0$, since the opposite case is similar. Since $f(b) \leq b$, the continuity of f tells us that there exists a fixed point in the interval $(x_0, b]$. Also from continuity, there exists the least fixed point, denoted by p, in this interval (otherwise, x_0 would be itself a fixed point since the set of fixed points is closed). Let us observe that

$$x_1 = (1 - \lambda)x_0 + \lambda f(x_0) > x_0.$$

We want to show, by induction, that the sequence is increasing and for every $n \in \mathbb{N}^*$, we have $x_n < p$ and $x_n < f(x_n)$. Suppose that these relations hold up to rank n, and we show it for rank $n + 1$. Suppose, by contradiction, that $p < x_{n+1}$. Then $x_n < p < x_{n+1}$, whence

$$0 < p - x_n < x_{n+1} - x_n = \lambda(f(x_n) - x_n),$$

from where

$$0 < \frac{1}{\lambda} |x_n - p| = (L + 1) |x_n - p|$$
$$< |f(x_n) - x_n| \leq |f(x_n) - f(p)| + |p - x_n|,$$

whence

$$L |x_n - p| < |f(x_n) - f(p)|$$

which contradicts the Lipschitz property of f. So $x_{n+1} < p$. Moreover,

$$x_{n+1} = (1 - \lambda)x_n + \lambda f(x_n) > x_n.$$

If we would have $f(x_{n+1}) < x_{n+1}$, then between x_n and x_{n+1}, a fixed point would exist, but since $x_n > x_0$ and $x_{n+1} < p$, this would contradict the choice of p. Therefore, the claims are proved. Now, since (x_n) is monotone and bounded, it converges towards a point $\overline{x} \in [a, b]$. We have

$$|\overline{x} - f(\overline{x})| \leq |\overline{x} - x_n| + |x_n - f(x_n)| + |f(x_n) - f(\overline{x})|$$
$$= |\overline{x} - x_n| + \frac{1}{\lambda} |x_n - x_{n+1}| + |f(x_n) - f(\overline{x})|.$$

At this moment, it is obvious that the right-hand side goes to 0 for $n \to \infty$ and we get $\overline{x} = f(\overline{x})$. $\qquad\square$

7.3 Smooth Optimization

Exercise 7.48. *Find the local extrema of the following functions:*
   (i) $f : \mathbb{R}^2 \to \mathbb{R}$, $f(x_1, x_2) = 6x_1^2 x_2 + 2x_2^3 - 45x_1 - 51x_2 + 7$;

(ii) $f : \mathbb{R}^3 \setminus \{(0, 0, 0)\} \to \mathbb{R}, f(x_1, x_2, x_3) = \frac{x_1}{x_2} + \frac{x_2}{4} + \frac{x_3}{x_1} + \frac{1}{x_3}$;

(iii) $f : \mathbb{R}^2 \to \mathbb{R}, f(x_1, x_2) = x_1 x_2 (x_1^2 + x_2^2 - 4)$;

(iv) $f : \mathbb{R}^3 \to \mathbb{R}, f(x_1, x_2, x_3) = x_1^4 + x_2^3 + x_3^3 + 4x_1 x_3 - 3x_2 + 2$;

(v) $f : \mathbb{R}^2 \to \mathbb{R}, f(x_1, x_2) = x_1^4 + x_2^4$;

(vi) $f : \mathbb{R}^2 \to \mathbb{R}, f(x_1, x_2) = x_1^2 + x_2^3$;

(vii) $f : \mathbb{R}^2 \to \mathbb{R}, f(x_1, x_2) = x_1 x_2^2 e^{x_1 - x_2}$.

Solution We have to deal with nonlinear optimization problems without restrictions. The general method for finding the extrema is as follows. One finds the stationary (critical) points by solving the equation $\nabla f(x) = 0$. In everyone of these points one computes $\nabla^2 f(\overline{x})$, which in fact identifies to Hessian matrix.

- If $\nabla^2 f(\overline{x})$ is positive definite, then $\overline{x}$ is a local minimum;
- if $\nabla^2 f(\overline{x})$ is negative definite, then $\overline{x}$ is a local maximum;
- if $\nabla^2 f(\overline{x})$ is indefinite, then $\overline{x}$ is not a local extremum point.

In order to verify these aspects, in some cases, one can use the method described after Corollary 3.1.29:

- if the determinants of the matrices $\left(\frac{\partial^2 f}{\partial x^i \partial x^j}(\overline{x}) \right)_{i,j \in \overline{1,k}}$, $k \in \overline{1, p}$ are strictly positive, then $\overline{x}$ is a local minimum;
- if the determinants of the matrices $\left(\frac{\partial^2 f}{\partial x^i \partial x^j}(\overline{x}) \right)_{i,j \in \overline{1,k}}$, $k \in \overline{1, p}$ are nonzero and alternate their signs starting as negative, then $\overline{x}$ is a local maximum;
- if the determinants of the matrices $\left(\frac{\partial^2 f}{\partial x^i \partial x^j}(\overline{x}) \right)_{i,j \in \overline{1,k}}$, $k \in \overline{1, p}$ are nonzero, then every other configuration of signs apart from those described above implies that the point is not an extremum.

If no one of the above conclusions apply, then one should consider every case in its particularities in order to decide the nature of the critical point.

(i) We solve the system coming from relation $\nabla f(x) = 0$ in order to find the critical points. We obtain the system

$$12x_1 y_1 = 45$$
$$6x_1^2 + 6x_2^2 = 51$$

which have the solutions $\left(\frac{3}{2}, \frac{5}{2} \right), \left(\frac{5}{2}, \frac{3}{2} \right), \left(-\frac{3}{2}, -\frac{5}{2} \right), \left(-\frac{5}{2}, -\frac{3}{2} \right)$.

Therefore, by the application of the above method, we obtain the conclusions: $\left(\frac{3}{2}, \frac{5}{2} \right)$ is a local minimum, $\left(\frac{5}{2}, \frac{3}{2} \right), \left(-\frac{5}{2}, -\frac{3}{2} \right)$ are not local extrema and $\left(-\frac{3}{2}, -\frac{5}{2} \right)$ is a local maximum.

The item (ii) is similar.

(v) The only critical point is $(0, 0)$, but the determinants given by the Hessian matrix are zero, so we cannot decide on this basis. Nevertheless, it is easy to observe that

$$f(x_1, x_2) \geq 0 = f(0, 0), \ \forall (x_1, x_2) \in \mathbb{R}^2,$$

so $(0, 0)$ is a global minimum.

(vi) Again, $(0, 0)$ is the unique critical point, but we cannot decide its nature on the above theory. Observe that $f(0, 0) = 0$, but for the sequence $x_n = (\frac{1}{n}, 0) \to (0, 0)$, $f(x_n) > 0$, while for $y_n = (0, -\frac{1}{n}) \to (0, 0)$, $f(y_n) < 0$. So, in every neighborhood of $(0, 0)$ there exist points where the objective function takes greater or smaller values. Therefore, the point is not a local extremum.

For the other items one proceeds similarly: there exist critical points where the above method works and critical points where we should use the structure of the problem. There are as well situations when one cannot decide if $\nabla^2 f(\overline{x})$ is positive (negative) definite or not, by using the direct calculus of it, the Sylvester Criterion being not applicable. □

Problem 7.49. *(i) Let us consider a differentiable function $f : \mathbb{R} \to \mathbb{R}$. Show that if it has only one critical (stationary) point $\overline{x}$, which turns to be local extremum, then it is necessarily a global extremum.*

(ii) By the study of the function $f : \mathbb{R}^p \to \mathbb{R}$ $(p \geq 2)$ defined by

$$f(x) = (1 + x_p)^3 \sum_{k=1}^{p-1} x_k^2 + x_p^2,$$

show that the assertion of (i) is no longer true in the case of several variables.

Solution (i) Suppose that $\overline{x}$ is a local minimum and it is not a global minimum point. Then, it would exist $u \in \mathbb{R}$ with $f(u) < f(\overline{x})$. One can suppose that $u < \overline{x}$. But, from the local minimality of $\overline{x}$, there exists $v \in \mathbb{R}$ with $u < v < \overline{x}$ and $f(\overline{x}) < f(v)$ (otherwise, f would be constant on an interval $(\overline{x} - \varepsilon, \overline{x})$ and therefore $\overline{x}$ would fail to be the only critical point). Therefore, $f(u) < f(\overline{x}) < f(v)$, whence the value $f(\overline{x})$ is attained inside the interval (u, v), at a point which we denote by w. By virtue of Rolle Theorem applied to f on $[w, \overline{x}]$, there exists $t \in (w, \overline{x})$ with $f'(t) = 0$, which is a contradiction.

(ii) We show that $0 \in \mathbb{R}^p$ is the only critical point of f, and it is a local strict minimum (of order $\alpha = 2$), but it is not a global minimum point (an illustration of the case $p = 2$ is given in the figure below).

We have an optimization problem without restrictions. After computation,

$$\frac{\partial f}{\partial x_k}(x) = 2x_k(1 + x_p)^3, \ \forall k \in \overline{1, p-1},$$

$$\frac{\partial f}{\partial x_p}(x) = 3(1 + x_p)^2 \sum_{k=1}^{p-1} x_k^2 + 2x_p,$$

and the only critical point $\overline{x}$ (i.e. $\nabla f(\overline{x}) = 0$, that is $\frac{\partial f}{\partial x_k}(\overline{x}) = 0$ for $k \in \overline{1, p}$) is $\overline{x} = 0$. An easy calculus shows that the Hessian matrix of f at $\overline{x}$ is the square matrix of dimensions $p \times p$ having the number 2 on the main diagonal and 0 in all the other positions, whence it is positive definite. According to Corollary 3.1.29, we deduce that

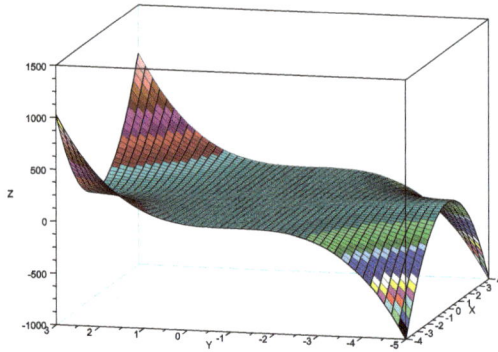

Figure 7.1: $f(x) = (1 + x_2)^3 x_1^2 + x_2^2$.

$\bar{x}$ is a local strict solution of order two. Let us observe that $f(1, 1, \ldots, x_p) = (p - 1)(1 + x_p)^3 + x_p^2$ is a third degree polynomial expression which attains, when $x_p \in \mathbb{R}$, all the real values. It follows that f cannot have a global minimum. □

Exercise 7.50. *Let* $f : \mathbb{R}^2 \to \mathbb{R}$, $f(x_1, x_2) = 3x_1^4 - 4x_1^2 x_2 + x_2^2$. *Find the minimum points of the function* f.

Solution Let us compute the critical points. The system

$$\begin{cases} \frac{\partial f}{\partial x_1}(x_1, x_2) = 0 \\ \frac{\partial f}{\partial x_2}(x_1, x_2) = 0 \end{cases}$$

has as a unique solution the element $\bar{x} = (0, 0)$. However, the sufficient optimality condition of order two is not satisfied, since the Hessian matrix of f at $\bar{x}$ is the matrix $\begin{pmatrix} 0 & 0 \\ 0 & 2 \end{pmatrix}$. Therefore, we cannot decide, on the basis of Corollary 3.1.29, if $\bar{x}$ is a minimum point. In such cases, we use the structure of the problem to get the conclusion. In our specific case, we observe that $f(x_1, x_2) = (x_1^2 - x_2)(3x_1^2 - x_2)$, and for the sequence $x_k = (\sqrt{k^{-1}}, -k^{-1}) \to \bar{x}$

$$f(x_k) = \frac{8}{k^2} > f(\bar{x})$$

while, for the sequence $x_k' = (\sqrt{(2k)^{-1}}, k^{-1}) \to \bar{x}$,

$$f(x_k') = -\frac{1}{4k^2} < f(\bar{x}).$$

Then, $\bar{x}$ is not an extremum point of f. The picture of the localization of the graph of f around $\bar{x}$ shows what happens there. □

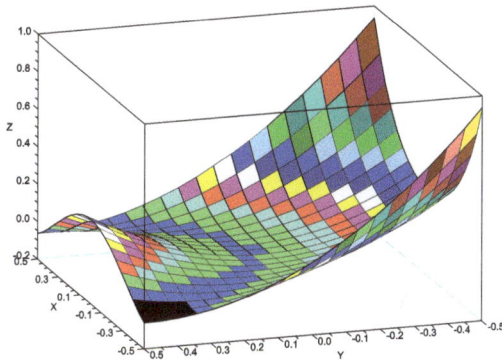

Figure 7.2: $f(x) = 3x_1^4 - 4x_1^2x_2 + x_2^2$.

Exercise 7.51. *Consider the Rosenbrock function* $f : \mathbb{R}^2 \to \mathbb{R}$, $f(x_1, x_2) = (1 - x_1)^2 + 100(x_2 - x_1^2)^2$. *Find the minima of this function.*

Solution The graph of f is depicted below.

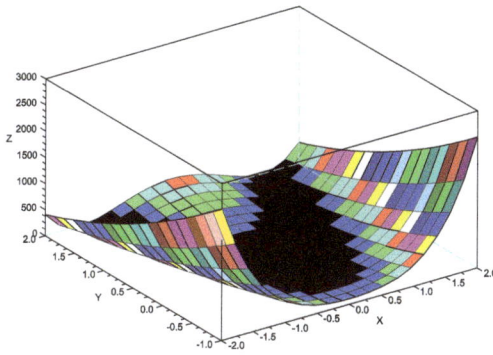

Figure 7.3: The Rosenbrock function.

The system

$$\begin{cases} \frac{\partial f}{\partial x_1}(x_1, x_2) = 0 \\ \frac{\partial f}{\partial x_2}(x_1, x_2) = 0 \end{cases}$$

is equivalent to

$$\begin{cases} -2(1 - x_1) - 400x_1(x_2 - x_1^2) = 0 \\ 200(x_2 - x_1^2) = 0, \end{cases}$$

and its unique solution is $(x_1, x_2) = (1, 1)$. One could observe that $f(1, 1) = 0$, and $f(x_1, x_2) \geq 0$ for every $(x_1, x_2) \in \mathbb{R}^2$, whence $(1, 1)$ is a global minimum. In fact, the Hessian matrix at this point

$$\begin{pmatrix} 802 & -400 \\ -400 & 200 \end{pmatrix}$$

is positive definite, whence $(1, 1)$ is a local strict solution of order two.

This function is called the Rosenbrock function and it is used in order to test numerical algorithms for the approximation of solutions. Generally speaking, because of the fact that the minimum point is situated in a relatively planar region (see the graph), this is not easy to approximate numerically. □

Exercise 7.52. *Find the global extrema of* $f : \mathbb{R}^3 \to \mathbb{R}$, $f(x_1, x_2, x_3) = x_1^3 + x_2^3 + x_3^3$ *on the sphere* $x_1^2 + x_2^2 + x_3^2 = 4$.

Solution We have a problem with a restriction given as an equality $h(x) = 0$, where

$$h : \mathbb{R}^3 \to \mathbb{R}, \quad h(x_1, x_2, x_3) = x_1^2 + x_2^2 + x_3^2 - 4.$$

Let us observe that for every $x \neq 0$, $\nabla h(x) \neq 0$. Since $x = 0$ is not feasible, the linear independence condition holds for every feasible point.

On the other hand, since the sphere is compact and f is continuous, from Weierstrass Theorem, the global extrema of the problem do exist. We observe as well that since we have only equalities (in fact a single one) as restrictions the necessary optimality conditions from Karush-Kuhn-Tucker Theorem look the same for both minima and maxima. Applying these conditions, we get the system

$$x_1(3x_1 + 2\mu) = 0$$
$$x_2(3x_2 + 2\mu) = 0$$
$$x_3(3x_3 + 2\mu) = 0$$
$$x_1^2 + x_2^2 + x_3^2 = 4,$$

where μ is a real number. We distinguish between several situations.

- If $x_1 = x_2 = x_3 = 0$, then the point is not feasible.
- If $x_1 = x_2 = 0$ and $x_3 \neq 0$, we get $x_3 = \pm 2$ (μ is not important at this stage).
- If $x_1 = 0$, $x_2, x_3 \neq 0$, we get $x_2 = x_3 = \pm\sqrt{2}$.
- If $x_1, x_2, x_3 \neq 0$, we get $x_1 = x_2 = x_3 = \pm\frac{2}{\sqrt{3}}$.

Taking into account the fact that the other cases are symmetric to these ones, we finally get the points

$$(\pm 2, 0, 0); (0, \pm 2, 0); (0, 0, \pm 2);$$

$$(0, \sqrt{2}, \sqrt{2}); (0, -\sqrt{2}, -\sqrt{2}); (\sqrt{2}, 0, \sqrt{2});$$
$$(-\sqrt{2}, 0, -\sqrt{2}); (\sqrt{2}, \sqrt{2}, 0); (\sqrt{2}, \sqrt{2}, 0);$$
$$\left(\frac{2}{\sqrt{3}}, \frac{2}{\sqrt{3}}, \frac{2}{\sqrt{3}} \right); \left(-\frac{2}{\sqrt{3}}, -\frac{2}{\sqrt{3}}, -\frac{2}{\sqrt{3}} \right).$$

By direct computation of the function values, we infer that the maximum value is 8 and it is attained at $(2, 0, 0)$, $(0, 2, 0)$, $(0, 0, 2)$, and the minimum value is -8, and it is attained at $(-2, 0, 0)$, $(0, -2, 0)$, $(0, 0, -2)$. □

Exercise 7.53. *Find the global extrema of $f : \mathbb{R}^3 \to \mathbb{R}$, $f(x_1, x_2, x_3) = x_1^3 + x_2^3 + x_3^3$ on the set of points which satisfy $x_1^2 + x_2^2 + x_3^2 = 4$ and $x_1 + x_2 + x_3 = 1$.*

Solution The points we are looking for do exist by the same reasons as before. We have a restriction of the form $h(x) = 0$, where

$$h : \mathbb{R}^3 \to \mathbb{R}^2, \quad h(x_1, x_2, x_3) = (x_1^2 + x_2^2 + x_3^2 - 4, x_1 + x_2 + x_3 - 1).$$

The linear independence condition is satisfied for all triples (x_1, x_2, x_3) for which at least two components are different. Since there are no feasible points with $x_1 = x_2 = x_3$, we deduce that the qualification condition do hold, so the extreme points are among the critical points of the Lagrangian. We obtain the system

$$3x_1^2 + 2\mu_1 x_1 + \mu_2 = 0$$
$$3x_2^2 + 2\mu_1 x_2 + \mu_2 = 0$$
$$3x_3^2 + 2\mu_1 x_3 + \mu_2 = 0$$
$$x_1^2 + x_2^2 + x_3^2 = 4$$
$$x_1 + x_2 + x_3 = 1,$$

where $\mu \in \mathbb{R}$. In order to have a compatible system in the variables μ_1, μ_2 from the first three equations, it is necessary and sufficient that

$$\begin{vmatrix} 3x_1^2 & 2x_1 & 1 \\ 3x_2^2 & 2x_2 & 1 \\ 3x_3^2 & 2x_3 & 1 \end{vmatrix} = 0,$$

whence

$$(x_1 - x_2)(x_1 - x_3)(x_2 - x_3) = 0$$
$$x_1^2 + x_2^2 + x_3^2 = 4$$
$$x_1 + x_2 + x_3 = 1.$$

We obtain

$$\left(\frac{1}{3} + \frac{\sqrt{22}}{6}, \frac{1}{3} + \frac{\sqrt{22}}{6}, \frac{1}{3} - \frac{\sqrt{22}}{3} \right)$$

$$\left(\frac{1}{3} - \frac{\sqrt{22}}{6}, \frac{1}{3} - \frac{\sqrt{22}}{6}, \frac{1}{3} + \frac{\sqrt{22}}{3} \right)$$

and their permutations. It is not difficult to verify that the first point (and its permutations) corresponds to the maximum, and the second one to the minimum. □

Exercise 7.54. *Let $n \geq 3$ and $a_i > 0$, for every $i \in \overline{1, n}$. Determine the minimum of*

$$\sum_{i=1}^{n} a_i x_i^2$$

under the constraint

$$\sum_{i=1}^{n} x_i = c,$$

where c is a given constant. What is the maximum of the above expression under the same constraint?

Solution Let us observe that if we denote by $M := \{x \in \mathbb{R}^n \mid \sum_{i=1}^n x_i = c\}$ the set of feasible points, and by $f : \mathbb{R}^n \to \mathbb{R}$, $f(x) = \sum_{i=1}^n a_i x_i^2$ the objective function, then for $v \geq \sum_{i=1}^n \frac{a_i c^2}{n^2}$, the set $M \cap N_v f$ is nonempty (it contains, for instance, the element $(cn^{-1}, ..., cn^1) \in \mathbb{R}^n$) and bounded. Hence, according to Theorem 3.1.7, there exists the minimum of the proposed problem. Notice that the restriction function is affine, so the quasiregularity qualification condition is fulfilled. The Karush-Kuhn-Tucker conditions say that there exists $\mu \in \mathbb{R}$ such that

$$2a_i x_i + \mu = 0, \ \forall i \in \overline{1, n},$$

whence

$$x_i = -\frac{\mu}{2a_i}.$$

We replace this into restriction and we get

$$c = -\sum_{i=1}^{n} \frac{\mu}{2a_i},$$

so

$$\mu = -\frac{2c}{\sum_{i=1}^{n} \frac{1}{a_i}}.$$

Therefore,

$$x_i = \frac{c}{a_i \sum_{i=1}^{n} \frac{1}{a_i}}, \ \forall i \in \overline{1, n}.$$

The given expression does not admit a maximum: we observe that for the sequence of feasible points $(p, -p, c, 0..., 0)_{p \in \mathbb{N}}$, the value of f goes to $+\infty$. □

Exercise 7.55. *Minimize* $f : \mathbb{R}^3 \to \mathbb{R}$,

$$f(x_1, x_2, x_3) = x_3 + \frac{1}{2}\left(x_1^2 + x_2^2 + \frac{x_3^2}{10}\right)$$

under constraints $x_1 + x_2 + x_3 = r \ (r > 0)$, $x_1 \geq 0$, $x_2 \geq 0$, $x_3 \geq 0$.

Solution The existence of minimum is ensured by Weierstrass Theorem. The constraints are linear, so we do not have to verify the qualification conditions. In order to bring the problem to the standard form, the inequalities restrictions are written as $-x_1 \leq 0, -x_2 \leq 0, -x_3 \leq 0$. Karush-Kuhn-Tucker conditions ensure that at a minimum point x there are $\lambda_1, \lambda_2, \lambda_3 \geq 0$ and $\mu \in \mathbb{R}$ such that

$$x_1 - \lambda_1 + \mu = 0$$
$$x_2 - \lambda_2 + \mu = 0$$
$$1 + \frac{x_3}{10} - \lambda_3 + \mu = 0$$
$$\lambda_1 x_1 = 0$$
$$\lambda_2 x_2 = 0$$
$$\lambda_3 x_3 = 0$$
$$x_1 + x_2 + x_3 = r.$$

After the study of all the possibilities, we find the solution
- for $r \leq 2$,

$$(x_1, x_2, x_3) = \left(\frac{r}{2}, \frac{r}{2}, 0\right) \text{ and } (\lambda_1, \lambda_2, \lambda_3, \mu) = \left(0, 0, 1 - \frac{r}{2}\right);$$

- for $r > 2$,

$$(x_1, x_2, x_3) = \left(\frac{r+10}{12}, \frac{r+10}{12}, \frac{5(r-2)}{6}\right) \text{ and }$$
$$(\lambda_1, \lambda_2, \lambda_3, \mu) = \left(0, 0, 0, -\frac{10+r}{12}\right).$$

The problem is solved. □

Exercise 7.56. *Maximize* $f : \mathbb{R}^3 \to \mathbb{R}$, $f(x) = x_1 x_2 x_3$ *under the restrictions* $x_1, x_2, x_3 \geq 0$, $2x_1 + 2x_2 + 4x_3 \leq a$, *where* $a > 0$.

Solution It is clear that the maximum is attained (Weierstrass Theorem) and it is strictly positive. For standardization, we deal with the problem

$$\min \ -x_1 x_2 x_3$$

$-x_1, -x_2, -x_3 \leq 0$, $2x_1 + 2x_2 + 4x_3 \leq a$. Following the theory for a minimum point x, there exist $\lambda_1, \lambda_2, \lambda_3, \lambda_4 \geq 0$ such that

$$-x_2 x_3 - \lambda_1 + 2\lambda_4 = 0$$

$$-x_1x_3 - \lambda_2 + 2\lambda_4 = 0$$
$$-x_1x_2 - \lambda_3 + 4\lambda_4 = 0$$
$$\lambda_1 x_1 = 0$$
$$\lambda_2 x_2 = 0$$
$$\lambda_3 x_3 = 0$$
$$(2x_1 + 2x_2 + 4x_3 - a)\lambda_4 = 0.$$

We infer that

$$x_1(-x_2x_3 + 2\lambda_4) = 0$$
$$x_2(-x_1x_3 + 2\lambda_4) = 0$$
$$x_3(-x_1x_2 + 4\lambda_4) = 0$$
$$(2x_1 + 2x_2 + 4x_3 - a)\lambda_4 = 0$$

We add the first three equations and we get $-3x_1x_2x_3 + \lambda_4(2x_1+2x_2+4x_3) = 0$, therefore

$$-3x_1x_2x_3 + a\lambda_4 = 0,$$

so $\lambda_4 = \frac{3x_1x_2x_3}{a}$. Replacing in $x_1(-x_2x_3 + 2\lambda_4) = 0$, since $x_1x_2x_3 \neq 0$, we find $x_1 = \frac{a}{6}$. Analogously, $x_2 = \frac{a}{6}$, $x_3 = \frac{a}{12}$. The multipliers are $(0, 0, 0, \frac{a^2}{144})$. So, the solution of the problem is $(\frac{a}{6}, \frac{a}{6}, \frac{a}{12})$. $\qquad\square$

Exercise 7.57. *Find the extrema of $f : \mathbb{R}^3 \to \mathbb{R}$, $f(x) = x_1x_2x_3$ under the restriction $h(x) = 0$, where $h : \mathbb{R}^3 \to \mathbb{R}^2$, $h(x) = (x_1x_2 + x_1x_3 + x_2x_3 - 8, x_1 + x_2 + x_3 - 5)$.*

Solution We show that the set of feasible points is bounded (hence compact).

Using the equality $x_3 = 5 - x_1 - x_2$, we can eliminate x_3 from the first restriction, that is

$$x_1^2 + x_2^2 + x_1x_2 - 5x_1 - 5x_2 + 8 = 0,$$

whence

$$\left(\frac{x_1}{\sqrt{2}} + \frac{x_2}{\sqrt{2}}\right)^2 + \left(\frac{x_1}{\sqrt{2}} - \frac{5}{\sqrt{2}}\right)^2 + \left(\frac{x_2}{\sqrt{2}} - \frac{5}{\sqrt{2}}\right)^2 = 17.$$

We deduce that x_1 and x_2 lie in a bounded set, as the x_3 as well. Therefore, the set of feasible points is compact and both minimization and maximization problems have solutions.

We verify the linear independence qualification condition: we ask if there exist two real numbers α_1, α_2 with $(\alpha_1, \alpha_2) \neq (0, 0)$ such that

$$\alpha_1(x_2 + x_3) + \alpha_2 = 0$$
$$\alpha_1(x_1 + x_3) + \alpha_2 = 0$$
$$\alpha_1(x_1 + x_2) + \alpha_2 = 0.$$

Since we are interested only on the set of feasible points, we deduce

$$\alpha_1(5 - x_1) + \alpha_2 = 0$$
$$\alpha_1(5 - x_2) + \alpha_2 = 0$$
$$\alpha_1(5 - x_3) + \alpha_2 = 0.$$

Then, we find that $x_1 = x_2 = x_3$, which cannot be true on the set of feasible points. Therefore, the linear independence qualification condition holds on the set of feasible points. The application of Theorem 3.2.6 (and the remarks afterwards) leads to the conclusion that if $\bar{x}$ is a minimum or maximum point of the problem, then there exist $\mu_1, \mu_2 \in \mathbb{R}$ such that

$$\bar{x}_2\bar{x}_3 + \mu_1(\bar{x}_2 + \bar{x}_3) + \mu_2 = 0$$
$$\bar{x}_1\bar{x}_3 + \mu_1(\bar{x}_1 + \bar{x}_3) + \mu_2 = 0$$
$$\bar{x}_1\bar{x}_2 + \mu_1(\bar{x}_1 + \bar{x}_2) + \mu_2 = 0$$
$$\bar{x}_1\bar{x}_2 + \bar{x}_1\bar{x}_3 + \bar{x}_2\bar{x}_3 = 8$$
$$\bar{x}_1 + \bar{x}_2 + \bar{x}_3 = 5.$$

Obviously, μ_1, μ_2 cannot be simultaneously 0. After easy manipulations we get

$$(\mu_1\bar{x}_3 + \mu_2)(\bar{x}_1 - \bar{x}_2) = 0$$
$$(\mu_1\bar{x}_2 + \mu_2)(\bar{x}_1 - \bar{x}_3) = 0$$
$$(\mu_1\bar{x}_1 + \mu_2)(\bar{x}_2 - \bar{x}_3) = 0$$
$$\bar{x}_1\bar{x}_2 + \bar{x}_1\bar{x}_3 + \bar{x}_2\bar{x}_3 = 8$$
$$\bar{x}_1 + \bar{x}_2 + \bar{x}_3 = 5.$$

If $\mu_1 = 0$, then $\bar{x}_1 = \bar{x}_2 = \bar{x}_3$, which is not possible. Then $\mu_1 \neq 0$ and since $\bar{x}_1, \bar{x}_2, \bar{x}_3$ cannot be equal, we find $\bar{x} = (2, 2, 1)$, $\bar{x} = (1, 2, 2)$, $\bar{x} = (2, 1, 2)$ and $\bar{x} = (\frac{4}{3}, \frac{4}{3}, \frac{7}{3})$, $\bar{x} = (\frac{7}{3}, \frac{4}{3}, \frac{4}{3})$, $\bar{x} = (\frac{4}{3}, \frac{7}{3}, \frac{4}{3})$. By easy comparison of function values we find that the first ones are maxima, while the last ones, minima. $\qquad\square$

Exercise 7.58. *Find the global minima of $f : \mathbb{R}^3 \to \mathbb{R}$, $f(x_1, x_2, x_3) = x_1^3 + x_2^3 + x_3^3$ on the set of points satisfying $x_1^2 + x_2^2 + x_3^2 \le 4$ and $x_1 + x_2 + x_3 \le 1$.*

Solution Once again, the existence of minimum is ensured by Weierstrass Theorem. It is the same objective function as in Exercise 7.53, but this time the constraints are written as inequalities. Observe that the restrictions are convex and, moreover, the Slater condition holds. So, the minima are to be found among the critical points of the Lagrangian. We have

$$3x_1^2 + 2\lambda_1 x_1 + \lambda_2 = 0$$
$$3x_2^2 + 2\lambda_1 x_2 + \lambda_2 = 0$$
$$3x_3^2 + 2\lambda_1 x_3 + \lambda_2 = 0$$
$$\lambda_1(x_1^2 + x_2^2 + x_3^2 - 4) = 0$$
$$\lambda_2(x_1 + x_2 + x_3 - 1) = 0$$

$$x_1^2 + x_2^2 + x_3^2 \le 4$$
$$x_1 + x_2 + x_3 \le 1$$
$$\lambda_1, \lambda_2 \ge 0.$$

Again, we distinguish several situations.

- If $\lambda_1 = \lambda_2 = 0$, then $x_1 = x_2 = x_3 = 0$.
- If $\lambda_1 = 0$, $x_1 + x_2 + x_3 = 1$, we deduce

$$\lambda_2 = -3x_1^2 = -3x_2^2 = -3x_3^2,$$

so the unique solution $\bar{x}$ is $(\frac{1}{3}, \frac{1}{3}, \frac{1}{3})$ which gives $\lambda_2 = -\frac{1}{3} < 0$, and this is not convenient.

- If $\lambda_2 = 0$, $x_1^2 + x_2^2 + x_3^2 = 4$, we have

$$12 + 2(x_1 + x_2 + x_3)\lambda_1 = 0,$$

that is

$$\lambda_1 = -\frac{6}{x_1 + x_2 + x_3}.$$

Replacing in the first three equations, we obtain the solutions

$$(-2, 0, 0), (0, -2, 0), (0, 0, -2)$$
$$(-\sqrt{2}, -\sqrt{2}, 0), (-\sqrt{2}, 0, -\sqrt{2}), (0, -\sqrt{2}, -\sqrt{2})$$
$$\left(-\frac{2}{\sqrt{3}}, -\frac{2}{\sqrt{3}}, -\frac{2}{\sqrt{3}}\right).$$

- If $x_1 + x_2 + x_3 = 1$, $x_1^2 + x_2^2 + x_3^2 = 4$, we are again in the situation from Exercise 7.53.

After the computation of the function values we get that the minimum is attained in $(-2, 0, 0), (0, -2, 0), (0, 0, -2)$, so, finally, the situation is different from that in Exercise 7.53. □

Problem 7.59. *Obtain the mean inequality from Karush-Kuhn-Tucker conditions applied for an appropriate optimization problem.*

Solution Let $f, g : \mathbb{R}^n \to \mathbb{R}$,

$$f(x) = \frac{x_1 + x_2 + \ldots + x_n}{n}$$
$$h(x) = 1 - x_1 x_2 \ldots x_n.$$

We study the problem of minimization of f under the restrictions

$$h(x) = 0, \quad x_i > 0, \quad \forall i \in \overline{1, n}.$$

Firstly, we observe that f has a minimum on the compact set $\{x \in \mathbb{R}^n \mid h(x) = 0, x_i \ge 0, \forall i \in \overline{1, n}\} \cap N_v f$ (where $v > 0$). Since the point with at least one zero component is not in this set, we deduce that the minimum belongs to the set of feasible points of the problem under consideration. On the other hand, the linear independence condition is satisfied at every feasible point. So, the Karush-Kuhn-Tucker condition means that for a solution x of the problem, there exists $\mu \in \mathbb{R}$ such that

$$0 = \frac{1}{n} - \frac{\mu}{x_i}, \ \forall i \in \overline{1, n}.$$

Therefore, $x_1 = x_2 = \ldots = x_n = 1$. So, $f(x) \geq f(1, \ldots, 1) = 1$ for every feasible point x.

Let $a_1, a_2, \ldots, a_n > 0$. Denote $G = \sqrt[n]{a_1 \ldots a_n}$ and $x_i = \frac{a_i}{G} > 0$, for every n. Then $h(x) = 0$, therefore

$$\frac{x_1 + x_2 + \ldots + x_n}{n} \geq 1,$$

that is

$$\frac{1}{n} \left(\frac{a_1}{G} + \ldots + \frac{a_n}{G} \right) \geq 1,$$

and the conclusion follows. □

Exercise 7.60. *Find the global extrema for $f : \mathbb{R}^2 \to \mathbb{R}$, $f(x_1, x_2) = -2x_1^2 + 4x_1x_2 + x_2^2$ on the unit circle.*

Solution The existence of solutions is ensured by the Weierstrass Theorem. We take $h : \mathbb{R}^2 \to \mathbb{R}$, $h(x_1, x_2) = 1 - x_1^2 - x_2^2$ and we have to study the problem of extrema of f under the constraint $h(x) = 0$. The linear independence qualification condition holds and for both minima and maxima we have to find the critical points of the Lagrangian. We have

$$(-2 - \mu)x_1 + 2x_2 = 0$$
$$2x_1 + (1 - \mu)x_2 = 0$$
$$x_1^2 + x_2^2 = 1.$$

We infer that

$$\begin{vmatrix} -2 - \mu & 2 \\ 2 & 1 - \mu \end{vmatrix} = 0,$$

so $\mu = 2$ or $\mu = -3$ and, correspondingly, $(x_1, x_2) = \left(-\frac{2}{\sqrt{5}}, \frac{1}{\sqrt{5}} \right)$, $(x_1, x_2) = \left(\frac{2}{\sqrt{5}}, -\frac{1}{\sqrt{5}} \right)$ or $(x_1, x_2) = \left(\frac{1}{\sqrt{5}}, -\frac{2}{\sqrt{5}} \right)$, $\left(-\frac{1}{\sqrt{5}}, \frac{2}{\sqrt{5}} \right)$. By direct computation of the function f values at these points, we find that the first two points are maxima, while the last two are minima. □

Exercise 7.61. *Show that $f : \mathbb{R}^3 \to \mathbb{R}$, $f(x_1, x_2, x_3) = x_1^2 + x_2^2 + x_3^2 - x_1 - x_2 - x_3$ is convex. Find its minimum value under the restrictions*

$$x_1^2 + x_2^2 = 4, \ -1 \leq x_3 \leq 1.$$

Solution The convexity of f follows from the fact that $\nabla^2 f(x)$ is positive definite for every $x \in \mathbb{R}^3$. Obviously, the set M of feasible points is not convex. So, a point $x \in M$ is a minimum if

$$-\nabla f(x) \in N(M, x) = \{u \in \mathbb{R}^3 \mid \langle u, c - x \rangle \le 0, \ \forall c \in M\},$$

i.e.,

$$\nabla f(x)(c - x) \ge 0, \ \forall c \in M.$$

The expression of the normal cone leads to the result, but this can be complicated. For instance, for a feasible point x with $x_3 \in (-1, 1)$,

$$N(M, x) = \mathbb{R}\{(x_1, x_2, 0)\},$$

and the condition becomes

$$-(2x_1 - 1, 2x_1 - 1, 2x_1 - 1) \in \mathbb{R}\{(x_1, x_2, 0)\},$$

and we get $x_3 = \frac{1}{2}$ and $x_1 = x_2 = \pm\sqrt{2}$. Therefore $\left(-\sqrt{2}, -\sqrt{2}, \frac{1}{2}\right)$ and $\left(\sqrt{2}, \sqrt{2}, \frac{1}{2}\right)$ are minima. The global minimum is attained at $\left(\sqrt{2}, \sqrt{2}, \frac{1}{2}\right)$. The remaining reasoning is similar.

A useful remark which can lead to the solution in a simple manner is as follows: the cost function on the feasible set is $f(x_1, x_2, x_3) = 4 - x_1 - x_2 + x_3^2 - x_3$, and the variables are now separated since it is sufficient to find minima of $x_3^2 - x_3$ on $[-1, 1]$ and the minima of $4 - x_1 - x_2$ under the restriction $x_1^2 + x_2^2 = 4$. These two problems are much more easy to handle than the initial one (the former being elementary). We obtain the minimum of the initial problem at $\left(\sqrt{2}, \sqrt{2}, \frac{1}{2}\right)$. □

Exercise 7.62. *Consider the region in $\mathbb{R}^2$ defined by*

$$x_1^2 - x_2^2 \le 1, \ x_1^2 + x_2^2 \le 4.$$

Find the extreme values on this region for

$$f : \mathbb{R}^2 \to \mathbb{R}, \ f(x_1, x_2) = x_1^2 + 2x_2^2 + x_1 x_2.$$

Solution The feasible set is compact. The linear independence condition means that $x_1 x_2 \ne 0$. If $x_1 x_2 = 0$, then we have two situations: for $x_1 = 0$, we get $x_2 \in [-2, 2]$, and for those points the maximum of f is 8, while the minimum is 0; for $x_2 = 0$, we obtain $x_1 \in [-1, 1]$, and the maximum of f is 1, while the minimum is 0.

After these cases, we can suppose that $x_1 x_2 \ne 0$ and we treat separately the minimization and the maximization problems. For minimization we have the conditions

$$2x_1 + x_2 + 2\lambda_1 x_1 + 2\lambda_2 x_1 = 0$$

$$x_1 + 4x_2 - 2\lambda_1 x_2 + 2\lambda_2 x_2 = 0$$
$$\lambda_1(x_1^2 - x_2^2 - 1) = 0$$
$$\lambda_2(x_1^2 + x_2^2 - 4) = 0$$
$$\lambda_1 \geq 0$$
$$\lambda_2 \geq 0.$$

The rest of the calculation is not very involved, and at the end we compare the objective function values. For maximization, the reasoning is similar. □

Exercise 7.63. *Find the closest point to the origin on the surfaces:*
(i) $x_1 x_2 + x_1 x_3 + x_2 x_3 = 1$;
(ii) $x_1^2 + x_2^2 - x_3^2 = 1$.

Solution In both cases, the objective function is $f : \mathbb{R}^3 \to \mathbb{R}$,

$$f(x_1, x_2, x_3) = x_1^2 + x_2^2 + x_3^2,$$

and on the set of feasible points the linear independence qualification condition holds.

For instance, at (i), for $v > 1$, the set $M \cap N_v f$ is nonempty (it contains, for example, the point $(1, 0, 0)$) and bounded. According to Theorem 3.1.7 there exists the global minimum of the proposed problem. For for a local minimum point, there exists $\mu \in \mathbb{R}$ such that

$$2x_1 + \mu(x_2 + x_3) = 0$$
$$2x_2 + \mu(x_1 + x_3) = 0$$
$$2x_3 + \mu(x_1 + x_2) = 0$$
$$x_1 x_2 + x_1 x_3 + x_2 x_3 = 1.$$

For the first three equations we cannot have $x_1 = x_2 = x_3 = 0$ (because of the fourth equation), so we infer that

$$\begin{vmatrix} 2 & \mu & \mu \\ \mu & 2 & \mu \\ \mu & \mu & 2 \end{vmatrix} = 0,$$

and calculations lead to

$$x = \left(\frac{1}{\sqrt{3}}, \frac{1}{\sqrt{3}}, \frac{1}{\sqrt{3}} \right) \text{ and } x = \left(-\frac{1}{\sqrt{3}}, -\frac{1}{\sqrt{3}}, -\frac{1}{\sqrt{3}} \right),$$

which both are minima.

Let us observe that there is no maximum point, since for every $n \in \mathbb{N}^*$ the point $\left(n, n, \frac{1-n^2}{2n} \right)$ is feasible, but $f\left(n, n, \frac{1-n^2}{2n} \right) \overset{n \to \infty}{\longrightarrow} \infty$. □

Exercise 7.64. *Let $f : \mathbb{R}^2 \to \mathbb{R}$,*

$$f(x) = -x_1 - 2x_2 - 2x_1x_2 + \frac{x_1^2}{2} + \frac{x_2^2}{2}$$

and the set of feasible points

$$M := \left\{ x \in \mathbb{R}^2 \mid x_1 + x_2 \leq 1,\ x_1 \geq 0,\ x_2 \geq 0 \right\}.$$

Solve the problem (P) of minimizing f on M.

Solution Let us remark that for every $x \in \mathbb{R}^2$,

$$\nabla^2 f(x) = \begin{pmatrix} 1 & -2 \\ -2 & 1 \end{pmatrix}.$$

Since this matrix is not positive definite (the determinant is negative), the function is not convex. If it would exist a minimum point $\bar{x}$ lying in the interior of M, then that point would be a minimum point without restrictions (according to Remark 3.1.2), whence, from Fermat Theorem, $\nabla f(\bar{x}) = 0$. But

$$\nabla f(\bar{x}) = (-1 + \bar{x}_1 - 2\bar{x}_2, -2 - 2\bar{x}_1 + \bar{x}_2),$$

and the resulting system has the solution $\bar{x} = (-\frac{5}{3}, -\frac{4}{3})$ which actually does not belong to M. Hence, the problem has no solutions in $\mathrm{int}\, M$. However f is continuous, M is compact, so the problem (P) has at least one solution.

We can approach the problem in two ways.

The first one takes advantage of the fact that the geometrical image of the set M is a simple one (a triangle with the vertices at $(0, 0)$, $(0, 1)$, and $(1, 0)$). It is not difficult to compute the Bouligand tangent and normal cones to M at its boundary points and then to verify the necessary optimality condition: $-\nabla f(\bar{x}) \in N(M, \bar{x})$ (see Theorem 3.1.18).

Therefore, if the point $\bar{x}$ is:

– on the open segment joining the vertices $(0, 1)$, $(1, 0)$:

$$T(M, \bar{x}) = \{u \in \mathbb{R}^2 \mid u_1 + u_2 \leq 0\};\ N(M, \bar{x}) = \mathbb{R}_+\{(1, 1)\};$$

– on the open segment joining the vertices $(0, 0)$, $(0, 1)$:

$$T(M, \bar{x}) = \{u \in \mathbb{R}^2 \mid u_1 \geq 0\};\ N(M, \bar{x}) = \mathbb{R}_+\{(-1, 0)\};$$

– on the open segment joining the vertices $(0, 0)$, $(1, 0)$:

$$T(M, \bar{x}) = \{u \in \mathbb{R}^2 \mid u_2 \geq 0\};\ N(M, \bar{x}) = \mathbb{R}_+\{(0, -1)\};$$

– exactly $(0, 1)$:

$$T(M, \bar{x}) = \{u \in \mathbb{R}^2 \mid u_1 + u_2 \leq 0,\ u_1 \geq 0\}$$

$$N(M, \bar{x}) = \{a(1, 1) + b(-1, 0) \mid a, b \geq 0\};$$

– exactly $(1, 0)$:

$$T(M, \bar{x}) = \{u \in \mathbb{R}^2 \mid u_1 + u_2 \leq 0, \ u_2 \geq 0\}$$
$$N(M, \bar{x}) = \{a(1, 1) + b(0, -1) \mid a, b \geq 0\};$$

– exactly $(0, 0)$:

$$T(M, \bar{x}) = \{u \in \mathbb{R}^2 \mid u_1 \geq 0, \ u_2 \geq 0\};$$
$$N(M, \bar{x}) = \{a(-1, 0) + b(0, -1) \mid a, b \geq 0\}.$$

A direct computation shows that there is only one point which satisfies the necessary optimality condition, and this is $\bar{x} = \left(\frac{1}{3}, \frac{2}{3}\right)$. Therefore, according to the preceding remark, this is the only minimum point of the problem.

One can also consider the problem of finding the global maximum of f on M (the existence of such a maximum is ensured by the Weierstrass' Theorem), which is equivalent to the finding the global minimum of $-f$ on M. With the same arguments as above, we find two points which verify the necessary optimality condition (i.e., $\nabla f(\bar{x}) \in N(M, \bar{x})$): $\bar{x} = (0, 0)$ and $\bar{x} = (1, 0)$. But $f(0, 0) = 0$, while $f(1, 0) = -2^{-1}$, so $(0, 0)$ is the global maximum point.

The second possible solution is to consider the function

$$g : \mathbb{R}^2 \to \mathbb{R}^3, \ g(x) := (x_1 + x_2 - 1, -x_1, -x_2)$$

and to reinterpret the problem as a problem with three functional inequalities given by $g(x) \leq 0$. Since all the functions g_i are linear, it is not necessary to verify qualification conditions (according to Theorem 3.2.24). Then, if $\bar{x} \notin \text{int } M$ is a solution of the problem, there exists $(\lambda_1, \lambda_2, \lambda_3) \in \mathbb{R}_+^3$ such that

$$\begin{cases} \nabla f(\bar{x}) + \lambda_1 \nabla g_1(\bar{x}) + \lambda_2 \nabla g_2(\bar{x}) + \lambda_3 \nabla g_3(\bar{x}) = 0 \\ \lambda_i g_i(\bar{x}) = 0, \ i \in \overline{1, 3}. \end{cases}$$

Now, again, the discussion should be divided into six cases which mirror those from before. For instance, if $\bar{x}$ is on the open segment joining $(0, 1)$ and $(1, 0)$, then $g_2(\bar{x}) < 0$, $g_3(\bar{x}) < 0$, whence $\lambda_2 = \lambda_3 = 0$, and the above system reduces to

$$\begin{cases} -1 - 2\bar{x}_2 + \bar{x}_1 + \lambda_1 = 0 \\ -2 - 2\bar{x}_1 + \bar{x}_2 + \lambda_1 = 0 \\ x_1 + x_2 - 1 = 0, \end{cases}$$

which gives the solution $\lambda_1 = 2$, $\bar{x} = \left(\frac{1}{3}, \frac{2}{3}\right)$. In the same way, in the other situations, the Karush-Kuhn-Tucker system has no solution and the conclusion is the same as in the first approach..

$\square$

Problem 7.65. *Let $f : \mathbb{R}^p \to \mathbb{R}$ be a convex and differentiable function. Write the optimality conditions for the minimization of f on the unit simplex.*

Solution According to Example 2.1.16 and Proposition 3.1.25, $\overline{x} \in M$ is a minimum point of f on M if and only if $-\nabla f(\overline{x}) \in N(M, \overline{x})$. From the particular form of $N(M, \overline{x})$, one infers that this condition become

$$\frac{\partial f}{\partial x_i}(\overline{x}) = c, \text{ (constant)}, \forall i \notin I(\overline{x})$$

$$\frac{\partial f}{\partial x_i}(\overline{x}) \geq c, \ \forall i \in I(\overline{x}). \qquad \square$$

Exercise 7.66. *Let us consider the objective function $f : \mathbb{R}^2 \to \mathbb{R}$, $f(x_1, x_2) = x_1 + x_2^2$ and a function which defines a equality constraint as $h : \mathbb{R}^2 \to \mathbb{R}$, $h(x_1, x_2) = x_1^3 - x_2^2$. Find the solution of the minimization of f under $h(x) = 0$.*

Solution We have $M = \{x \in \mathbb{R}^2 \mid h(x) = 0\}$. In order to verify the linear independence qualification condition at a point $\overline{x} \in M$, it is necessary and sufficient to have $\nabla h(\overline{x}) \neq 0$, and this happens for all points in M, but $\overline{x} = (0, 0)$. For the moment, we avoid this point. If $\overline{x} \in M \setminus \{(0, 0)\}$ is a minimum point of (P), then, according to Theorem 3.2.6, there exists $\mu \in \mathbb{R}$ such that

$$\nabla f(\overline{x}) + \mu \nabla h(\overline{x}) = 0.$$

A simple calculation shows that the resulting system has no solution, whence (P) has no solution different from $(0, 0)$. Let us remark that $\overline{x} = (0, 0)$ is a solution (even a global one) since $f(\overline{x}) = 0$, and for every $x \in M$, $x_1^3 = x_2^2 \geq 0$, so $f(x) = x_1 + x_2^2 \geq 0$. $\square$

Exercise 7.67. *Let $n_1, ..., n_p \in \mathbb{N}^*$, and let $f : \mathbb{R}^p \to \mathbb{R}$, $f(x) = -x_1^{n_1} x_2^{n_2} ... x_p^{n_p}$. Minimize this function on the unit simplex of $\mathbb{R}^p$.*

Solution Clearly, the problem has a solution, since f is continuos and M is compact. Since f vanishes if at least one of the components of the argument is zero, it is clear that the solutions will actually be from

$$\left\{ x \in \mathbb{R}^p \mid x_i > 0, \ \forall i \in \overline{1, p}, \ \sum_{i=1}^{p} x_i = 1 \right\}.$$

With the notation from Example 2.1.16, this means that $I(\overline{x}) = \emptyset$. Firstly, the necessary optimality condition, $-\nabla f(\overline{x}) \in N(M, \overline{x})$, could be written, with the expression of the normal cone to the unit simplex in mind (Example 2.1.16), as

$$\frac{n_i}{\overline{x}_i} f(\overline{x}) = c, \text{ (constant)} \ \forall i \in \overline{1, p},$$

that is

$$\frac{n_i}{\overline{x}_i} = c', \text{ (constant)} \ \forall i \in \overline{1, p}.$$

Because $\sum_{i=1}^{p} \bar{x}_i = 1$, by denoting $N := \sum_{i=1}^{p} n_i$, we find

$$\bar{x}_i = \frac{n_i}{N}, \ \forall i \in \overline{1, p}.$$

Since the problem admits a solution and only one point satisfies the necessary condition, we deduce that this point is the solution we are looking for.

Another approach consists of transforming the geometrical restriction into a functional one. Let $h : \mathbb{R}^p \to \mathbb{R}$, $h(x) = \sum_{i=1}^{p} x_i - 1$. Clearly, $M = \{x \in \mathbb{R}^p \mid h(x) = 0\}$. Let $\bar{x}$ be a solution of the problem. As $\nabla h(\bar{x}) \neq 0$, we can apply Theorem 3.2.6 in order to deduce the existence of a number $\mu \in \mathbb{R}$ such that

$$\nabla f(\bar{x}) + \mu \nabla h(\bar{x}) = 0,$$

that is

$$-\frac{n_i}{\bar{x}_i} f(\bar{x}) = \mu, \text{ (constant) } \forall i \in \overline{1, p}.$$

So, as expected, the same conclusion follows. □

Exercise 7.68. *Let $a \in \mathbb{R}$, $f, h : \mathbb{R}^2 \to \mathbb{R}$, $f(x) = (x_1 - 1)^2 + x_2^2$, $h(x) = -x_1 + ax_2^2$. Decide if $\bar{x} = (0, 0)$ is a minimum point for f under the restriction $h(x) = 0$.*

Solution We have $\nabla h(\bar{x}) = (-1, 2a\bar{x}_2) \neq (0, 0)$, so the linear independence qualification condition is fulfilled. We study the necessary optimality condition from Theorem 3.2.6 for $\bar{x} = (0, 0)$. There exists $\mu \in \mathbb{R}$ with

$$\nabla f(\bar{x}) + \mu \nabla h(\bar{x}) = 0,$$

and we obtain $\mu = -2$. We discuss now the second-order necessary optimality condition (Theorem 3.3.2). It is easy to see that $T_B(M, \bar{x}) = \{0\} \times \mathbb{R}$, and for $u = (0, u_2) \in T_B(M, \bar{x})$,

$$\nabla_{xx}^2 L(\bar{x}, \mu)(u, u) = 2(1 - 2a)u_2^2.$$

If $a > \frac{1}{2}$, the condition from Theorem 3.3.2 is not satisfied, so the point is not a minimum.

If $a < \frac{1}{2}$, the sufficient condition from Theorem 3.3.3 is satisfied, so the point is even a strict solution of second order.

For $a = \frac{1}{2}$, only necessary optimality condition is satisfied. In this case we observe that for $x \in M$, $x_1 = 2^{-1} x_2^2$, whence

$$f(x_1, x_2) = (2^{-1} x_2^2 - 1)^2 + x_2^2 = 4^{-1} x_2^4 + 1 \geq 1 = f(0, 0),$$

so $(0, 0)$ is a minimum point. □

Exercise 7.69. *Solve the problem of minimization of $f : \mathbb{R}^3 \to \mathbb{R}$, $f(x) = -x_1 - x_2 - x_2 x_3 - x_1 x_3$ under the affine constraint $h(x) = 0$, where $h : \mathbb{R}^3 \to \mathbb{R}$, $h(x) = x_1 + x_2 + x_3 - 3$.*

Solution Let $\bar{x} \in \mathbb{R}^3$ be a feasible point, i.e., $h(\bar{x}) = 0$. We verify if the necessary optimality condition in Theorem 3.2.6 holds. There should exist some $\mu \in \mathbb{R}$ with

$$\nabla f(\bar{x}) + \mu \nabla h(\bar{x}) = 0.$$

We get the linear system

$$\begin{cases} \bar{x}_1 + \bar{x}_2 + \bar{x}_3 = 3 \\ -1 - \bar{x}_3 = -\mu \\ -\bar{x}_1 - \bar{x}_2 = -\mu \end{cases}$$

which gives $\mu = 2$, $\bar{x} = (\bar{x}_1, 2 - \bar{x}_1, 1)$, $\bar{x}_1 \in \mathbb{R}$.

Let us observe that these points also verify the second-order necessary optimality conditions (see Theorem 3.3.2 and Remark 3.3.1). It is clear (as in the case of unit simplex) that $T(M, \bar{x}) = \{u \in \mathbb{R}^3 \mid u_1 + u_2 + u_3 = 0\}$. Then for every $u \in T(M, \bar{x})$,

$$\nabla^2_{xx} L(\bar{x}, \mu)(u, u) = (u_1, u_2, u_3) \begin{pmatrix} 0 & 0 & -1 \\ 0 & 0 & -1 \\ -1 & -1 & 0 \end{pmatrix} (u_1, u_2, u_3)^t$$

$$= -2(u_1 + u_2)u_3 = 2(u_1 + u_2)^2 \geq 0.$$

One can observe that for $u = (u_1, -u_1, 0)$ with $u_1 \neq 0$, the sufficient second-order optimality condition in Theorem 3.3.3 is not satisfied, so we should decide if the above points are solutions using different tools. More precisely, will try to exploit the particular form of the problem. We observe that for every $\bar{x}_1 \in \mathbb{R}$, $f(\bar{x}_1, 2 - \bar{x}_1, 1) = -4$, and for an arbitrary $x \in M$, $f(x) + 4 = (x_1 + x_2 - 2)^2 \geq 0$, so all the points which satisfy the necessary optimality conditions are solutions for our problem. □

Exercise 7.70. *Let* $a > 4^{-1}$, $f, g : \mathbb{R}^2 \to \mathbb{R}$, $f(x) = x_1^2 + ax_2^2 + x_1x_2 + x_1$ *and* $g(x) = x_1 + x_2 - 1$. *Solve the problem* (P), *with the usual notations.*

Solution Let us observe that

$$\nabla^2 f(x) = \begin{pmatrix} 2 & 1 \\ 1 & 2a \end{pmatrix}$$

is positive definite, whence f is convex (Theorem 2.2.10). Since the remaining problem data is convex (affine, in fact), according to Theorems 3.2.6 and 3.2.8, a point $\bar{x}$ is solution of (P) if and only is there exists $\lambda \geq 0$ such that

$$\begin{cases} \nabla f(\bar{x}) + \lambda \nabla g(\bar{x}) = 0 \\ \lambda g(\bar{x}) = 0. \end{cases}$$

This system can be written as $\bar{x} \in \text{int } M$ (i.e., $g(\bar{x}) < 0$) or $\bar{x} \in \text{bd } M$ (i.e., $g(\bar{x}) = 0$) in one of the following forms:

$$\begin{cases} \bar{x}_1 + \bar{x}_2 - 1 < 0 \\ 2\bar{x}_1 + \bar{x}_2 + 1 = 0 \\ \bar{x}_1 + 2a\bar{x}_2 = 0 \end{cases}$$

and

$$\begin{cases} \overline{x}_1 + \overline{x}_2 - 1 = 0 \\ 2\overline{x}_1 + \overline{x}_2 + 1 + \lambda = 0 \\ \overline{x}_1 + 2a\overline{x}_2 + \lambda = 0 \\ \lambda \geq 0, \end{cases}$$

respectively. The former one admits solution for $a > 3^{-1}$, and this is $\overline{x} = \left(-\frac{2a}{4a-1}, \frac{1}{4a-1}\right)$. The latter one has solution for $a \in (4^{-1}, 3^{-1}]$, and this is $\overline{x} = \left(1 - \frac{1}{a}, \frac{1}{a}\right)$, $\lambda = \frac{1-3a}{a}$. $\square$

Exercise 7.71. *Let $f, g : \mathbb{R}^2 \to \mathbb{R}$, $f(x) = x_1^3 + x_2^2$ and $g(x) = x_1^2 + x_2^2 - 9$. Study the minima of f under the restriction $g(x) \leq 0$.*

Solution We remark from the beginning that the set of feasible points is compact. Since f is continuous, the problem admits a solution. Let us remark that g is convex, since its Hessian matrix is positive definite at every point. But $g(0,0) < 0$ and the Slater condition holds, so a solution $\overline{x}$ of (P) verifies Karush-Kuhn-Tucker conditions that is, there exists $\lambda \geq 0$ such that

$$\begin{cases} \nabla f(\overline{x}) + \lambda \nabla g(\overline{x}) = 0 \\ \lambda g(\overline{x}) = 0. \end{cases}$$

As before, we have two distinct situations and we get the solutions

$$\overline{x} = (0, 0), \lambda = 0 \text{ and } \overline{x} = (-3, 0), \lambda = \frac{9}{2}.$$

But $(0, 0)$ is not a solution, because it is enough to consider the sequences $(x_n) = (\frac{1}{n}, 0) \to (0, 0)$ and $(y_n) = (-\frac{1}{n}, 0) \to (0, 0)$ for which

$$f(y_n) < f(0, 0) < f(x_n), \forall n \in \mathbb{N}^*.$$

The conclusion is that $\overline{x} = (-3, 0)$ is the only solution of the problem. $\square$

Exercise 7.72. *Let A be a symmetric square matrix of dimension p. We consider the application $f : \mathbb{R}^p \to \mathbb{R}$, $f(x) = \left\langle (Ax^t)^t, x \right\rangle$. Solve the problems of minimization and maximization of this function on the unit sphere of $\mathbb{R}^p$.*

Solution Clearly, in both cases there is a solution. We observe that, following Remark 3.2.21, at a point x of the sphere, the Bouligand tangent cone to the sphere is $\{u \in \mathbb{R}^p \mid \langle x, u \rangle = 0\}$, whence the corresponding normal cone is $\{ax \mid x \in \mathbb{R}\}$. Then, from Theorem 3.1.18, a point $\overline{x}$ on the unit sphere is a solution for minimization or maximization problems if there exists $\mu \in \mathbb{R}$ with $(A\overline{x}^t)^t = \mu\overline{x}$, that is $\mu = \left\langle (A\overline{x}^t)^t, \overline{x} \right\rangle$. We deduce that the points which satisfy the optimality conditions correspond to some eigenvectors of A, while the values of f in those points are eigenvalues for f. We conclude that the greater eigenvalue of A is $\lambda_1 := \max_{\|x\|=1} \left\langle (Ax^t)^t, x \right\rangle$, while the smallest is $\lambda_p := \min_{\|x\|=1} \left\langle (Ax^t)^t, x \right\rangle$. Therefore

$$\lambda_1 = \max_{x \in \mathbb{R}^p \setminus \{0\}} \frac{\left\langle (Ax^t)^t, x \right\rangle}{\|x\|^2} \text{ and } \lambda_p = \min_{x \in \mathbb{R}^p \setminus \{0\}} \frac{\left\langle (Ax^t)^t, x \right\rangle}{\|x\|^2}. \qquad \square$$

Exercise 7.73. *Let us consider the function $f : \mathbb{R}^p \to \mathbb{R}$ given by $f(x) = \frac{1}{2} \langle (Qx^t)^t, x \rangle + \langle c, x \rangle$, where Q is a symmetric positive definite square matrix of dimension p. We consider as well the restriction $Ax^t = b^t$ where A is a matrix of dimensions $q \times p$ of rank q, where $1 \leq q \leq p$ and $b \in \mathbb{R}^q$. Solve the problem of the minimization of f subject to the given restriction.*

Solution First of all, we see that f is strictly convex ($\nabla^2 f(x) = Q$ for every $x \in \mathbb{R}^p$), $\lim_{\|x\| \to \infty} f(x) = +\infty$ and the restriction is affine. Therefore, there is a unique minimum point $\bar{x}$ fully characterized by the relations

$$A\bar{x}^t = b^t$$
$$\exists \mu \in \mathbb{R}^q \text{ such that } \nabla f(\bar{x}) + \mu A = 0.$$

So

$$\begin{cases} A\bar{x}^t = b^t \\ (Q\bar{x}^t)^t + c + \mu A = 0, \end{cases}$$

and

$$\begin{cases} A\bar{x}^t = b^t \\ Q\bar{x}^t + c^t + A^t \mu^t = 0. \end{cases}$$

Since Q is invertible, we get

$$\begin{cases} A\bar{x}^t = b^t \\ \bar{x}^t + Q^{-1}c^t + Q^{-1}A^t \mu^t = 0 \end{cases}$$

and, by multiplication by A in the second relation,

$$\begin{cases} A\bar{x}^t = b^t \\ b + AQ^{-1}c^t + AQ^{-1}A^t \mu^t = 0. \end{cases}$$

But $AQ^{-1}A^t$ is a symmetric square matrix of dimension q and for every $y \in \mathbb{R}^q$,

$$\left\langle (AQ^{-1}A^t y^t)^t, y \right\rangle = \left\langle (Q^{-1}A^t y^t)^t, (A^t y^t)^t \right\rangle \geq 0.$$

The equality in the above relation holds if and only if $A^t y^t = 0$. Since the linear function associated with A is surjective (from the rank condition), we infer that the linear function associated to A^t is injective, so $A^t y^t = 0$ if and only if $y = 0$. Therefore $AQ^{-1}A^t$ is positive definite, whence invertible. We obtain

$$\mu^t = -(AQ^{-1}A^t)^{-1}(b + AQ^{-1}c^t)$$

and

$$\bar{x}^t = -Q^{-1}c^t + Q^{-1}A^t(AQ^{-1}A^t)^{-1}(b + AQ^{-1}c^t). \qquad \square$$

Exercise 7.74. *Solve the problem to minimize and to maximize the objective function $f : \mathbb{R}^3 \to \mathbb{R}$, $f(x) = x_1 x_2 x_3$ under the constraint $h(x) = 0$, where $h : \mathbb{R}^3 \to \mathbb{R}$, $h(x) = x_1 x_2 + x_1 x_3 + x_2 x_3 - 8$.*

Solution We remark that these problems do not admit global solutions because, for instance, the points

$$\left(n, n, \frac{8 - n^2}{2n}\right), \left(-n, -n, -\frac{8 - n^2}{2n}\right), \ n \in \mathbb{N}^*$$

are feasible, but

$$\lim_n f\left(n, n, \frac{8 - n^2}{2n}\right) = \lim_n n\frac{8 - n^2}{2} \to -\infty$$

$$\lim_n f\left(-n, -n, -\frac{8 - n^2}{2n}\right) = \lim_n n\frac{-8 + n^2}{2} \to +\infty,$$

whence f is neither lower, nor upper bounded on the feasible set. So, we are looking for local solutions.

It is easy to see that the linear independence qualification condition holds in every feasible point. The application of Theorem 3.2.6 leads to the conclusion that if $\overline{x}$ is solution of one of the problems, then there exists $\mu \in \mathbb{R}$ such that

$$\begin{cases} \overline{x}_2\overline{x}_3 + \mu(\overline{x}_2 + \overline{x}_3) = 0 \\ \overline{x}_1\overline{x}_3 + \mu(\overline{x}_1 + \overline{x}_3) = 0 \\ \overline{x}_1\overline{x}_2 + \mu(\overline{x}_1 + \overline{x}_2) = 0 \\ \overline{x}_1\overline{x}_2 + \overline{x}_1\overline{x}_3 + \overline{x}_2\overline{x}_3 = 8. \end{cases}$$

Solving this system (by appropriate multiplication and subtraction of the equations), we get $\overline{x} = \left(-\frac{2\sqrt{6}}{3}, -\frac{2\sqrt{6}}{3}, -\frac{2\sqrt{6}}{3}\right)$, $\mu = \frac{\sqrt{6}}{2}$ and $\overline{x} = \left(\frac{2\sqrt{6}}{3}, \frac{2\sqrt{6}}{3}, \frac{2\sqrt{6}}{3}\right)$, $\mu = -\frac{\sqrt{6}}{2}$. We need to verify second-order optimality conditions. The set of critical directions in both cases is $\{u \in \mathbb{R}^3 \mid u_1 + u_2 + u_3 = 0\}$. For the first point, $\nabla^2_{xx}L(\overline{x}, \mu)(u, u) = \frac{-\sqrt{6}}{3}(u_1u_2 + u_1u_3 + u_2u_3)$. Since

$$0 = (u_1 + u_2 + u_3)^2 = u_1^2 + u_2^2 + u_3^2 + 2(u_1u_2 + u_1u_3 + u_2u_3)$$

we deduce that $u_1u_2 + u_1u_3 + u_2u_3 < 0$ for any nonzero critical direction, i.e., $\nabla^2_{xx}L(\overline{x}, \mu)(u, u) > 0$ for such a direction, whence the sufficient optimality condition in Theorem 3.3.3 do hold, and therefore the reference point is a local (strict) minimum. The same holds for the second point and the corresponding sufficient condition for maximization. Hence the second point is the solution of the maximization problem.□

Exercise 7.75. *The Karush-Kuhn-Tucker conditions tell us that a solution of (P) is a critical point with respect to x for the Lagrangian function.*

(i) Show that a solution of (P) is not necessarily a minimum point of $x \mapsto L(x, (\lambda, \mu))$. For this, consider $f : \mathbb{R}^2 \to \mathbb{R}$, $f(x_1, x_2) = x_1^2 - x_2^2 - 3x_2$ and $h : \mathbb{R}^2 \to \mathbb{R}$, $h(x_1, x_2) = x_2$.

(ii) Show that there exist as well situations when a solution of (P) is a minimum point of $x \mapsto L(x, (\lambda, \mu))$. For this, consider $f : \mathbb{R}^2 \to \mathbb{R}$, $f(x_1, x_2) = 5x_1^2 + 4x_1x_2 + x_2^2$ and $h : \mathbb{R}^2 \to \mathbb{R}$, $h(x_1, x_2) = 3x_1 + 2x_2 + 5$.

Solution (i) It is clear that $(0, 0)$ is a minimum point of f under linear restriction $h(x_1, x_2) = 0$. Implementing Karush-Kuhn-Tucker conditions, we get the multiplier $\mu = 3$, but $(0, 0)$ is not a minimum point for $f(x_1, x_2) + 3h(x_1, x_2) = x_1^2 - x_2^2$.

(ii) Clearly, $f(x_1, x_2) = (2x_1 + x_2)^2 + x_1^2$ satisfies the conditions of Proposition 3.1.8, whence the problem has solutions. Since the restriction $h(x) = 0$ is affine, we can apply Theorem 3.2.6 and we get the linear system

$$\begin{cases} 10x_1 + 4x_2 + 3\mu = 0 \\ 4x_1 + 2x_2 + 2\mu = 0 \\ 3x_1 + 2x_2 + 5 = 0, \end{cases}$$

which has the solution $(x_1, x_2, \mu) = (1, -4, 2)$. Then Lagrangian function in (x_1, x_2) is $l : \mathbb{R}^2 \to \mathbb{R}$,

$$l(x_1, x_2) = 5x_1^2 + 4x_1x_2 + x_2^2 + 6x_1 + 4x_2 + 10$$
$$= (2x_1 + x_2 + 2)^2 + (x_1 - 1)^2 + 5,$$

and $(1, -4)$ is a global minimum without restrictions. □

Exercise 7.76. *Find the local minima of the function $f(x_1, x_2) = x_1^2 + (x_2 - 1)^2$ under the constraints $x_2 \le x_1^2$ and $x_2 \le x_1$.*

Solution Consider the functions $g_1(x_1, x_2) = x_2 - x_1^2$, $g_2(x_1, x_2) = x_2 - x_1$. The the constraints set is

$$M = \left\{ (x_1, x_2) \in \mathbb{R}^2 \mid g_i(x_1, x_2) \le 0, \ i = 1, 2 \right\}.$$

By considering the gradients of the functions g_i, equal to $(-2x, 1)$, $(-1, 1)$, respectively, observe that the linear independence condition is not satisfied (for $x = \frac{1}{2}$). To find the minima, we write then the Fritz John necessary conditions (i.e., Theorem 3.2.1): we must find the scalars $\lambda_0, \lambda_1, \lambda_2 \ge 0$, not all 0, such that

$$\begin{cases} \lambda_0 \nabla f(x_1, x_2) + \lambda_1 \nabla g_1(x_1, x_2) + \lambda_2 \nabla g_2(x_1, x_2) = 0 \\ \lambda_1 \cdot g_1(x_1, x_2) = 0, \ \lambda_2 \cdot g_2(x_1, x_2) = 0, \\ g_1(x_1, x_2) \le 0, g_2(x_1, x_2) \le 0 \end{cases}$$

hence we arrive at

$$\begin{cases} 2\lambda_0 x_1 - 2\lambda_1 x_1 + \lambda_2 = 0 \\ 2\lambda_0(x_2 - 1) + \lambda_1 + \lambda_2 = 0 \\ \lambda_1(x_2 - x_1^2) = 0 \\ \lambda_2(x_2 - x_1) = 0 \\ x_2 - x_1^2 \le 0, \lambda_1 \ge 0 \\ x_2 - x_1 \le 0, \lambda_2 \ge 0. \end{cases} \tag{7.3.1}$$

Suppose $\lambda_0 = 0$. Then we get from the second relation $\lambda_1 + \lambda_2 = 0$, and since $\lambda_1 \ge 0, \lambda_2 \ge 0$, it follows $\lambda_1 = \lambda_2 = 0$. We obtain the contradiction that not all scalars are 0, hence $\lambda_0 > 0$, and we may suppose without loss of generality $\lambda_0 = 1$.

Consider then the Lagrangian

$$L(x, x_2; \mu_1, \mu_2) = f(x_1, x_2) + \lambda_1 g_1(x_1, x_2) + \lambda_2 g_2(x_1, x_2)$$
$$= x_1^2 + (x_2 - 1)^2 + \lambda_1(x_2 - x_1^2) + \lambda_2(x_2 - x_1).$$

The Karush-Kuhn-Tucker necessary conditions (see Theorem 3.2.6) will be then

$$\begin{cases} 2x_1 - 2\lambda_1 x_1 + \lambda_2 = 0 \\ 2(x_2 - 1) + \lambda_1 + \lambda_2 = 0 \\ \lambda_1(x_2 - x_1^2) = 0 \\ \lambda_2(x_2 - x_1) = 0 \\ x_2 - x_1^2 \leq 0, \lambda_1 \geq 0 \\ x_2 - x_1 \leq 0, \lambda_2 \geq 0. \end{cases} \tag{7.3.2}$$

For $\lambda_1 = \lambda_2 = 0$, we get from the first two relations the solution $(0, 1)$, which does not satisfy $x_2 - x_1 \leq 0$.

For $\lambda_1, \lambda_2 > 0$, we have $x_2 - x_1^2 = 0$ and $x_2 - x_1 = 0$, with solutions $(0, 0)$ and $(1, 1)$. For $(0, 0)$, we obtain from the first relation $\lambda_2 = 0$, and then $\lambda_1 = 2$. For $(1, 1)$, we get from the second relation $\lambda_1 + \lambda_2 = 0$, which is impossible because $\lambda_1, \lambda_2 > 0$.

For $\lambda_1 = 0, \lambda_2 > 0$, we must solve the system

$$2x_1 + \lambda_2 = 0$$
$$2(x_2 - 1) + \lambda_2 = 0$$
$$x_2 - x_1 = 0.$$

Subtracting the first two relations and adding the third, one gets the contradiction $-1 = 0$.

For $\lambda_1 > 0, \lambda_2 = 0$, we must solve the system

$$x - \lambda_1 x_1 = 0$$
$$2(x_2 - 1) + \lambda_1 = 0$$
$$x_2 - x_1^2 = 0.$$

From the first relation, if $x_1 = 0$, one gets $x_2 = 0$ and $\lambda_1 = 2$, that is, one obtains the above solution. Suppose $x_1 \neq 0$. Then $\lambda_1 = 1$, and $x_2 = \frac{1}{2}$, which gives $x_1 = \pm\frac{1}{\sqrt{2}}$. From the obtained solutions, the only feasible one is $\left(\frac{1}{\sqrt{2}}, \frac{1}{2}\right)$.

For concluding, we have got the following Karush-Kuhn-Tucker points and associated multipliers:

$$a = (0, 0), \quad \lambda_a = (2, 0)$$
$$b = \left(\frac{1}{\sqrt{2}}, \frac{1}{2}\right), \quad \lambda_b = (1, 0).$$

We calculate $\nabla^2_{xx}L(x_1, x_2, \lambda_1, \lambda_2)$. Because

$$\frac{\partial^2 L}{\partial x_1^2}(x_1, x_2, \lambda_1, \lambda_2) = 2 - 2\lambda_1, \quad \frac{\partial^2 L}{\partial x_2^2}(x_1, x_2, \lambda_1, \lambda_2) = 2, \quad \frac{\partial^2 L}{\partial x_1 \partial x_2}(x_1, x_2, \lambda_1, \lambda_2) = 0,$$

one gets

$$\nabla^2_{xx}L(x_1, x_2, \lambda_1, \lambda_2)(h_1, h_2) = (2 - 2\lambda_1)h_1^2 + 2h_2^2.$$

Using Theorem 3.3.3, we must verify that

$$\nabla^2_{xx}L(a, \lambda_a)(h_1, h_2) = -2h_1^2 + 2h_2^2$$

is positive definite on the linear subspace which is orthogonal to the gradients of those restrictions which are active at a, for which the associated multiplier is strictly positive.

At a, both restrictions are active, but just $\lambda_1 > 0$ (i.e., $A(a) = \{1\}$). We have $\nabla g_1(a) = (0, 1)$, so we will consider the orthogonal subspace to this vector, that is

$$\left\{(u, v) \in \mathbb{R}^2 \mid \langle \nabla g_1(a), (u, v) \rangle = 0 \right\} = \{(u, v) \mid v = 0\}$$
$$= \{(u, 0) \mid u \in \mathbb{R}\}.$$

Since

$$\nabla^2_{xx}L(a, \lambda_a)(u, 0) = -2u^2$$

is not positive definite, we cannot apply Theorem 3.3.3 to decide if a is a local minimum.

Observe, though, that in a both restrictions are active, and the gradients $(0, 1)$, $(-1, 1)$ are linearly independent. We can apply then Theorem 3.3.2 for $A(a) = \{1\}$, and observe that the necessary optimality condition is not satisfied, hence a is not a local minimum.

For b, we have

$$\nabla^2_{xx}L(b, \lambda_b)(h_1, h_2) = 2h_2^2,$$

and again the first restriction is active. We consider the orthogonal subspace on $\nabla g_1(b) = \left(-\sqrt{2}, 1\right)$, that is

$$\left\{(u, v) \in \mathbb{R}^2 \mid -\sqrt{2}u + v = 0\right\} = \left\{(u, \sqrt{2}u) \mid u \in \mathbb{R}\right\}.$$

$\nabla^2_{xx}L(b, \lambda_b)$ is positive definite on this subspace, hence b is a local minimum.

We end by mentioning that in $(0, 0)$ the restriction g_2 is active, and the associated multiplier is 0. $\qquad \square$

Problem 7.77. *Let $u \in \mathbb{R}^p \setminus \{0_{\mathbb{R}^p}\}$ and $a \in \mathbb{R}$. We consider the set (called hyperplane)*

$$M := \{x \in \mathbb{R}^p \mid \langle u, x \rangle = a\}$$

and a point $v \in \mathbb{R}^p \setminus M$. Show that M is a convex, closed set, and determine the explicit expression of the projection of v on M, as well as the value of the distance from v to M.

Solution The convexity and the closedness of M are obvious. So, from Theorem 2.1.5, there exists a unique projection of v on M, which we denote here by v_a.

Then v_a is the unique solution (see Theorem 2.1.5) of the problem of minimizing the function

$$f(x) = \frac{1}{2} \|x - v\|^2 ,$$

for $x \in M$. The choice of the objective function above has the same motivation as in the case of the least squares method. Since the constraint (in functional interpretation) is affine and f is convex, the element v_a is characterized by

$$\begin{cases} \langle u, v_a \rangle = a \\ \exists \mu \in \mathbb{R}, \nabla f(v_a) + \mu u = 0. \end{cases}$$

Then, by the differentiation of f,

$$\begin{cases} \langle u, v_a \rangle = a \\ v_a - v + \mu u = 0. \end{cases}$$

The second relation gives

$$\langle v_a - v, u \rangle + \mu \|u\|^2 = 0,$$

so

$$\mu = \frac{\langle u, v \rangle - a}{\|u\|^2}.$$

Therefore, we finally get

$$v_a = v - \frac{\langle u, v \rangle - a}{\|u\|^2} u.$$

To end, the distance from v to M has the value.

$$\|v - v_a\| = \frac{|\langle u, v \rangle - a|}{\|u\|}. \qquad \square$$

Problem 7.78. *(i) Consider a function $f : \mathbb{R} \to \mathbb{R}$ of class C^2, and let $\bar{x}$ be a simple root of f (i.e. $f(\bar{x}) = 0$, $f'(\bar{x}) \neq 0$). Deduce the algorithm of Newton's method from the Banach Principle, applied to an appropriate contraction.*

(ii) Show that, if the root is not simple, then the order of the convergence is not quadratic. In the case that the order of the root is known, appropriately modify the iterations such that a quadratic convergence holds.

Solution (i) Recall that the Newton iterations are given by:

$$x_{k+1} = x_k - \frac{f(x_k)}{f'(x_k)}.$$

Let put now this into a rigorous perspective. Let $L \in (0, 1)$. Let us denote by V a closed interval centered at $\bar{x}$ for which $f'(x) \neq 0$ for any $x \in V$ and

$$\left| \frac{f(x)f''(x)}{f'(x)^2} \right| < L, \quad \forall x \in V. \tag{7.3.3}$$

The above relation is possible exactly because $f(\overline{x}) = 0$. Take now the function $g : V \to \mathbb{R}$

$$g(x) = x - \frac{f(x)}{f'(x)}.$$

Clearly, this function is well defined on V and since

$$g'(x) = 1 - \frac{f'(x)^2 - f(x)f''(x)}{f'(x)^2} = \frac{f(x)f''(x)}{f'(x)^2},$$

from the choice of V, we deduce that g is a contraction. On the other hand, $g(\overline{x}) = \overline{x}$, so, in particular,

$$\left|g(x) - \overline{x}\right| \le L\left|x - \overline{x}\right|, \ \forall x \in V$$

which means that g applies V in V. Now, if we start with $x_0 \in V$, let us observe that the Newton iteration is in fact a Picard iteration associated to g. On the basis of the theory previously developed, the Newton iteration converges (for any initial date $x_0 \in V$) to the fixed point of which is exactly the root of f in V, that is $\overline{x}$. As shown before,

$$g'(\overline{x}) = 0$$

and we can apply Remark 6.1.1 in order to deduce that one has a quadratic convergence, that is

$$\lim_k \frac{x_{k+1} - \overline{x}}{(x_k - \overline{x})^2} = \lim_{x \to \overline{x}} \frac{g(x) - \overline{x}}{(x - \overline{x})^2}$$

$$= \lim_{x \to \overline{x}} \frac{1}{f'(x)} \frac{xf'(x) - f(x) - \overline{x}f'(x)}{(x - \overline{x})^2} = \frac{f''(\overline{x})}{2f'(\overline{x})} \in \mathbb{R}.$$

(ii) The above discussion emphasizes that in order to be sure that the Newton iteration converges quadratically to the underlying root, it is necessary that x_0 to be in a neighborhood V of the solution where the derivative do not vanish and condition (7.3.3) takes place. Therefore, for functions with several roots, depending on initial data, we can find different solutions.

Let us suppose that f is as $f(x) = (x - \overline{x})^q u(x)$, where $q > 1$ is a natural number, u is of class C^2 and $u(\overline{x}) \ne 0$. Then g is

$$g(x) = x - \frac{(x - \overline{x})u(x)}{qu(x) + (x - \overline{x})u'(x)}$$

and, after calculation,

$$g'(x) = \frac{\left(1 - \frac{1}{q}\right) + (x - \overline{x})\frac{2u'(x)}{qu(x)} + (x - \overline{x})^2 \frac{u''(x)}{q^2 u(x)}}{\left[1 + (x - \overline{x})\frac{u'(x)}{qu(x)}\right]^2}.$$

For values close enough to $\overline{x}$, $\left|g'(x)\right| < 1$, so the Picard iterations converge to the fixed point of g which is exactly $\overline{x}$ (in the neighborhood one considers). On the other hand,

$g'(\overline{x}) = 1 - \frac{1}{q} \neq 0$, so the convergence is only linear. Therefore, the Newton procedure converges quadratically only for simple roots. If for a given root $\overline{x}$ the order of multiplicity q is known, then one can consider the function

$$g_q(x) = x - q\frac{f(x)}{f'(x)} = x - q\frac{(x - \overline{x})u(x)}{qu(x) + (x - \overline{x})u'(x)},$$

and a similar calculation shows that $g_q'(\overline{x}) = 0$, so, by means of the above reasonings, one gets again the quadratic convergence. $\qquad\square$

7.4 Nonsmooth Optimization

Problem 7.79. *Let $g, h : \mathbb{R}^p \to \mathbb{R}$ be convex functions, and g of class C^1. We denote $f : \mathbb{R}^p \to \mathbb{R}, f = g + h$. The next assertions are equivalent:*

    (i) $\overline{x}$ is a minimum point for f;

    (ii) for every $x \in \mathbb{R}^p$,

$$\nabla g(\overline{x})(x - \overline{x}) + h(x) - h(\overline{x}) \geq 0;$$

    (iii) for every $x \in \mathbb{R}^p$,

$$\nabla g(x)(x - \overline{x}) + h(x) - h(\overline{x}) \geq 0.$$

Solution $(i) \Rightarrow (ii)$ Let $\overline{x}$ be a minimum point of f on $\mathbb{R}^p$, let $x \in \mathbb{R}^p$ and $\lambda \in (0, 1)$. Then

$$f(\overline{x}) \leq f(\lambda x + (1 - \lambda)\overline{x}),$$

that is,

$$g(\overline{x}) + h(\overline{x}) \leq g(\lambda x + (1 - \lambda)\overline{x}) + h(\lambda x + (1 - \lambda)\overline{x})$$
$$\leq g(\overline{x} + \lambda(x - \overline{x})) + \lambda h(x) + (1 - \lambda)h(\overline{x}).$$

We obtain

$$0 \leq g(\overline{x} + \lambda(x - \overline{x})) - g(\overline{x}) + \lambda(h(x) - h(\overline{x})),$$

from where, if we divide by λ,

$$0 \leq \frac{g(\overline{x} + \lambda(x - \overline{x})) - g(\overline{x})}{\lambda} + h(x) - h(\overline{x}).$$

Making $\lambda \to 0$, we get $0 \leq \nabla g(\overline{x})(x - \overline{x}) + h(x) - h(\overline{x})$.

    $(ii) \Rightarrow (i)$ The convexity of g gives

$$g(x) \geq g(\overline{x}) + \nabla g(\overline{x})(x - \overline{x}),$$

which together with the hypothesis leads us to

$$g(x) + h(x) \geq g(\overline{x}) + h(\overline{x}).$$

$(ii) \Rightarrow (iii)$ Again, the convexity of g allows us to write

$$\left(\nabla g(x) - \nabla g(\bar{x})\right)(x - \bar{x}) \geq 0, \ \forall x \in \mathbb{R}^p,$$

which together with the hypothesis leads to the conclusion.

$(iii) \Rightarrow (ii)$ Let $\bar{x} \in \mathbb{R}^p$ with

$$\nabla g(x)(x - \bar{x}) + h(x) - h(\bar{x}) \geq 0, \ \forall x \in \mathbb{R}^p.$$

Let $x \in \mathbb{R}^p$ and $\lambda \in (0, 1)$. Then

$$\lambda \nabla g((1 - \lambda)\bar{x} + \lambda x)(x - \bar{x}) + h((1 - \lambda)\bar{x} + \lambda x) - h(\bar{x}) \geq 0,$$

and using the convexity of h,

$$\lambda \nabla g((1 - \lambda)\bar{x} + \lambda x)(x - \bar{x}) + \lambda \left[h(x) - h(\bar{x})\right] \geq 0,$$

that is

$$\nabla g((1 - \lambda)\bar{x} + \lambda x)(x - \bar{x}) + h(x) - h(\bar{x}) \geq 0.$$

But ∇g is a continuous application, so, for $\lambda \to 0$, we get the conclusion. $\square$

Problem 7.80. *Show that every sublinear function is convex. Show that for any $x \in \mathbb{R}^p$,*

$$\partial f(x) = \{u \in \partial f(0) \mid \langle x, u \rangle = f(x)\}.$$

Solution The first part is obvious by direct calculation: for any $\alpha \in (0, 1)$ and any $x, x_2 \in \mathbb{R}^p$, the sublinearity of f yields

$$f(\alpha x + (1 - \alpha)y) \leq f(\alpha x) + f((1 - \alpha)y) = \alpha f(x) + (1 - \alpha)f(y),$$

so f is convex.

Let us prove the second affirmation. Before that, notice that $f(0) = 0$. Take $u \in \partial f(0)$ and $x \in \mathbb{R}^p$ with $\langle x, u \rangle = f(x)$. Then for every $y \in \mathbb{R}^p$, $\langle y, u \rangle \leq f(y)$, whence

$$\langle y - x, u \rangle = \langle y, u \rangle - \langle x, u \rangle \leq f(y) - f(x).$$

This shows that $u \in \partial f(x)$. Conversely, consider $x \in \mathbb{R}^p$ and suppose that $u \in \partial f(x)$. So,

$$\langle y - x, u \rangle \leq f(y) - f(x), \ \forall y \in \mathbb{R}^p$$

which for $y = 0$ gives $\langle x, u \rangle \geq f(x)$ and for $y = x + tz$ with $t > 0$ and $z \in \mathbb{R}^p$ gives

$$t \langle z, u \rangle \leq f(x + tz) - f(x) \leq f(x) + tf(z) - f(x) = tf(z).$$

So, for every $z \in \mathbb{R}^p$, $\langle z, u - 0 \rangle \leq f(z) - f(0)$, i.e., $u \in \partial f(0)$. In particular, we also get $\langle x, u \rangle \leq f(x)$, so, in fact, coupled with the opposite inequality proved before, it follows that $\langle x, u \rangle = f(x)$. $\square$

Problem 7.81. *Find the subdifferential of the norm.*

Solution The norm is a differentiable function away from 0. So, for any $x \in \mathbb{R}^p \setminus \{0\}$,

$$\partial \|\cdot\| (x) = \{\nabla \|\cdot\| (x)\} = \{(x_1 \cdot \|x\|^{-1}, x_2 \cdot \|x\|^{-1}, \ldots, x_p \cdot \|x\|^{-1})\}.$$

Let us find the subdifferential of the norm at 0. Following the definition,

$$u \in \partial \|\cdot\| (0) \Leftrightarrow \langle u, x \rangle \leq \|x\|, \ \forall x \in \mathbb{R}^p$$
$$\Leftrightarrow \langle u, x \rangle \leq 1, \ \forall x \in S(0, 1) \Leftrightarrow |\langle u, x \rangle| \leq 1, \ \forall x \in S(0, 1) \Leftrightarrow \|u\| \leq 1.$$

Hence, $\partial \|\cdot\| (0) = D(0, 1)$. Notice that the norm is a sublinear function, so $\partial \|\cdot\| (x)$ at $x \in \mathbb{R}^p \setminus \{0\}$ can be deduced as well from Problem 7.80.

Observe as well that, due to the convexity of the norm, one could also apply Example 5.1.5. $\qquad \square$

Problem 7.82. *Let $K \subset \mathbb{R}^p$ be a closed convex cone with nonempty interior and let $e \in \mathrm{int}\, K$. Consider the function $s_e : \mathbb{R}^p \to \mathbb{R}$ given by*

$$s_e(x) := \inf\{t \in \mathbb{R} \mid x \in te - K\}. \tag{7.4.1}$$

Show that:

(i) s_e is well defined, i.e., the set in the right-hand side cannot be $\emptyset$ and the infimum cannot be $-\infty$.

(ii) for every $\lambda \in \mathbb{R}$ and $v \in \mathbb{R}^p$ one has

$$\{x \in \mathbb{R}^p \mid s_e(x) \leq \lambda\} = \lambda e - K, \tag{7.4.2}$$

$$\{x \in \mathbb{R}^p \mid s_e(x) < \lambda\} = \lambda e - \mathrm{int}\, K, \tag{7.4.3}$$

$$\{x \in \mathbb{R}^p \mid s_e(x) = \lambda\} = \lambda e - \mathrm{bd}\, K, \tag{7.4.4}$$

and

$$s_e(v + \lambda e) = s_e(v) + \lambda; \tag{7.4.5}$$

(iii) s_e is sublinear;

(iv) s_e is strictly $\mathrm{int}\, K$–monotone, i.e., for all $x_1, x_2 \in \mathbb{R}^p$ with $x_2 - x_1 \in \mathrm{int}\, K$, one has $s_e(x_1) < s_e(x_2)$.

(v) the subdifferential of s_e at a point $x \in \mathbb{R}^p$ is

$$\partial s_e(x) = \{u \in -K^- \mid u(e) = 1, \langle u, x \rangle = s_e(x)\}.$$

Solution (i) The first item follows from Problem 7.30 *(iii)* and *(iv)*.

(ii) Fix $\lambda \in \mathbb{R}$. The inclusion

$$\lambda e - K \subset \{x \in \mathbb{R}^p \mid s_e(x) \leq \lambda\}$$

follows directly from the definition of s_e. Take now an element x from the second term of the above inclusion. Then for every $n \geq 1$, one has that $s_e(x) < \lambda + n^{-1}$, therefore there exists $t_n \in \mathbb{R}$ such that $t_n < \lambda + n^{-1}$ and $x \in t_n e - K$. Then, taking into account Problem 7.30 (i),

$$x \in \left(\lambda + n^{-1}\right) e + \left(t_n - \lambda - n^{-1}\right) e - K$$
$$\subset \left(\lambda + n^{-1}\right) e - K, \ \forall n \in \mathbb{N}^*.$$

Making $n \to \infty$, the closedness of K yields the conclusion. So, we have proved that

$$\lambda e - K = \{x \in \mathbb{R}^p \mid s_e(x) \leq \lambda\}.$$

Notice that, in particular, a similar type of argument show that the infimum in definition of s_e is actually attained, i.e., for any $x \in \mathbb{R}^p$, $x \in s_e(x)e - K$. We prove now the relation concerning the strict level sets. To this end, take first $x \in \lambda e - \operatorname{int} K$. Then there exists $\varepsilon > 0$ such that $x \in \lambda e - \varepsilon e - \operatorname{int} K$, whence $s_e(x) \leq \lambda - \varepsilon < \lambda$. Take now $\lambda \in \mathbb{R}$ and $x \in \mathbb{R}^p$ with $s_e(x) < \lambda$. Then, by use of Problem 7.30 (ii),

$$x \in s_e(x)e - K \subset \lambda e - (\lambda - s_e(y))e - K$$
$$\subset \lambda e - \operatorname{int} K,$$

and the converse inclusion follows. Then (7.4.3) holds. Now, (7.4.2) shows that s_e is lower semicontinuous, while (7.4.3) shows that it is upper semicontinuous. Then s_e is continuous. Again, using in conjunction (7.4.2) and (7.4.3), one obtains (7.4.4).

Let us prove the last equality of (ii). Fix $\lambda \in \mathbb{R}$ and $v \in \mathbb{R}^p$. There exists a sequence (t_n) such that $t_n \to s_e(v + \lambda e)$ and for every $n \in \mathbb{N}^*$,

$$v + \lambda e \in t_n e - K.$$

This means that

$$v \in (t_n - \lambda)e - K$$

and from (7.4.2), it follows that

$$s_e(v) \leq t_n - \lambda.$$

Passing to the limit as $n \to \infty$, one gets $s_e(v) + \lambda \leq s_e(v + \lambda e)$. Conversely, one has from (7.4.2) that

$$v \in s_e(v)e - K,$$

i.e.,

$$v + \lambda e \in (s_e(v) + \lambda)e - K,$$

whence

$$s_e(v + \lambda e) \leq s_e(v) + \lambda.$$

This final inequality completes the proof of (ii).

(*iii*) The relation (7.4.2) equally shows that

$$\text{epi}\, s_e = \{(x, t) \in \mathbb{R}^p \times \mathbb{R} \mid x \in te - K\}, \tag{7.4.6}$$

and this is clearly a closed convex cone. So, s_e is sublinear (see Problem 7.31).

(*iv*) Take $x_1, x_2 \in \mathbb{R}^p$ with $x_2 - x_1 \in \text{int}\, K$. Then

$$y_2 \in s_e(y_2)e - K,$$

whence

$$y_1 \in y_2 - \text{int}\, K \subset s_e(y_2)e - K - \text{int}\, K$$
$$\subset s_e(y_2)e - \text{int}\, K.$$

Therefore, $s_e(y_1) < s_e(y_2)$ and the conclusion follows.

(*v*) Taking into account Problem 7.80, it is enough to prove that the subdifferential of s_e at 0 is

$$\partial s_e(0) = \{u \in -K^- \mid u(e) = 1\}. \tag{7.4.7}$$

Notice that $s_e(0) = 0$, so an element $u \in \mathbb{R}^p$ is in $\partial s_e(0)$ if and only if

$$s_e(y) \geq \langle u, y \rangle, \ \forall x \in \mathbb{R}^p.$$

This means that for all $y \in \mathbb{R}^p$ and for all $\lambda \in \mathbb{R}$ with $\lambda \geq s_e(y)$, one has $\lambda \geq \langle u, y \rangle$. Consequently, taking into account (7.4.2), for all $y \in \lambda e - K$ one has $\lambda \geq \langle u, y \rangle$, i.e.,

$$\lambda \geq \lambda \langle u, e \rangle - \langle u, k \rangle, \ \forall k \in K.$$

Since the above inequality holds for all λ, making $k = 0$, one deduces that $\langle u, e \rangle = 1$. On the other hand, $\langle u, k \rangle \geq 0$ for all $k \in K$, whence $u \in -K^-$. The first inclusion in relation (7.4.7) is proved. For the converse, take $u \in -K^-$ such that $\langle u, e \rangle = 1$. Fix $y \in \mathbb{R}^p$ and take $\lambda \geq s_e(y)$. Then there exists $k \in K$ such that $y = \lambda e - k$. Accordingly,

$$\langle u, y \rangle = \lambda \langle u, e \rangle - \langle u, k \rangle \leq \lambda.$$

Since $\lambda \geq s_e(y)$ was arbitrarily chosen, one has

$$\langle u, y \rangle \leq s_e(y), \ \forall y \in \mathbb{R}^p,$$

i.e., $u \in \partial s_e(0)$. $\qquad\qquad\qquad\qquad\qquad\qquad\qquad\qquad\qquad\qquad\qquad\square$

Remark 7.4.1. *The mapping studied in the previous problem is the Gerstewitz (Tammer) scalarization functional which successfully used in optimization problems with vector-valued functions. For more details, the reader is invited to consult (Göpfert et al., 2003).*

Up until now, we have avoided speaking about functions which can take $+\infty$ as a value. For the end of the discussion about convex functions we shall say something about this subject. We consider a function with real-extended values $f : \mathbb{R}^p \to \mathbb{R} \cup \{+\infty\}$. For such a function the domain is

$$\operatorname{dom} f := \{x \in \mathbb{R}^p \mid f(x) < +\infty\}.$$

The definition of convexity for f is formally the usual one (for any $x, y \in \mathbb{R}^p$ and $\alpha \in [0, 1]$, $f(\alpha x + (1 - \alpha)y) \le \alpha f(x) + (1 - \alpha)f(y)$) with the convention $+\infty + r = +\infty$ and $s(+\infty) = +\infty$ for any $r \in \mathbb{R} \cup \{+\infty\}$ and $s \in [0, \infty]$. It is clear that $x, y \in \operatorname{dom} f$ implies that $\alpha x + (1 - \alpha)y \in \operatorname{dom} f$, so $\operatorname{dom} f$ is a convex set. Moreover, it is enough to have the inequality in the definition of convexity fulfilled only for $x, y \in \operatorname{dom} f$. Many of the results given for convex functions with real values are still valid for functions with real-extended values. It is a useful exercise for the reader to verify how this adaptation can be made.

Problem 7.83. *Let $f, g : \mathbb{R}^p \to \mathbb{R} \cup \{+\infty\}$ be convex functions. Suppose that the function $f \square g : \mathbb{R}^p \to \mathbb{R} \cup \{+\infty\}$,*

$$(f \square g)(x) := \inf\{f(x - y) + g(y) \mid y \in \mathbb{R}^p\}$$

(called the convolution of f and g) is well defined (i.e., the infimum is not $-\infty$). Show that $\operatorname{dom} f \square g = \operatorname{dom} f + \operatorname{dom} g$, $f \square g = g \square f$, and $f \square g$ is convex.

Solution Take $x \in \operatorname{dom} f \square g$, that is $\inf\{f(x - y) + g(y) \mid y \in \mathbb{R}^p\} < +\infty$. Then there exists $y \in \mathbb{R}^p$ such that $f(x - y), g(y) \in \mathbb{R}$, so $x = (x - y) + y \in \operatorname{dom} f + \operatorname{dom} g$. Conversely, take $x \in \operatorname{dom} f + \operatorname{dom} g$. Then $x = x_1 + x_2$ with $x_1 \in \operatorname{dom} f$ and $x_2 \in \operatorname{dom} g$. We have

$$(f \square g)(x) = \inf\{f(x - y) + g(y) \mid y \in \mathbb{R}^p\} \le f(x_1) + f(x_2) \in \mathbb{R},$$

so $x \in \operatorname{dom} f \square g$.

The equality $f \square g = g \square f$ is obvious. We show that $f \square g$ is convex. Consider $x_1, x_2 \in \operatorname{dom} f \square g$ and $\alpha \in (0, 1)$. Take $t_1, t_2 \in \mathbb{R}$ with $t_1 > (f \square g)(x_1)$ and $t_2 > (f \square g)(x_2)$. Then there exist $y_1, y_2 \in \mathbb{R}^p$ with

$$f(x_1 - y_1) + g(y_1) < t_1$$
$$f(x_2 - y_2) + g(y_2) < t_2.$$

Then

$$
\begin{aligned}
(f \square g)(\alpha x_1 + (1 - \alpha)x_2) &\le f(\alpha x_1 + (1 - \alpha)x_2 - \alpha y_1 - (1 - \alpha)y_2) + g(\alpha y_1 + (1 - \alpha)y_2) \\
&= f(\alpha(x_1 - y_1) + (1 - \alpha)(x_2 - y_2)) + g(\alpha y_1 + (1 - \alpha)y_2) \\
&\le \alpha f(x_1 - y_1) + (1 - \alpha)f(x_2 - y_2) + \alpha g(y_1) + (1 - \alpha)g(y_2) \\
&= \alpha(f(x_1 - y_1) + g(y_1)) + (1 - \alpha)(f(x_2 - y_2) + g(y_2))
\end{aligned}
$$

$$< \alpha t_1 + (1 - \alpha)t_2.$$

Since t_1, t_2 are chosen arbitrarily such that $t_1 > (f \Box g)(x_1)$ and $t_2 > (f \Box g)(x_2)$, we infer that

$$(f \Box g)(\alpha x_1 + (1 - \alpha)x_2) \le \alpha f \Box g)(x_1) + (1 - \alpha)(f \Box g)(x_2).$$

So, $f \Box g$ is convex. $\square$

Problem 7.84. *If $f : \mathbb{R} \to \mathbb{R}$ is strictly convex and differentiable, compute $f \Box f$.*

Solution To compute a convolution means in fact to solve an optimization problem (in order to determine the infimum from the definition of the convolution). In our specific case, consider, for fixed x, $h : \mathbb{R} \to \mathbb{R}$, $h(y) = f(x - y) + f(y)$. The derivative of h is $h'(y) = -f'(x - y) + f'(y)$. Since f is strictly convex, its derivative is injective (being strictly increasing), so $\bar{y} = 2^{-1}x$ is the only critical point. A study of the monotonicity of h reveals that $\bar{y}$ is a global minimum point for h, so, $(f \Box f)(x) = f(x - 2^{-1}x) + f(2^{-1}x) = 2f(2^{-1}x)$. $\square$

Problem 7.85. *Let $A \subset \mathbb{R}^p$, $A \ne \mathbb{R}^p$ be a nonempty convex set. Show that the function $\mu : \mathbb{R}^p \to \mathbb{R} \cup \{+\infty\}$ given by*

$$\mu(y) := \begin{cases} -d_{\mathbb{R}^p \setminus A}(y), & y \in A \\ +\infty, & y \notin A. \end{cases}$$

is convex.

Solution Take $y_1, y_2 \in A = \operatorname{dom} \mu$ and observe that $D(y_1, d_{\mathbb{R}^p \setminus A}(y_1)) \subset \operatorname{cl} A$, and similarly for y_2. Then,

$$B := \operatorname{conv} \left[D(y_1, d_{\mathbb{R}^p \setminus A}(y_1)) \cup D(y_2, d_{\mathbb{R}^p \setminus A}(y_2)) \right] \subset \operatorname{cl} A.$$

For every $\lambda \in [0, 1]$,

$$D(\lambda y_1 + (1 - \lambda)y_2, \lambda d_{\mathbb{R}^p \setminus A}(y_1) + (1 - \lambda)d_{\mathbb{R}^p \setminus A}(y_2)) \subset B,$$

whence

$$d_{\mathbb{R}^p \setminus A}(\lambda y_1 + (1 - \lambda)y_2) \ge \lambda d_{\mathbb{R}^p \setminus A}(y_1) + (1 - \lambda)d_{\mathbb{R}^p \setminus A}(y_2)$$

which is exactly the desired property. $\square$

Problem 7.86. *Let $A \subset \mathbb{R}^p$, $A \ne \mathbb{R}^p$ be a nonempty set. The oriented distance function associated to A is $\Delta_A : \mathbb{R}^p \to \mathbb{R}$, given as*

$$\Delta_A(y) := d_A(y) - d_{\mathbb{R}^p \setminus A}(y).$$

Show that:

(i) Δ_A *is real-valued and 1–Lipschitz;*

(ii) $\Delta_A(y) < 0$ *for every* $y \in \operatorname{int} A$, $\Delta_A(y) = 0$ *for every* $y \in \operatorname{bd} A$ *and* $\Delta_A(y) > 0$ *for every* $y \in \operatorname{int}(\mathbb{R}^p \setminus A)$;

(iii) *if* A *is closed, then* $A = \{y \in \mathbb{R}^p \mid \Delta_A(y) \le 0\}$;

(iv) *if* A *is convex, then* Δ_A *is convex;*

(v) *if* A *is a cone, then* Δ_A *is positively homogeneous;*

(vi) *if* A *is a closed convex cone, then* $-\Delta_A$ *is* A–*monotone;*

(vii) *if* A *is a closed convex cone with nonempty interior, then* $-\Delta_A$ *is strictly* $\operatorname{int} A$–*monotone;*

(viii) *if* A *is a closed convex cone, then for every* $y \in \mathbb{R}^p$, $\partial \Delta_{-A}(y) \subset -A^-$.

Solution (i) Let $y_1, y_2 \in \mathbb{R}^p$. If $y_1, y_2 \in A$ or $y_1, y_2 \in \mathbb{R}^p \setminus A$, the inequality

$$\left| \Delta_A(y_1) - \Delta_A(y_2) \right| \le \| y_1 - y_2 \|$$

follows from the similar property of the distance to a set function. The same if at least one of the points is on the boundary of A. Suppose now that $y_1 \in \operatorname{int} A$ and $y_2 \in \operatorname{int}(\mathbb{R}^p \setminus A)$. Then it exists $\lambda \in (0, 1)$ such that $y := \lambda y_1 + (1 - \lambda) y_2 \in \operatorname{bd} A$. Then

$$\left| \Delta_A(y_1) - \Delta_A(y_2) \right| = d_{\mathbb{R}^p \setminus A}(y_1) + d_A(y_2) \le \| y_1 - y \| + \| y_2 - y \| = \| y_1 - y_2 \|$$

and the proof of the property is complete.

(ii) Clearly, if $y \in \operatorname{int} A$, then $d_A(y) = 0$ and $d_{\mathbb{R}^p \setminus A}(y) < 0$, and similar for the last situation. If $y \in \operatorname{bd} A$, since $\operatorname{bd} A = \operatorname{bd}(\mathbb{R}^p \setminus A)$, then both $d_A(y)$ and $d_{\mathbb{R}^p \setminus A}(y)$ are zero.

(iii) Suppose that A is closed. If $y \in A$, then $d_A(y) = 0$ whence $\Delta_A(y) \le 0$. Conversely, if $\Delta_A(y) \le 0$, then supposing that $y \notin A$ we would have $y \in \operatorname{int}(\mathbb{R}^p \setminus A)$ whence $\Delta_A(y) > 0$. The conclusion follows.

(iv) Observe now that

$$\Delta_A(y) = \inf\{ \| y - x \| + \mu(x) \mid x \in \mathbb{R}^p \}, \tag{7.4.8}$$

where μ is defined in the previous problem. Indeed, if $y \in A$, $\Delta_A(y) = -d_{\mathbb{R}^p \setminus A}(y)$, while

$$\inf\{ \| y - x \| + \mu(x) \mid x \in \mathbb{R}^p \} = \inf\{ \| y - x \| - d_{\mathbb{R}^p \setminus A}(x) \mid x \in A \} \le -d_{\mathbb{R}^p \setminus A}(y)$$

is obvious. Now observe that for every $x \in A$, $\| y - x \| + d_{\mathbb{R}^p \setminus A}(y) \ge d_{\mathbb{R}^p \setminus A}(x)$, which means that

$$\| y - x \| + \mu(x) \ge -d_{\mathbb{R}^p \setminus A}(y)$$

so $\inf\{ \| y - x \| + \mu(x) \mid x \in \mathbb{R}^p \} \ge -d_{\mathbb{R}^p \setminus A}(y)$ and in this case (7.4.8) follows. If $y \in \mathbb{R}^p \setminus A$, $\Delta_A(y) = d_A(y)$. Clearly,

$$\inf\{ \| y - x \| + \mu(x) \mid x \in \mathbb{R}^p \} \le \inf\{ \| y - x \| \mid x \in A \} = d_A(y).$$

In order to prove the reverse inequality, observe first that there always exists a sequence $(y_n) \subset A$ with $\| y - y_n \| \to d_A(y)$. We claim that $d_{\mathbb{R}^p \setminus A}(y_n) \to 0$. Indeed, in the opposite case we can suppose, without loss of generality that there exists $\varepsilon > 0$ such that for every $n \in \mathbb{N}$, $B(y_n, \varepsilon) \subset A$. Take now $\delta \in (0, 1)$ such that $\delta \|y - y_n\| < \varepsilon$ for every n (note that such a δ does exist). Consider $z_n := y + (1 - \delta)(y - y_n)$. One has, on one hand, that

$$\|y_n - z_n\| = \delta \|y - y_n\| < \varepsilon,$$

hence $z_n \in A$ and, on the other hand,

$$d_A(y) \leq \|y - z_n\| = (1 - \delta) \|y - y_n\| \to (1 - \delta)d_A(y) < d_A(y).$$

This contradiction can be eliminated only if we admit that $d_{\mathbb{R}^p \setminus A}(y_n) \to 0$. Therefore,

$$\|y - y_n\| - d_{\mathbb{R}^p \setminus A}(y_n) \geq \inf\{\|y - x\| + \mu(x) \mid x \in \mathbb{R}^p\},$$

whence, passing to the limit, we get

$$d_A(y) \geq \inf\{\|y - x\| + \mu(x) \mid x \in \mathbb{R}^p\},$$

so (7.4.8) is finally completely proven. So, according to relation (7.4.8), the function Δ_A appears as a convolution of two convex function, so it is itself a convex function.

(v) If A is a cone, then $\mathbb{R}^p \setminus A$ shares the property to be closed at multiplication with positive scalars, whence the property of Δ_A we are looking for comes from the similar property of the distance to a set function.

(vi) If A is a convex cone then, from (iv) and (v), we deduce that Δ_A is subadditive since

$$\Delta_A(y_1 + y_2) = 2\Delta_A(2^{-1}y_1 + 2^{-1}y_2) \leq 2\Delta_A(2^{-1}y_1) + 2\Delta_A(2^{-1}y_2)$$
$$= \Delta_A(y_1) + \Delta_A(y_2).$$

Now, if $y_2 - y_1 \in -A$, we can write successively

$$0 \geq \Delta_A(y_1 - y_2) \geq \Delta_A(y_1) - \Delta_A(y_2).$$

(vii) The last assertion is similar, taking into account (ii).

(viii) Let $u \in \partial \Delta_{-A}(y)$. One has

$$\langle u, z - y \rangle \leq \Delta_{-A}(z) - \Delta_{-A}(y), \quad \forall z \in \mathbb{R}^p. \tag{7.4.9}$$

From (vi), it follows $\Delta_{-A}(a + y) \leq \Delta_{-A}(y)$ for every $a \in -A$ and whence, from (7.4.9), $\langle u, a \rangle \leq 0$. This implies that for every $y \in \mathbb{R}^p$

$$\partial \Delta_{-A}(y) \subset -K^-.$$

The solution is complete. □

Problem 7.87. *Let $f : \mathbb{R}^p \to \mathbb{R} \cup \{+\infty\}$ be a function. One defines the conjugate of f as $f^* : \mathbb{R}^p \to \mathbb{R} \cup \{+\infty\}$,*

$$f^*(u) = \sup\{\langle x, u \rangle - f(x) \mid x \in \mathbb{R}^p\}.$$

Show that f^* is convex.

Solution Take $u_1, u_2 \in \operatorname{dom} f^*$ and $\alpha \in (0, 1)$. Consider $t \in \mathbb{R}$ with $t < f^*(\alpha u_1 + (1 - \alpha)u_2)$. Then there exists $x \in \mathbb{R}^p$ such that $t < \langle x, \alpha u_1 + (1 - \alpha)u_2 \rangle - f(x)$. Thus,

$$t < \alpha \langle x, u_1 \rangle - \alpha f(x) + (1 - \alpha) \langle x, u_2 \rangle - (1 - \alpha)f(x)$$
$$\leq \alpha f^*(u_1) + (1 - \alpha)f^*(u_2).$$

Since $t < f^*(\alpha u_1 + (1 - \alpha)u_2)$ was chosen arbitrarily, we deduce that

$$f^*(\alpha u_1 + (1 - \alpha)u_2) \leq \alpha f^*(u_1) + (1 - \alpha)f^*(u_2),$$

so f^* is convex. Note that this is a direct proof. □

Exercise 7.88. *Compute the conjugate for the following functions:*
(i) $f : \mathbb{R} \to \mathbb{R}$, $f(x) = \frac{1}{p}|x|^p$ where $p > 1$;
(ii) $f : \mathbb{R} \to \mathbb{R} \cup \{+\infty\}$,

$$f(x) = \begin{cases} -\ln x, & \text{if } x > 0 \\ +\infty, & \text{if } x \leq 0; \end{cases}$$

(iii) $f : \mathbb{R} \to \mathbb{R}$, $f(x) = e^x$;
(iv) $f : \mathbb{R} \to \mathbb{R} \cup \{+\infty\}$,

$$f(x) = \begin{cases} 0, & \text{if } |x| \leq 1 \\ +\infty, & \text{if } x > 1. \end{cases}$$

Solution As in the case of convolution, to compute a conjugate is equivalent to solving an optimization problem (in order to determine the supremum in the definition of the conjugate). For our functions, it is a routine to compute the derivatives for $h(x) = xu - f(u)$ and to get the variation of this function and, from that, the conclusions. We obtain:

(i) $f^* : \mathbb{R} \to \mathbb{R}$, $f^*(u) = \frac{1}{q}|u|^q$, where $\frac{1}{p} + \frac{1}{q} = 1$;
(ii) $f^* : \mathbb{R} \to \mathbb{R} \cup \{+\infty\}$,

$$f^*(u) = \begin{cases} -\ln(-u), & \text{if } u < 0 \\ +\infty, & \text{otherwise}; \end{cases}$$

(iii) $f^* : \mathbb{R} \to \mathbb{R} \cup \{+\infty\}$,

$$f^*(u) = \begin{cases} u\ln(u) - u, & \text{if } u > 0 \\ 0, & \text{if } u = 0 \\ +\infty, & \text{if } u < 0; \end{cases}$$

(iv) $f^* : \mathbb{R} \to \mathbb{R}$, $f^*(u) = |u|$. □

Problem 7.89. *Let* $f : \mathbb{R}^k \to \mathbb{R}$ *be a convex function. Show that for every* $x, u \in \mathbb{R}^k$,

$$\langle u, x \rangle \leq f(x) + f^*(u)$$

with equality if and only if $u \in \partial f(x)$.
Write the above inequality for the convex function $f : \mathbb{R} \to \mathbb{R}, f(x) = \frac{1}{p}|x|^p$, *where* $p > 1$.
Find $\partial f(x)$.

Solution From the definition of the conjugate, for every $u \in \mathbb{R}^k$,

$$f^*(u) = \sup\{\langle x, u \rangle - f(x) \mid x \in \mathbb{R}^k\} \geq \langle u, x \rangle - f(x), \ \forall x \in \mathbb{R}^k,$$

so the inequality holds. The equality means that the supremum is attained at x, so for any $y \in \mathbb{R}^k$,

$$\langle x, u \rangle - f(x) \geq \langle y, u \rangle - f(y),$$

which means that $u \in \partial f(x)$.

Taking into account (*i*) in the preceding problem, the inequality becomes the inequality of Young already studied before (see Subsection 2.2.3), namely

$$xu \leq \frac{|x|^p}{p} + \frac{|u|^q}{q}, \ \forall x, u \in \mathbb{R}, \ \text{where } p, q > 1, \ \frac{1}{p} + \frac{1}{q} = 1.$$

We have seen that

$$ab \leq \frac{a^p}{p} + \frac{b^q}{q}, \ \forall p, q > 1, \ \frac{1}{p} + \frac{1}{q} = 1, \ a, b \geq 0,$$

with the equality if and only if $a^p = b^q$. Thus, for every $x, u \in \mathbb{R}$,

$$xu = \langle x, u \rangle \leq |x| \, |u| \leq \frac{|x|^p}{p} + \frac{|u|^q}{q},$$

with the equality if and only of $xu = |x|^p = |u|^q$, whence $\partial f(x) = \{u \in \mathbb{R} \mid xu = |x|^p = |u|^q\}$. $\square$

Problem 7.90. *Let* $C \subset \mathbb{R}^p$ *be a nonempty closed convex set. Show that for every* $x \in$ bd C, $N(C, x) \neq \{0\}$. *Deduce that the distance function* d_C *is not differentiable at the points of* bd C.

Solution Without loss of generality, we can suppose that $x = 0 \in$ bd C. There exists a sequence $(x_k) \subset \mathbb{R}^p \setminus C$ such that $x_k \to 0$. Take, for every k, y_k as the projection of x_k on C. From the continuity of the projection (Proposition 4.1.2), one deduces that $y_k \to 0$. Consider $z_k := x_k - y_k \neq 0$ (for every k) and the limit $\mu \neq 0$ of a convergent subsequence of the bounded sequence $\|z_k\|^{-1} z_k$. Take $x \in C$. Since for any $\alpha \in (0, 1)$, $\alpha x + (1 - \alpha)y_k \in C$, one deduces that $\|y_k - x_k\| \leq \|\alpha x + (1 - \alpha)y_k - x_k\|$, so

$$\|y_k - x_k\|^2 \leq \|\alpha x + (1 - \alpha)y_k - x_k\|^2.$$

After some calculations, one gets that

$$2 \langle z_k, x - y_k \rangle - \alpha \|x - y_k\|^2 \leq 0.$$

Making $\alpha \to 0$, one gets $\langle z_k, x - y_k \rangle \le 0$ for any k. In particular,

$$\left\langle \|z_k\|^{-1} z_k, x - y_k \right\rangle \le 0$$

for any k. Passing to the limit, we obtain that $\langle \mu, x \rangle \le 0$ for all $x \in C$. Then $\mu \in N(C, 0)$, so the conclusion.

If d_C would be differentiable at a point $\bar{x}$ in its boundary, then $\partial d_C(\bar{x})$ reduces to a singleton (Proposition 4.2.3). But, as shown in Proposition 4.3.3, $N(C, \bar{x}) \cap D(0, 1) \subset \partial d_C(\bar{x})$. In view of the above proved fact, $N(C, \bar{x}) \cap D(0, 1)$ cannot be a singleton, so a contradiction arises. Therefore, d_C cannot be differentiable at $\bar{x}$.

Remark as well that a solution of the problem could be easily deduced from the more general result of Corollary 5.2.29. □

Problem 7.91. *Let $C \subset \mathbb{R}^p$ be a nonempty closed, convex set. Show that d_C^2 is differentiable and for every $x \in \mathbb{R}^p$, and $\nabla d_C^2(x) = 2(x - \mathrm{pr}_C x)$.*

Solution Let $x \in \mathbb{R}^p$ and $h \in \mathbb{R}^p \setminus \{0\}$. Therefore,

$$\begin{aligned}
d_C^2(x + h) - d_C^2(x) &\ge d_C^2(x + h) - \left\| x - \mathrm{pr}_C(x + h) \right\|^2 \\
&= \left\| \mathrm{pr}_C(x + h) - (x + h) \right\|^2 - \left\| x - \mathrm{pr}_C(x + h) \right\|^2 \\
&= 2 \left\langle x - \mathrm{pr}_C(x + h), h \right\rangle + \|h\|^2,
\end{aligned}$$

so, using Proposition 4.1.2,

$$\begin{aligned}
d_C^2(x + h) - d_C^2(x) - 2 \left\langle x - \mathrm{pr}_C x, h \right\rangle &\ge 2 \left\langle x - \mathrm{pr}_C(x + h), h \right\rangle - 2 \left\langle x - \mathrm{pr}_C x, h \right\rangle + \|h\|^2 \\
&\ge -2 \left\| \mathrm{pr}_C x - \mathrm{pr}_C(x + h) \right\| \|h\| + \|h\|^2 \\
&\ge - \|h\|^2.
\end{aligned}$$

On the other hand,

$$\begin{aligned}
d_C^2(x + h) - d_C^2(x) &\le \left\| (x + h) - \mathrm{pr}_C x \right\|^2 - \left\| x - \mathrm{pr}_C x \right\|^2 \\
&= 2 \left\langle x - \mathrm{pr}_C x, h \right\rangle + \|h\|^2.
\end{aligned}$$

Finally, we can write

$$- \|h\|^2 \le d_C^2(x + h) - d_C^2(x) - 2 \left\langle x - \mathrm{pr}_C x, h \right\rangle \le \|h\|^2,$$

so,

$$- \|h\| \le \|h\|^{-1} \left(d_C^2(x + h) - d_C^2(x) - 2 \left\langle x - \mathrm{pr}_C x, h \right\rangle \right) \le \|h\|,$$

which confirms that d_C^2 is differentiable at every $x \in \mathbb{R}^p$ and $\nabla d_C^2(x) = 2(x - \mathrm{pr}_C x)$. □

Problem 7.92. *Let $a_1, a_2, a_3 \in \mathbb{R}^p$ non colinear points. Show that there exists a unique point $\bar{x} \in \mathbb{R}^p$ which minimize on $\mathbb{R}^p$ the function $f : \mathbb{R}^p \to \mathbb{R}$,*

$$f(x) = \|x - a_1\| + \|x - a_2\| + \|x - a_3\|.$$

Then prove that $\overline{x} \in \text{conv}\{a_1, a_2, a_3\}$. Deduce that if $\overline{x} \notin \{a_1, a_2, a_3\}$, then the angle between $\overline{x} - a_i$ and $\overline{x} - a_j$ is $\frac{2\pi}{3}$ for all $i, j \in \overline{1, 3}$, $i \neq j$. (The point $\overline{x}$ is called the Torricelli point of the triangle of vertices a_1, a_2, a_3.)

Solution It is easy to observe that f is convex and coercive, so it has a global minimum on $\mathbb{R}^p$. Moreover, f is strictly convex. Indeed, for $x, y \in \mathbb{R}^p$, the equality $\|x\| + \|y\| = \|x + y\|$ is possible if and only if x, y are on a half-line passing through the origin. If there exist $x, y \in \mathbb{R}^p$, $x \neq y$ and $\alpha \in (0, 1)$ with $f(\alpha x + (1 - \alpha)y) = \alpha f(x) + (1 - \alpha)f(y)$, then a_1, a_2, a_3 must be on the same straight-line determined by x and y, which is a contradiction. Therefore, there exists a unique element $\overline{x} \in \mathbb{R}^p$, global minimum of f. Thus, from Theorem 4.3.1, $0 \in \partial f(\overline{x})$. But, by use of Theorem 4.2.7 and Problem 7.81,

$$\partial f(x) = \begin{cases} \sum_{i=1}^{3} \|x - a_i\|^{-1} (x - a_i), & \text{if } x \notin \{a_1, a_2, a_3\} \\ \sum_{i=1, i \neq j}^{3} \|x - a_i\|^{-1} (x - a_i) + D(0, 1), & \text{if } x = a_j. \end{cases}$$

If $\overline{x}$ is one of the points a_1, a_2, a_3, then surely belongs to $\text{conv}\{a_1, a_2, a_3\}$. Suppose that $\overline{x} \notin \{a_1, a_2, a_3\}$. Then

$$\sum_{i=1}^{3} \|\overline{x} - a_i\|^{-1} (\overline{x} - a_i) = 0,$$

whence

$$\overline{x} = \left(\sum_{i=1}^{3} \|\overline{x} - a_i\|^{-1} \right)^{-1} \sum_{i=1}^{3} \|\overline{x} - a_i\|^{-1} a_i \in \text{conv}\{a_1, a_2, a_3\}.$$

For the last part, if $\overline{x} \notin \{a_1, a_2, a_3\}$, again from $\sum_{i=1}^{3} \|\overline{x} - a_i\|^{-1} (\overline{x} - a_i) = 0$ we deduce that $\left\langle \|\overline{x} - a_i\|^{-1} (\overline{x} - a_i), \|\overline{x} - a_j\|^{-1} (\overline{x} - a_j) \right\rangle = 2^{-1}$ for all $i, j \in \overline{1, 3}$, $i \neq j$. Consequently, the angle between $\overline{x} - a_i$ and $\overline{x} - a_j$ is always $\frac{2\pi}{3}$. This ends the solution. □

Problem 7.93. *Consider the Fermat's principle concerning the propagation of light: "in an inhomogeneous medium a ray of light travels between two points along the path requiring the shortest time". Prove the law of refraction: when light passes from one medium to another, the directions of the light satisfy $\frac{\sin \alpha_1}{v_1} = \frac{\sin \alpha_2}{v_2}$, where α_1, α_2 are the angles between the directions of the ray and the normal to the surface which separates the two media, and v_1, v_2 are the speeds of the light in the two media, respectively.*

Solution Suppose that the light travels from the point $(0, a)$ in the first medium to the point (b, d), $b > 0$ in the second one (see the figure below).

For easy calculus, suppose that the surface which separates the two media is the Ox axis. Let v_1, v_2 be the speeds of the light in the two media, respectively. Suppose that the light passes from the first medium to the second one at the point $(x, 0)$, where x is to be determined following Fermat's principle. The amount of time spent by the

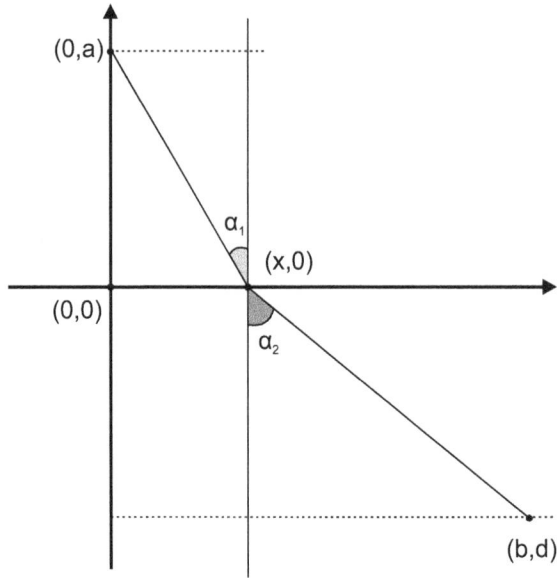

Figure 7.4: Propagation of light.

light in the first and the second media are, respectively,

$$t_1 = \frac{\sqrt{a^2 + x^2}}{v_1}$$

$$t_2 = \frac{\sqrt{(b - x)^2 + c^2}}{v_2},$$

so the total time to minimize is $\frac{\sqrt{a^2+x^2}}{v_1} + \frac{\sqrt{(b-x)^2+c^2}}{v_2}$. Take $f : \mathbb{R} \to \mathbb{R}$,

$$f(x) = \frac{\sqrt{a^2 + x^2}}{v_1} + \frac{\sqrt{(b - x)^2 + c^2}}{v_2}.$$

Now the problem is to minimize f on $\mathbb{R}$. The derivative of f is

$$f'(x) = \frac{x}{v_1\sqrt{a^2 + x^2}} + \frac{x - b}{v_2\sqrt{(x - b)^2 + c^2}}.$$

Since $f'(0) = \frac{-b}{v_2\sqrt{b^2+c^2}} < 0$ and $f'(b) = \frac{b}{v_1\sqrt{a^2+b^2}} > 0$ and

$$f''(x) = \frac{1}{v_1} \cdot \frac{a^2}{(a^2 + x^2)^{\frac{3}{2}}} + \frac{1}{v_2} \cdot \frac{c^2}{((x - b)^2 + c^2)^{\frac{3}{2}}} > 0, \ \forall x \in \mathbb{R},$$

there exists a unique critical point of f situated in the interval $[0, b]$. Denote by $\bar{x}$ this critical point. The variation of f shows that $\bar{x}$ is a minimum point. Then

$$\frac{\bar{x}}{v_1\sqrt{a^2 + \bar{x}^2}} = \frac{b - \bar{x}}{v_2\sqrt{(\bar{x} - b)^2 + c^2}}.$$

But,

$$\frac{\overline{x}}{\sqrt{a^2 + \overline{x}^2}} = \sin \alpha_1$$

and

$$\frac{b - \overline{x}}{\sqrt{(\overline{x} - b)^2 + c^2}} = \sin \alpha_2,$$

so,

$$\frac{\sin \alpha_1}{v_1} = \frac{\sin \alpha_2}{v_2}.$$

and the light refraction law in proved. □

Exercise 7.94. *Show that $f : (0, \infty) \times (0, \infty) \to \mathbb{R}$,*

$$f(x_1, x_2) = \frac{1}{x_1} + \frac{1}{x_2} - \frac{1}{x_1 + x_2}$$

is convex.

Solution The calculus of second-order partial derivatives leads to the following form of the Hessian matrix at a point x in the domain of f :

$$\begin{pmatrix} \frac{2}{x_1^3} - \frac{2}{(x_1+x_2)^3} & -\frac{2}{(x_1+x_2)^3} \\ -\frac{2}{(x_1+x_2)^3} & \frac{2}{x_2^3} - \frac{2}{(x_1+x_2)^3} \end{pmatrix}.$$

On this basis, it is easy to verify that $\nabla^2 f(x)$ is positive definite for every $x \in (0, \infty) \times (0, \infty)$, so, according to Theorem 2.2.10, the function f is convex. □

Problem 7.95. *Let $a, b, c, d \in \mathbb{R}$ with $a < b$, $c < d$, and $f : [a, b] \times [c, d] \to \mathbb{R}$. Define $\varphi : [a, b] \to \mathbb{R}$,*

$$\varphi(x) = \inf\{f(x, y) \mid y \in [c, d]\}.$$

Show that φ is well defined and continuous.

Solution The mapping

$$[c, d] \ni y \longmapsto f(x, y)$$

is continuous and hence attains its minimum on $[c, d]$. Then there exist $y_x \in [c, d]$ such that $\varphi(x) = f(x, y_x)$.

From Cantor Theorem (Theorem 1.2.24), f is uniformly continuous on the compact set $[a, b] \times [c, d]$: for every $\varepsilon > 0$, there exists $\delta_\varepsilon > 0$, such that for every (x', y'), $(x'', y'') \in [a, b] \times [c, d]$ with $\|(x', y') - (x'', y'')\| < \delta_\varepsilon$, one has $|f(x', y') - f(x'', y'')| < \varepsilon$.

Let $x', x'' \in [a, b]$ with $|x' - x''| \leq \delta_\varepsilon$. Then

$$\varphi(x') - \varphi(x'') = f(x', y_{x'}) - f(x'', y_{x''})$$
$$\leq f(x', y_{x''}) - f(x'', y_{x''}) < \varepsilon.$$

Changing the roles of x' and x'', we get

$$\varphi(x'') - \varphi(x') < \varepsilon,$$

from where the conclusion. □

Problem 7.96. *Let $f : \mathbb{R}^p \to \mathbb{R}$ be a function and $M \subset \mathbb{R}^p$ be a closed set. Consider the problem*

$$\min f(x), \ x \in M.$$

Suppose that $\bar{x}$ is a solution of this problem and that f is locally Lipschitz of constant $L \geq 0$ around $\bar{x}$. Let $\varphi : \mathbb{R}^p \to [0, \infty)$ with $\varphi(\bar{x}) = 0$ a lower semicontinuous function. Show that one and only one of the following holds:
  (i) there exists $\lambda > 0$ such that $\bar{x}$ is a local minimum without constraints for $f + \lambda\varphi$;
  (ii) there exists $(z_n)_{n \in \mathbb{N}^*} \subset \mathbb{R}^p \setminus M$, $z_n \to \bar{x}$ such that for every $n \in \mathbb{N}^*$, the mapping

$$x \mapsto \varphi(x) + n^{-1} \|x - z_n\|$$

attains its minimum at z_n.

Solution We can have one and only one of the following possibilities:
- there exists a neighborhood V of $\bar{x}$ and $a > 0$ such that for every $x \in V$,

$$a\varphi(x) \geq d(x, M);$$

- there exists $x_n \to \bar{x}$ such that

$$2n\varphi(x_n) < d(x_n, M).$$

In the first situation, let $\alpha > 0$ such that $B(\bar{x}, \alpha) \subset U \cap V$, where U is the neighborhood of $\bar{x}$ where the Lipschitz condition holds and where $\bar{x}$ is a minimum point.
  For every $x \in B(\bar{x}, \alpha/3)$ there exists $u \in M$ with

$$\|x - u\| \leq 2d(x, M) \leq 2 \|x - \bar{x}\|.$$

Hence, in particular,

$$\|u - \bar{x}\| \leq \|u - x\| + \|x - \bar{x}\| \leq 3 \|x - \bar{x}\| < \alpha.$$

Then

$$f(x) \geq f(u) - L \|u - x\| \geq f(\bar{x}) - 2La\varphi(x)$$

so we are in the first alternative of the conclusion.
  In the second situation, since φ has positive values, $(x_n) \subset \mathbb{R}^p \setminus M$ and

$$\varphi(x_n) \leq \inf_{x \in \mathbb{R}^p} \varphi(x) + 2^{-1}n^{-1}d(x_n, M).$$

We take $\varepsilon := 2^{-1}n^{-1}d(x_n, M) > 0$ and $\delta := 2^{-1}d(x_n, M)$, and we apply Ekeland Variational Principle to φ for the ε–minimum x_n. We infer that there exists $z_n \in \mathbb{R}^p$ with

$$\varphi(z_n) \le \varphi(x_n) < n^{-1}d(x_n, M)$$
$$\|z_n - x_n\| \le 2^{-1}d(x_n, M)$$
$$\varphi(z_n) \le \varphi(x) + n^{-1}\|x - z_n\|, \ \forall x \in \mathbb{R}^p.$$

The last relation shows that for every $n \in \mathbb{N}^*$ the mapping $x \mapsto \varphi(x) + n^{-1}\|x - z_n\|$ attains its minimum at z_n. Moreover,

$$\|z_n - \overline{x}\| \le \|z_n - x_n\| + \|x_n - \overline{x}\|$$
$$\le d(x_n, M) + \|x_n - \overline{x}\| \le 2\|x_n - \overline{x}\|,$$

whence $z_n \to \overline{x}$. If we would have $z_n \in M$, then

$$\|z_n - x_n\| \le 2^{-1}d(x_n, M) < d(x_n, M) \le \|z_n - x_n\|,$$

which is a contradiction. The solution is now complete. $\qquad\square$

Let us remark that Problem 7.96 is a generalization of Theorem 4.3.2.

Exercise 7.97. *Consider the function $f : \mathbb{R}^2 \to \mathbb{R}$ given by $f(x, y) := |x_1| - |x_2|$. Prove that*

$$\partial_C f(0, 0) = \mathrm{conv}\left\{(-1, -1), (-1, 1), (1, -1), (1, 1)\right\}.$$

Solution The formula can be deduced in a similar manner to Example 5.1.9, taking into account formula (5.1.8). $\qquad\square$

Other useful properties of the Clarke tangent and the normal cones are contained in the next exercise.

Problem 7.98. *Let $f, g : \mathbb{R}^p \to \mathbb{R}$ be locally Lipschitz functions around $\overline{x}$. The following hold:*

(i) $f \cdot g : \mathbb{R}^p \to \mathbb{R}$, given by $(f \cdot g)(x) := f(x) \cdot g(x)$ for every x is locally Lipschitz around $\overline{x}$, and

$$\partial_C(f \cdot g)(\overline{x}) \subset g(\overline{x})\partial_C f(\overline{x}) + f(\overline{x})\partial_C g(\overline{x}). \tag{7.4.10}$$

If, moreover, $f(x) \ge 0$ and $g(x) \ge 0$ for every x, and f and g are both regular, then fg is regular and equality holds in (7.4.10).

(ii) Suppose $g(x) \ne 0$ for every x. Then the function $\dfrac{f}{g} : \mathbb{R}^p \to \mathbb{R}$, given by $\left(\dfrac{f}{g}\right)(x) := \dfrac{f(x)}{g(x)}$ for every x is locally Lipschitz around $\overline{x}$ and

$$\partial_C\left(\frac{f}{g}\right)(\overline{x}) \subset \frac{g(\overline{x})\partial_C f(\overline{x}) - f(\overline{x})\partial_C g(\overline{x})}{g^2(\overline{x})}. \tag{7.4.11}$$

If, moreover, $f(x) \ge 0$ and $g(x) > 0$ for every x, and f and $-g$ are both regular, then $\dfrac{f}{g}$ is regular and equality holds in (7.4.11).

Solution For (i), take $h : \mathbb{R}^2 \to \mathbb{R}$ given by $h(x_1, x_2) := x_1 \cdot x_2$. Then Theorem 5.1.18 applies, and one has the conclusion. The proof of (ii) is similar. $\qquad\square$

Problem 7.99. *Let $f_1, \ldots, f_k : \mathbb{R}^p \to \mathbb{R}$ be locally Lipschitz functions around x. Then the function $h : \mathbb{R}^p \to \mathbb{R}$ given as*

$$h(x) := \max\{f_1(x), \ldots, f_k(x)\}$$

is locally Lipschitz around x and

$$\partial_C h(x) \subset \operatorname{conv} \bigcup_{i \in A(x)} \partial_C f_i(x),$$

where, $A(x) = \left\{ i \in \overline{1, k} \mid h(x) = f_i(x) \right\}$ denotes the set of active indices at x.

Solution Observe that $h = g \circ f$, where $f : \mathbb{R}^p \to \mathbb{R}^k$ is $f(x) = (f_1(x), \ldots, f_k(x))$ and $g : \mathbb{R}^k \to \mathbb{R}$ is $g(y) = \max\{y_1, \ldots, y_k\}$. Since f and g are locally Lipschitz, one may apply Theorem 5.1.18.

Observe moreover that, since g is convex, its Clarke subdifferential coincides with the convex subdifferential, i.e.,

$$\partial_C g(f(x)) = \partial g(f(x))$$
$$= \left\{ \eta \in \mathbb{R}_+^k \mid \eta_1 + \ldots + \eta_k = 1, \ \eta_1 f_1(x) + \ldots + \eta_k f_k(x) \geq h(x) \right\}$$
$$= \left\{ \eta \in \mathbb{R}_+^k \mid \eta_1 + \ldots + \eta_k = 1, \ \eta_1 f_1(x) + \ldots + \eta_k f_k(x) = h(x) \right\}$$
$$= \left\{ \eta \in \mathbb{R}_+^k \mid \eta_i = 0 \text{ if } i \notin A(x), \ \eta_1 + \ldots + \eta_k = 1 \right\}.$$

From (5.1.17), we get that

$$\partial_C h(x) \subset \operatorname{cl\,conv} \left\{ \partial_C \langle \eta, f \rangle(x) \mid \eta \in \mathbb{R}_+^k, \ \eta_i = 0 \text{ if } i \notin A(x), \ \sum_{i=1}^k \eta_i = 1 \right\}$$
$$= \operatorname{cl\,conv} \left\{ \partial_C \left(\sum_{i \in A(x)} \eta_i f_i(x) \right) \mid \eta_i \geq 0, \ \sum_{i \in A(x)} \eta_i = 1 \right\}$$
$$\subset \operatorname{cl\,conv} \left\{ \sum_{i \in A(x)} \eta_i \partial_C f_i(x) \mid \eta_i \geq 0, \ \sum_{i \in A(x)} \eta_i = 1 \right\}.$$

Since $\sum_{i \in A(x)} \eta_i \partial_C f_i(x)$ from above is a convex combination of the sets $\partial_C f_i(x)$ for $i \in A(x)$, each of one being convex and compact, one gets the desired conclusion. $\qquad\square$

Problem 7.100. *Suppose $A_1 \subset \mathbb{R}^p$ and $A_2 \subset \mathbb{R}^q$ are two sets, and take $x_1 \in A_1$, $x_2 \in A_2$. Prove that*

$$T_C(A_1 \times A_2, (x_1, x_2)) = T_C(A_1, x_1) \times T_C(A_2, x_2)$$

and

$$N_C(A_1 \times A_2, (x_1, x_2)) = N_C(A_1, x_1) \times N_C(A_2, x_2)$$

Solution The first equality follows easily from the characterization given by Theorem 5.1.25 (ii). Then the second equality follows by polarity. □

Problem 7.101. *Consider a nonempty set $A \subset \mathbb{R}^p$ and $\overline{x} \in A$. Define the indicator function of A as $\iota_A : \mathbb{R}^p \to \mathbb{R} \cup \{+\infty\}$,*

$$\iota_A(x) := \begin{cases} 0, & \text{if } x \in A \\ \infty, & \text{if } x \notin A. \end{cases}$$

Prove that

$$\partial_F \iota_A(\overline{x}) = N_F(A, \overline{x})$$

and

$$\partial_M \iota_A(\overline{x}) = \partial^\infty \iota_A(\overline{x}) = N_M(A, \overline{x}).$$

Solution It follows from the fact that epi $\iota_A = A \times [0, \infty)$, hence by Proposition 5.2.3 (vi)

$$N_F(\text{epi } \iota_A, (\overline{x}, 0)) = N_F(A, \overline{x}) \times (-\infty, 0]$$

and

$$N_M(\text{epi } \iota_A, (\overline{x}, 0)) = N_M(A, \overline{x}) \times (-\infty, 0]. \qquad \square$$

Exercise 7.102. *Consider the sets in $\mathbb{R}^3$ given by*

$$A = \left\{ (t, p, q) \mid (p, q) \in [(0, 0), (\cos t, \sin t)] \right\},$$

$$Q = \left\{ (t, p, q) \mid (p, q) \in \text{cone } D\left((\cos t, \sin t), \frac{\sqrt{2}}{2} \right) \right\}.$$

Compute the Fréchet and the Mordukhovich normal cones to these sets at $(0, 0, 0)$.

Solution Observe first that the set A can be equivalently written as

$$\begin{cases} x(u, v) = u \\ y(u, v) = v \cos u, & u \in \mathbb{R}, \ v \in [0, 1]. \\ z(u, v) = v \sin u \end{cases}$$

Our intention is to compute the Fréchet normal cone to the set A using its equality to $T_B(A, (x, y, z))^-$. Consider hence points (x_0, y_0, z_0) from A close to $(0, 0, 0)$. If (x_0, y_0, z_0) is such that $v \in (0, 1)$, then the Fréchet normal cone to this point is the cone generated by the normal vector to the surface, whose expression is $(-v, -\sin u, \cos u)$. In fact, it is the line

$$\left\{ (x, y, z) \in \mathbb{R}^3 \mid \frac{x - u}{-v} = \frac{y - v \cos u}{-\sin u} = \frac{z - v \sin u}{\cos u} \right\}. \qquad (7.4.12)$$

When $(x_n, y_n, z_n) \to (0, 0, 0)$, it means that the corresponding $(u_n, v_n) \to (0, 0)$, which gives, when passing to the limit when considering elements from the set (7.4.12), the line $\{0\} \times \{0\} \times \mathbb{R}$, i.e., the Oz axis. Remark also that, for the point $(x_0, y_0, z_0) \in A$, the Bouligand tangent cone is the tangent plane to the surface at (x_0, y_0, z_0) translated to $(0, 0, 0)$. This plane is

$$\left\{(x, y, z) \in \mathbb{R}^3 \mid -vx - y \sin u + z \cos u = 0\right\}. \tag{7.4.13}$$

Consider now $(x_0, y_0, z_0) \in A$ is such that $v = 0$, i.e., points of the type $(u, 0, 0)$. In this case, the Bouligand tangent cone is the half-plane obtained by taking $y \geq 0$ in the equation (7.4.13):

$$P := \left\{(x, y, z) \in \mathbb{R}^3 \mid -y \sin u + z \cos u = 0, \ y \geq 0\right\}.$$

In this case, the Fréchet normal cone is P^-, i.e.,

$$\left\{(x, y, z) \in \mathbb{R}^3 \mid x = 0, \ y \cos u + z \sin u = 0, \ y \leq 0\right\}.$$

When $(x_n, y_n, z_n) \to (0, 0, 0)$, one has again $(u_n, v_n) \to (0, 0)$, hence passing to the limit when taking elements from the previous set one obtains the half-plane

$$\{(x, y, z) \in \mathbb{R}^3 \mid x = 0, \ y \leq 0\}.$$

In conclusion,

$$N_M(A, (0, 0, 0)) = \{(x, y, z) \in \mathbb{R}^3 \mid x = 0, \ y \leq 0\}.$$

Remark also that the structure of Q and the calculus of $N_M(Q, (0, 0, 0))$ are somehow similar. This is because Q is bounded by two helicoids, given parametrically as

$$(H_1) : \begin{cases} x(u, v) = u \\ y(u, v) = v \cos\left(u + \dfrac{\pi}{4}\right), \\ z(u, v) = v \sin\left(u + \dfrac{\pi}{4}\right) \end{cases} \quad \text{and} \ (H_2) : \begin{cases} x(u, v) = u \\ y(u, v) = v \cos\left(u - \dfrac{\pi}{4}\right), \\ z(u, v) = v \sin\left(u - \dfrac{\pi}{4}\right) \end{cases}$$

both parametrized for $u \in \mathbb{R}$, $v \geq 0$.

Taking points from (H_1) or (H_2) such that $v > 0$, one gets the normal cones

$$\left\{(x, y, z) \in \mathbb{R}^3 \mid \frac{x - u}{-v} = \frac{y - v \cos\left(u \pm \dfrac{\pi}{4}\right)}{-\sin\left(u \pm \dfrac{\pi}{4}\right)} = \frac{z - v \sin\left(u \pm \dfrac{\pi}{4}\right)}{\cos\left(u \pm \dfrac{\pi}{4}\right)}\right\},$$

which tend when $(u, v) \to (0, 0)$ to the lines

$$\{(0, y, y) \mid y \in \mathbb{R}\} \ \text{and} \ \{(0, y, -y) \mid y \in \mathbb{R}\}.$$

When $(x_0, y_0, z_0) \in Q$ is such that $v = 0$, then

$$T_B(Q, (x_0, y_0, z_0)) = \left\{ (x, y, z) \in \mathbb{R}^3 \mid \begin{array}{c} -y \sin\left(u + \dfrac{\pi}{4}\right) + z \cos\left(u + \dfrac{\pi}{4}\right) \leq 0 \\ -y \sin\left(u - \dfrac{\pi}{4}\right) + z \cos\left(u - \dfrac{\pi}{4}\right) \geq 0 \\ y \geq 0 \end{array} \right\}$$

and

$$N_F(Q, (x_0, y_0, z_0)) = \left\{ (x, y, z) \in \mathbb{R}^3 \mid \begin{array}{c} x = 0 \\ y \cos\left(u + \dfrac{\pi}{4}\right) + z \sin\left(u + \dfrac{\pi}{4}\right) \leq 0 \\ y \cos\left(u - \dfrac{\pi}{4}\right) + z \sin\left(u - \dfrac{\pi}{4}\right) \geq 0 \\ y \leq 0 \end{array} \right\}.$$

Passing to the limit for $(u, v) \to (0, 0)$, one gets

$$N_M(Q, (0, 0, 0)) = \{(x, y, z) \in \mathbb{R}^3 \mid x = 0, \ y \leq 0, \ y \leq z \leq -y\}. \qquad \square$$

Problem 7.103. *Let $f, g : \mathbb{R}^p \to \mathbb{R}$ be Lipschitz around $\overline{x} \in \mathbb{R}^p$.*
(i) If $\partial_F(-f(\overline{x})g)(\overline{x}) \neq \emptyset$, then

$$\partial_F(f \cdot g)(\overline{x}) \subset \bigcap_{\xi \in \partial_F(-f(\overline{x})g)(\overline{x})} [\partial_F(g(\overline{x})f)(\overline{x}) - \xi],$$

which holds with equality if g is Fréchet differentiable at $\overline{x}$.
(ii) If $g(\overline{x}) \neq 0$, and if $\partial_F(f(\overline{x})g)(\overline{x}) \neq \emptyset$, then

$$\partial_F\left(\frac{f}{g}\right)(\overline{x}) \subset \bigcap_{\xi \in \partial_F(f(\overline{x})g)(\overline{x})} \frac{[\partial_F(g(\overline{x})f)(\overline{x}) - \xi]}{(g(\overline{x}))^2},$$

which holds with equality if g is Fréchet differentiable at $\overline{x}$.

Solution (i) Define $F : \mathbb{R}^p \to \mathbb{R}^2$ and $G : \mathbb{R}^2 \to \mathbb{R}$ by

$$F(x) := (f(x), g(x)) \quad \text{and} \quad G(y_1, y_2) := y_1 \cdot y_2.$$

Then $f \cdot g = G \circ F$, hence one can apply the chain rule from Theorem 5.2.36 to get that

$$\partial_F(f \cdot g)(\overline{x}) = \partial_F(G \circ F)(\overline{x}) = \partial_F \langle \xi, F \rangle (\overline{x}),$$

where $\xi \in \nabla G(\overline{y})$, since G is Fréchet differentiable at $\overline{y} := (f(\overline{x}), g(\overline{x}))$. Moreover, $\nabla G(\overline{y}) = (\overline{y}_2, \overline{y}_1)$, hence we have from above that

$$\partial_F(f \cdot g)(\overline{x}) = \partial_F(\overline{y}_2 f + \overline{y}_1 g)(\overline{x}) = \partial_F [g(\overline{x}) \cdot f - (-f(\overline{x})) \cdot g] (\overline{x}),$$

which gives, using Theorem 5.2.33, the conclusion.
The proof of (ii) is similar. $\qquad \square$

Problem 7.104. *Let $f_1, ..., f_k : \mathbb{R}^p \to \mathbb{R}$ be some functions, and $\overline{x} \in \mathbb{R}^p$. Then the function $g : \mathbb{R}^p \to \mathbb{R}$ given as*

$$g(x) := \min \{f_1(x), ..., f_k(x)\}, \ \forall x \in \mathbb{R}^p$$

satisfies

$$\partial_F g(\overline{x}) \subset \bigcap_{i \in A(\overline{x})} \partial_F f_i(\overline{x}),$$

where $A(\overline{x}) = \left\{i \in \overline{1, k} \mid g(\overline{x}) = f_i(\overline{x})\right\}$ denotes the set of active indices at $\overline{x}$.

Solution Take $\xi \in \partial_F g(\overline{x})$. Hence for any $\varepsilon > 0$, one can find $\delta > 0$ such that

$$\langle \xi, x - \overline{x} \rangle \leq g(x) - g(\overline{x}) + \varepsilon \|x - \overline{x}\|$$

whenever $\|x - \overline{x}\| < \delta$. For such x, for any $i \in A(\overline{x})$, one has

$$\begin{aligned}
\langle \xi, x - \overline{x} \rangle &\leq g(x) - g(\overline{x}) + \varepsilon \|x - \overline{x}\| \\
&= g(x) - f_i(\overline{x}) + \varepsilon \|x - \overline{x}\| \\
&\leq f_i(x) - f_i(\overline{x}) + \varepsilon \|x - \overline{x}\|,
\end{aligned}$$

which gives us that $\xi \in \partial_F f_i(\overline{x})$. $\qquad\square$

Bibliography

D. Bartl (2012). A very short algebraic proof of the Farkas Lemma, Mathematical Methods of Operations Research, 75, 101–104.

M. S. Bazaraa, H. D. Sherali, C. M. Shetty (2006). Nonlinear programming: theory and algorithms, John Wiley & Sons, New Jersey.

O. Cârjă (2003). Methods of nonlinear Functional Analysis, Editura Matrix Rom, Bucureşti (in Romanian).

F.H. Clarke (1983). Optimization and nonsmooth analysis, John Wiley & Sons, New York.

F.H. Clarke (2013). Functional analysis, calculus of variations and optimal control, Springer-Verlag, London.

A.L. Dontchev, R.T. Rockafellar (2009). Implicit functions and Solution Mappings, Springer, Berlin.

A. Forsgren, P.E. Gill, M.H. Wright (2002). Interior methods in nonlinear optimization, SIAM Review 44, 525–597.

A. Göpfert, H. Riahi, C. Tammer, C. Zălinescu (2003). Variational Methods in Partially Ordered Spaces, Springer, Berlin.

M.R. Hestenes (1975). Optimization Theory. The finite dimensional case, John Wiley & Sons, New York.

J.-B. Hiriart-Urruty (1983). A short proof of the variational principle for approximate solutions of a minimization problem, The American Mathematical Monthly, 90, 206–207.

J.-B. Hiriart-Urruty (2009). Optimization et analyse convexe, EDP Sciences, Paris.

J.-B. Hiriart-Urruty (2008). Les mathématiques du mieux faire, Volume 1, Premiers pas en optimization, Ellipses, Paris.

E. Isaacson, H.B. Keller (1966) Analysis of numerical methods, Dover Publications, New York.

D. Klatte, B. Kummer, (2002). Nonsmooth Equations in Optimization: Regularity, Calculus, Methods and Applications, Kluwer Academic Publishers, Dordrecht.

G. Lebourg (1975). Valeur moyenne pour un gradient généralisé, Comptes Rendus de l'Académie des Sciences, Série A, 281, 795–797.

B.S. Mordukhovich (2006). Variational Analysis and Generalized Differentiation, Vol. I: Basic Theory, Vol. II: Applications, Springer, Grundlehren der mathematischen Wissenschaften (A Series of Comprehensive Studies in Mathematics), Vol. 330 and 331, Berlin.

B.S. Mordukhovich, N.M. Nam, N.D. Yen (2006). Fréchet subdifferential calculus and optimality conditions in nondifferentiable programming, Optimization, 55, 685–708.

C. Niculescu, L.-E. Persson, (2006). Convex functions and their applications, Springer, New York.

J. Nocedal, S. J. Wright (2006). Numerical optimization, Springer, New York.

B.G. Pachpatte (2005). Mathematical inequalities, Elsevier, Amsterdam.

P. Pedregal (2004). Introduction to optimization, Springer-Verlag, New York.

T.L. Rădulescu, V. Rădulescu, T. Andreescu, (2009). Problems in Real Analysis: advanced calculus on the real axis, Springer, Dordrecht.

R.T. Rockafellar (1970). Convex Analysis, Princeton University Press.

R.T. Rockafellar (1985). Maximal monotone relations and the second derivatives of nonsmooth functions, Annales de l'Institut Henri Poincaré, Analyse non linéaire, 2, 167–184.

R.T. Rockafellar, R.J.-B. Wets (1998). Variational Analysis, Springer, Grundlehren der mathematischen Wissenschaften (A Series of Comprehensive Studies in Mathematics), Vol. 317, Berlin.

A. Quarteroni, F. Saleri (2006). Scientific computing with MATLAB and Octave, Springer, Milano.

C. Zălinescu (1998). Mathematical programming on infinite dimensional normed vector spaces, Editura Academiei, Bucureşti (in Romanian).

C. Zălinescu (2002). Convex Analysis in General Vector Spaces, World Scientific, River Edge, Singapore, 2002.

List of Notations

Operations and Symbols

:=	equal by definition
$\langle \cdot, \cdot \rangle$	scalar product
$\times$	cartesian product
$x \to \overline{x}$	x converges to $\overline{x}$
$x \xrightarrow{A} \overline{x}$	$x \to \overline{x}$ with $x \in A$
$x \xrightarrow{f} \overline{x}$	$x \to \overline{x}$ with $f(x) \to f(\overline{x})$
$x_n \to \overline{x}, (x_n) \to \overline{x}$	sequence (x_n) has the limit $\overline{x}$
$x_n \xrightarrow{A} \overline{x}$	$x \to \overline{x}$ with $(x_n) \subset A$
lim	limit (for sequences or functions)
lim inf	lower limit
lim sup	upper limit
$\|\cdot\|$	Euclidean norm of $\mathbb{R}^p$, $p > 1$
$|\cdot|$	modulus (on $\mathbb{R}$)
$\square$	end of proof/solution

Sets and Spaces

$\emptyset$	empty set
$\mathbb{N}$	set of natural numbers
$\mathbb{N}^*$	$\mathbb{N} \setminus \{0\}$
$\mathbb{Z}$	set of integers
$\mathbb{Q}$	set of rational numbers
$\mathbb{R}$	set of real numbers
$\mathbb{R}^p$, $p \in \mathbb{N}^*$	p-dimensional Euclidean space
$\mathbb{R}^p_+$	positive orthant of $\mathbb{R}^p$
$B(x, r)$	open ball of center x and radius r
$D(x, r)$	closed ball of center x and radius r
$S(x, r)$	sphere of center x and radius r
int A	interior of the set A
cl A	closure of the set A
bd A	boundary of the set A

conv A	convex hull of the set A
cone A	conic hull of the set A
A^-	polar of the set A
$T_B(A, \overline{x})$	Bouligand tangent cone to A at $\overline{x}$
$T_C(A, \overline{x})$	Clarke tangent cone to A at $\overline{x}$
$N(A, \overline{x})$	normal cone to the convex set A at $\overline{x}$
$N_B(A, \overline{x})$	Bouligand normal cone to A at $\overline{x}$
$N_C(A, \overline{x})$	Clarke normal cone to A at $\overline{x}$
$N_F(A, \overline{x})$	Fréchet normal cone to A at $\overline{x}$
$N_M(A, \overline{x})$	Mordukhovich normal cone to A at $\overline{x}$
$\partial f(\overline{x})$	convex subdifferential of f at $\overline{x}$
$\partial_C f(\overline{x})$	Clarke subdifferential of f at $\overline{x}$
$\partial_F f(\overline{x})$	Fréchet subdifferential of f at $\overline{x}$
$\partial_M f(\overline{x})$	Mordukhovich subdifferential of f at $\overline{x}$
$\partial^\infty f(\overline{x})$	singular subdifferential of f at $\overline{x}$
$\mathrm{pr}_A \overline{x}$	projection set of $\overline{x}$ on A
rank A	rank of the matrix A
A^t, x^t	transpose of the matrix A/vector x
dim X	dimension of the linear (sub)space X

Functions

$f : A \to B$	function from A to B
f^{-1}	inverse of f
$f \circ g$	composition of functions
gr f	graph of f
epi f	epigraph of f
$N_v f$	v-level set of f
dom f	domain of f
Ker T	kernel of the linear operator T
Im f	image of the mapping f
$\nabla f(\overline{x})$	gradient of f at $\overline{x}$
$\nabla^2 f(\overline{x})$	second order differential of f at $\overline{x}$
f^*	conjugate of f
$d_A, d(\cdot, A)$	distance function to the set A
Δ_A	oriented distance function to the set A
s_e	Gerstewitz (Tammer) scalarization functional

Index